Industrial Applications of Combinatorial Optimization

Applied Optimization

Volume 16

Series Editors:

Panos M. Pardalos
University of Florida, U.S.A.

Donald Hearn
University of Florida, U.S.A.

The titles published in this series are listed at the end of this volume.

Industrial Applications of Combinatorial Optimization

Edited by

Gang Yu

Department of Management Science & Information Systems,
and the Center for Management of Operations and Logistics,
University of Texas at Austin,
Austin, Texas, U.S.A.

KLUWER ACADEMIC PUBLISHERS

DORDRECHT / BOSTON / LONDON

A C.I.P. Catalogue record for this book is available from the Library of Congress.

ISBN 0-7923-5073-1

Published by Kluwer Academic Publishers,
P.O. Box 17, 3300 AA Dordrecht, The Netherlands.

Sold and distributed in North, Central and South America
by Kluwer Academic Publishers,
101 Philip Drive, Norwell, MA 02061, U.S.A.

In all other countries, sold and distributed
by Kluwer Academic Publishers,
P.O. Box 322, 3300 AH Dordrecht, The Netherlands.

Printed on acid-free paper

Printed in the Netherlands

CONTENTS

PREFACE

Industries rely more and more on advanced technology. Accelerated computer evolution makes large-scale computation practical. Many enterprises are beginning to benefit from more efficient allocation of resources and more effective planning, scheduling, manufacturing, and distribution by adopting state-of-the-art decision support systems. Academics increasingly emphasize application-driven research. All these forces have moved optimization from a pure classroom and textbook terminology to an accepted tool in today's business world.

This book chronicles and describes applications of combinatorial optimization in industry. A wide range of applications is included:

- manpower planning
- production planning
- job sequencing and scheduling
- manufacturing layout design
- facility planning
- vehicle scheduling and routing
- retail seasonal planning
- space shuttle scheduling, and
- telecommunication network design.

The applications covered in this book comprise a representative set of industry sectors including electronics, airlines, manufacturing, tobacco, retail, telecommunication, defense, and livestock. These examples should encourage operations researchers and applied mathematicians by pointing out how the importance and practicality of optimization is starting to be realized by the management of various organizations and how some pioneering developments in this field are beginning to bear fruit.

Since the birth nearly three decades ago of complexity theory which classified a majority of combinatorial optimization problems as NP-complete–a class of intractable problems–researchers and practitioners have never given up their efforts in exploring ways to tackle large-scale, complex, real-world combinatorial problems. Significant progress has been made in the areas of polyhedral theory and bound generations. More importantly, various effective heuristics such as simulated annealing, tabu search, GRASP, genetic algorithms, neural networks, etc. have been developed. These novel approaches, together with rapidly increasing computational speed, enable researchers and practitioners to solve real-world combinatorial optimization problem to an acceptable degree. This book demonstrates the results of such endeavors.

To put the future into perspective, however, not enough emphasis is placed on the integration and timeliness of systems used to support planning and operational tasks. Supply chain management promoting enterprise-wide optimization is a big step in this direction. Integrated, real-time decision support systems based on optimization technology promise of a whole new level of advancements. Real-time management and operations control poses its own challenges, because of the large scale of operations, the complexities in balancing resource utilization and regulatory concerns, the dynamic industrial operational environment, and the desire to provide high-quality, reliable, and consistent service to customers. The key to attaining smooth operational control, capturing a stable and profitable share of the competitive market, and meeting customers' rising demands and expectations is the deployment of advanced decision support systems which efficiently manage scarce and valuable resources.

Currently, most industries store data and develop systems through fragmented architecture. The systems don't communicate seamlessly with each other. Complete real-time data are lacking. Multiple data instances reside in different systems, leading to pitfalls in data integrity and synchronization. Graphical user interfaces don't have the same look and feel across the systems. Lastly, many managers still rely on computer printouts and manually- generated charts to plan and make daily operational decisions.

More important, even though some companies deploy isolated decision support information technology systems, these systems generate sub-optimal, localized, and uncoordinated solutions. For example, the system for generating production planning traditionally does not take into account the difficulty in compiling corresponding staffing assignments. The systems for generating production schedules and inventory policies tend to disregard the complexity of their consequent distribution issues. The system for manpower planning doesn't consider training scheduling and training resource use. The systems for planning decisions don't offer robust solutions needed for speedy recovery when there are disruptions. In all, this lack of integration among decision support systems leads to inferior responsiveness to change and imposes costs throughout the enterprise.

In addition to integration of systems, timeliness is another critical metric of decision effectiveness overlooked by many companies. Few companies have implemented or are even in the process of implementing real-time decision support technology to control daily operations. There is no doubt that the quality of a solution is important to revenue, cost, performance, and customer satisfaction. But the window of opportunity is extremely narrow for deploying the needed resources to implement real-time decisions. Decision support system solutions must be generated quickly to avoid complications associated with environment changes, resource availability modifications, and the introduction of new problems.

The rewards from applying integrated, real-time decision systems can be high. Several issues need to be noted.

- The integrated decision support framework must accommodate a wide range of scenarios and components.
- The focus needs to be on solution efficiency in order to achieve real-time advantages.
- The decision support system must provide for resource locking for multiple users responsible for the same set of resources.
- The system must be able to offer multiple solutions, allowing users to make choices in handling soft issues that are difficult to embed in the optimization model.
- There must be the capability to explore "what-ifs" which will facilitate interaction among decision makers and anticipate future problems.
- The new system must interface with information technology systems being phased out, with network communication, and with data integrity.
- The system's design and the graphic display must assist in user learning and acceptance.

It is my hope that these issues will be adequately addressed in future research and development. I am also looking forward to seeing future volumes of a similar theme that cover success stories on optimization applications in industry.

Gang Yu
The University of Texas at Austin
Austin, Texas, USA

ACKNOWLEDGMENTS

Those who contributed to this book deserve my highest respect and appreciation. Without their quality submissions, dedicated effort in making revisions, and patience while waiting for the birth of the book, publication of this volume would have been impossible. I owe a great deal to the anonymous referees who gave the reviewing tasks high priority and returned their reports faster than most of the journals I have submitted to. Working with the editor, John Martindale, whose positive attitude and accommodating manner relieved the pressure from my shoulder, has been a very pleasant process. My appreciation also goes to Caryn Cluiss, Jian Yang, and Xiaowei Wang who helped me with the tedious and time-consuming process of editing, formatting, and proofreading.

Having assumed too many responsibilities during the time this book was to be compiled, the completion was stretched beyond the targeted completion date. I am fortunate to have had my family's complete support and understanding, and I am especially indebted to my wife, Xiaomei Song. This book will leave a memorable mark on my life, since it began with the birth of my son and ended with the birth of my daughter.

Gang Yu
Editor
January 1998

CONTRIBUTORS

Michael F. Argüello
Center for Management of Operations and Logistics
The University of Texas at Austin
and CALEB Technologies Corp.

Dawei Bai
Department of Civil Engineering and Operations Research
Princeton University

Jonathan F. Bard
Department of Mechanical Engineering
The University of Texas at Austin

George A. Bohoris
School of Industrial Management
University of Pireaus
Pireaus, Greece

Shaukat A. Brah
Department of Decision Sciences
National University of Singapore
Singapore

Soo Y. Chang
Department of Industrial Engineering
Pohang University of Science and Technology
Kkyungbook, South Korea

Rahul Dé
Department of Business, Public Administration, and Economics
Bowie State University

Geon Cho
Department of Business Administration
Chonnan National University
Kwang Ju, Korea

Andreas Drexl
Institut für Betriebswirtschaftslehre
Christian-Albrechts-Universität zu Kiel
Kiel, Germany

Moshe Dror
Management Information Systems Department
University of Arizona

Stacy Dugan
CALEB Technologies Corp.

Fred Glover
The University of Colorado at Boulder

John L. Hunsucker
Department of Industrial Engineering
University of Houston

Hark-Chin Hwang
Department of Industrial Engineering
Pohang University of Science and Technology
Kkyungbook, South Korea

Alf Kimms
Institut für Betriebswirtschaftslehre
Christian-Albrechts-Universität zu Kiel
Kiel, Germany

Janny M. Y. Leung
Management Information Systems Department
University of Arizona

John Mulvey
Department of Civil Engineering and Operations Research
Princeton University

Katta G. Murty
University of Michigan

Miles G. Nicholls
School of Information Systems
Swinburne University of Technology
Victoria, Australia

George G. Polak
Department of Management Science and Information Systems
Wright State University

Edward C. Sewell
Department of Mathematics and Statistics
Southern Illinois University

Dong X. Shaw
School of Industrial Engineering
Purdue University

Stephen A. Smith
Santa Clara University

Yuri N. Sotskov
Institute of Engineering Cybernetics
Minsk, Belarus

J. C. Tan
Department of Systems
University of Pennsylvania

Vyacheslav S. Tanaev
Institute of Engineering Cybernetics
Minsk, Belarus

Michael T. Tapia
Department of Management Science and Information Systems
The University of Texas at Austin

Jane M. Thomas
Leicestershire, United Kingdom

Frank Werner
Otto-von-Guericke Universität Magdeburg
Facultät für Mathematik
Magdeburg, Germany

Gang Yu
Department of Management Science and Information Systems
and Center for Management of Operations and Logistics
The University of Texas at Austin

1

MOVING TOWARD AN INTEGRATED DECISION SUPPORT SYSTEM FOR MANPOWER PLANNING AT CONTINENTAL AIRLINES: OPTIMIZATION OF PILOT TRAINING ASSIGNMENTS

Gang Yu, Stacy Dugan,* and Mike Argüello**

Department of Management Science and Information Systems and
Center for Management of Operations and Logistics
The University of Texas at Austin
Austin, TX 78712
USA
**CALEB Technologies Corp.*
Kaleido II, Suite 420
9390 Research Blvd.
Austin, TX 78759
USA
***CALEB Technologies Corp.*
Kaleido II, Suite 420
9390 Research Blvd.
Austin, TX 78759
and
Center for Management of Operations and Logistics
The University of Texas at Austin
Austin, TX 78712
USA

ABSTRACT

Manpower planning and deployment is one of the most important and challenging issues faced by management in today's industry. This is particularly true in technology-

G. Yu (ed.), Industrial Applications of Combinatorial Optimization, 1–24.

driven industries such as the airline industry. Optimal long-term planning and daily management of crew resources are critical to the operational success of airlines. Due to the complexities associated with generating staffing plans which comply with business requirements and the multitude of data processing necessary to determine compliance with revenue-enhancing business opportunities, the application of an integrated decision support system is paramount to attain optimal operational and financial performance. In this paper, we discuss an integrated decision support system for manpower planning, and in particular, we layout a mathematical framework for a pilot training assignment module. A prototype based on heuristic solutions of this model has demonstrated over $6 million in annual savings for Continental Airlines due to better pilot training assignments alone. Based on a conservative estimate, an integrated DSS will save Continental over $10 million annually.

1 INTRODUCTION

Manpower planning and deployment is one of the most important and challenging issues faced by management in today's industry. Many chief executives in corporate America consider human resources to be the most valuable asset in their enterprise. This is particularly true in technology-driven industries such as airlines, semiconductors, electronics, and telecommunications.

An airline's product is measured by its timeliness, accuracy, functionality, quality, and price. For the air transportation customer, these criteria translate into flexible schedules, on-time flights, safety, satisfactory in-flight services, proper baggage handling, and convenient ticket purchases. Crew, aircraft, airport facilities, and maintenance facilities are the must-have components to provide the air transportation product. Among all the cost items, crew cost is the second largest expense in airline operations, fuel being the largest.

When conducting manpower planning, airlines must (1) accurately forecast staffing needs based on future business plan requirements, (2) identify personnel to fill these needs, and (3) schedule training. Due to the enormous data involved and the inter-dependencies in it, manpower planning is a time consuming and costly process when done manually. The number of people required to staff each position and the dates by which these positions need to be filled must be determined initially. Next, a hiring and training schedule must be generated to fill these open positions. Although new hires may be utilized, current employees are frequently advanced from within the organization to fill the required positions. Once these employees enter training, they are often unavailable for work until their training is complete. Thus, the business plan requirements and current staffing levels at each position must be explicitly considered when assigning employees to training.

An integrated manpower planning decision support system development project undertaken by the authors of this paper is in the process of being evaluated and implemented at Continental Airlines. This system contains pilot and flight attendant fleet optimization, monthly planning optimization, training optimization, staffing administration, training administration, vacation administration, and seniority administration. The training assignment problem discussed in this paper is a part of the fleet optimization module.

The training assignment problem requires the generation of a training schedule which meets the business plan set forth by the organization. Inputs consist of the identity and current and future assignments of each person moving to a new position, the training requirements for each person assigned to a new position, the dates by which each person must be available at their new position, the current staffing levels at each position, and the business plan requirements for each position. The objective function of this problem may vary according to the availability of the attributes that make one training schedule more attractive than another. For example, this function could be to maximize business plan compliance, maximize business rule compliance, minimize overall cost, minimize training cycles, or plan for the most efficient use of training resources (more discussion of objectives ensues in later sections). In the airline industry, pilots are one of the most expensive human resources. When new pilots are hired and/or existing pilots are moved from one position to another (due to advancement, crew base change, fleet change, etc.), extensive training is required.

Effective manpower planning is a key element for success in the highly competitive airline industry. Due to the dynamic nature of the business, airlines are constantly evaluating which markets to serve, the number of fleets and the number of corresponding aircraft to operate, the amount of time each fleet will be utilized, the number of pilots required to execute the flight schedule, and the locations where the pilots will be based. Several organizational departments such as marketing, finance, and flight scheduling cooperate in determining the business plan for the airline by setting the number of flight hours required to be staffed at each base for each fleet. Given the business plan for a specified period of time, the manpower planning managers must create a staffing plan to meet the business requirements. Projected staffing deficiencies at known positions may be opened up for bid by current pilots. The number of new hires required may also be calculated. After the identification of which pilots will be transferred to new positions and how many new hires will be implemented, the manpower planning managers are faced with the training assignment problem.

Currently, many airlines make their manpower planning decisions either manually or using fragmented systems. These systems do not communicate seamlessly with each other. Complete and up-to-date data is lacking. Multiple data instances reside in different systems which lead to pitfalls in data integrity and synchronization. Graphical user interfaces do not have the same look and feel across systems. Many managers still rely on spreadsheets and other data charts and tables to make human resource allocation and deployment decisions. More importantly, even though some companies deploy isolated decision support systems, these systems generate sub-optimal, localized, uncoordinated solutions. For example, the system for manpower hiring decisions does not consider subsequent training scheduling and training resource usage. The lack of integration among decision support systems leads to inferior solutions in terms of enterprise-wide cost and responsiveness to change.

Due to the enormous amount of data involved, long-term, global planning is difficult or even impossible to do manually. As a result, frequent short-term planning is often inevitable for managing the staffing levels and training schedules required to meet the business plan. Consequently, increased training costs and missed opportunity costs, as well as increased new hire requirements, are the reality for many airlines. Pilots are sent to training as needed in the short-term, instead of in groups to best utilize training resources and reduce the number of training cycles. Missed opportunity costs result from the inability to quickly and efficiently evaluate revenue enhancing opportunities such as the chance to purchase new aircraft, extend the use of older aircraft, or increase service at a new market. These opportunities require the evaluation of manpower feasibility. New hires provide a simple source for increasing capability at a position, without affecting the capability levels at alternate positions. For this reason, an increased use of new hires may result from short-term planning practices. An optimal training assignment schedule will smooth the training and staffing levels in order to increase compliance with the current business plan and reduce the number of new hires required to meet the business plan at each position.

Continental Airlines (including Continental Express and Continental Micronesia) currently serves 196 airports, has over 2,000 daily departures, operates 445 aircraft, and employs 38,700 people. Such a large operational scope, together with various crew legality rules enforced by the Federal Aviation Administration (FAA), union contracts, crew qualifications, and crew seniority policies, makes manpower planning a complex task.

The application of optimization models to manpower planning dates back to the 40's, and a large fraction of the earlier published work reviews military applications. Some of the models can be found in [9]. A significant amount of recent research is on personnel task assignment and scheduling. For example, the shift-scheduling problem is one that involves determining the number of people to be assigned to each specified shift and specifying the timing of their relief breaks. This problem has applications in a variety of service organizations such as telephone companies, airlines, hospitals, police departments, and banks. Research conducted in this area can be found in Segal [12], Keith [10], Glover et al. [8], Taylor and Huxley [13], Aykin [1], and Brusco and Jacobs [4]. Workforce scheduling also has been a much-researched topic. This problem can be generically described as seeking a workforce schedule that minimizes workforce size while satisfying staffing requirements in each shift and some work rules. Representative literature includes Baker [2], Brownell and Lowerre [3], Burns and Carter [5], and Burns and Koop [6]. Manpower planning research in a manufacturing environment can be found in Faaland and Schmitt [7], Treleven and Elvers [14], and Park and Bobrowski [11]. More relevant work is by Yu et al. [15] on Navy manpower planning and training scheduling. In this research, a network-based integer programming model is used to optimize the Navy's manpower and resource usage and to minimize the cost in the delayed-entry program and awaiting instructions.

This paper describes a mathematical programming model and a heuristic solution method designed to solve the pilot training assignment problem. This system was deployed at Continental Airlines as a PC-based application with a PowerBuilder interface. Two versions of this heuristic are described, the Limits version and the No Limits version.

The layout of the paper is as follows. Section 2 presents the mathematical model for the training assignment model. Section 3 describes the heuristic solution method developed. Section 4 presents the computational results based on data from Continental's April 1997–1998 Pilot System Bid. Conclusions and directions for future work are discussed in Section 5.

2 THE OPTIMIZATION MODEL

For an airline, the pilot training assignment problem can be stated as the process of generating a training schedule for pilots that meets the planned block hours (flying time required to accomplish the published flight schedule) set forth for each fleet at each base. A pilot's position is identified by sub-base, fleet, and seat. For example, a pilot with position IAH/727/CA implies that the pilot is a captain for a Boeing 727 and currently resides at the Houston crew base.

The training assignment problem requires the assignment of pilots to training in such a manner that the block hours established for each fleet and base are covered by qualified pilots. For example, an MD80 fleet first officer located in Newark may have been awarded a DC9 captain position in Houston. This pilot must be assigned to DC9 captain training in time to comply with the flight hours established for the DC9 fleet in Houston while not adversely affecting the ability of the MD80 first officer group in Newark to meet its flight hours. Pilots remain at their current position until their training is to begin. Once training begins, they are unavailable to fly until their training is complete and they have been advanced to their new position. Similarly, new hires become available only after their training is complete. Pilots are also assigned a due date with their new position that stipulates the latest date by which they may enter training without subjecting the airlines to variable monetary penalties. Thus, the objective of the pilot training assignment problem is to assign all pilots to training before their due dates while maximizing the compliance with the business plan (the stipulated block hours for each base-fleet).

Every training schedule generated must be feasible with respect to the following set of constraints:

- training course capacity - each class offered is restricted to a maximum number of students;
- fixed course duration - each course has a fixed duration which differs by the course type;
- maximum course frequency - the interval of two courses of the same type should be greater than a specified value (courses of the same type may overlap, but their starting dates must differ by the specified value);
- prevention of training assignments during known absences - the scheduled training time for a pilot should not overlap with the pilot's known absences such as vacations, military training, and company business (the known absences are inputs to the model);
- avoidance of the use of new hires for captain positions; and
- mandatory retirement or demotion to a second officer position for captains and first officers at age sixty.

Furthermore, there exists a set of soft constraints that reveals valuable attributes of a training schedule: (1) the assignment of pilots to training in proper seniority order–increasing for advancing moves and decreasing for lateral and downward moves and (2) assignment of pilots to training courses in captain/first-officer and second-officer/second-officer pairs. The second requirement is due to the fact that simulator training mimics the practical setting of a cockpit with two pilots in operation. If a pilot is not paired, an extra instructor is used to fill in, which incurs much higher cost. Additionally, such an instructor may not always be available.

The following are possible objective functions:

- Maximizing business plan compliance. The daily pilot capability at each base and for each fleet and seat should meet the business demand for as many weeks as possible.
- Minimizing training cycles. This will in turn reduce the pilot unavailability to service due to training.
- Minimizing overall cost. This includes: cost for training resource usage (instructors, facilities, devices, etc.), cost for missing training deadlines, cost for assigning pilots to training in improper seniority order, cost for not assigning pilots in pairs to classes, etc.

The training assignment model presented below captures the details of the training assignment problem in a concise and intuitive manner. The time horizon has been partitioned into weekly blocks for this model.

The following notation is used in the formulation.

Parameters

I = set of pilots who need to be trained, indexed by i

J = set of course types, indexed by j

K = set of weeks in the planning horizon, indexed by k

L = set of all possible positions described by (Base, Fleet, Seat), indexed by l

(a_i, b_i) = pilot i's beginning and ending positions

T_i = pilot i's course type

T_l = new hire for position l's course type

d_{lk} = demand of position l at week k, which can be obtained through preprocessing by using the required block hours transformed into needed pilots in each position and subtracting retirees and pilots whose known absences fall into the week

c_j = capacity for course type j

w_j = cost for offering a course of type j including expenses for instructors, equipment, facilities, etc.

q_l = cost for hiring a new pilot for position l

U_i = set of unavailable weeks for pilot i due to known absences such as vacation, planned military leaves, etc.

m_j = a parameter of type j indicating the minimum interval between courses

n_j = a parameter indicating the duration of the corresponding course of type j

M = a big number used to bound the total number of new hires

Variables

x_{ik} = 1 if pilot i is assigned to a class beginning at week k; 0 otherwise

y_{jk} = 1 if a class of type j is scheduled to begin at week k; 0 otherwise

z_{lk} = number of pilots at position l available for regular operation for week k

v_{lk} = number of new hires for position l available for training beginning at week k

Model

$$\text{(MP)} \qquad \min \sum_{k \in K} \sum_{j \in J} w_j y_{jk} + \sum_{k \in K} \sum_{l \in L} q_l v_{lk}$$

Subject to

$$\sum_{k \in K \setminus U_i} x_{ik} = 1 \qquad i \in I \tag{1.1}$$

$$\sum_{k \in U_i} x_{ik} = 0 \qquad i \in I \tag{1.2}$$

$$x_{ik} \leq y_{jk} \qquad j = T_i; i \in I; k \in K \tag{1.3}$$

$$v_{lk} \leq M y_{jk} \qquad j = T_l; l \in L; k \in K \tag{1.4}$$

$$\sum_{k'=k}^{k+m_j} y_{jk'} \leq 1 \qquad j \in J; k \in K \tag{1.5}$$

$$z_{lk} = z_{l,k-1} + \sum_{j:j=T_l} v_{l,k-n_j} + \sum_{\substack{i:b_i=l \\ j=T_i}} x_{i,k-n_j} - \sum_{i:a_i=l} x_{ik}$$

$$l \in L; k \in K \tag{1.6}$$

$$z_{lk} \geq d_{lk} \qquad l \in L; k \in K \tag{1.7}$$

$$\sum_{i:T_i=j} x_{ik} + v_{lk} \leq c_j \qquad j = T_l; j \in J \tag{1.8}$$

$$x_{ik} = \{0,1\} \qquad i \in I; k \in K \tag{1.9}$$

$$y_{jk} = \{0,1\} \qquad j \in J; k \in K \tag{1.10}$$

$$z_{lk} = \text{nonnegative integer} \qquad l \in L; k \in K \tag{1.11}$$

$$v_{lk} = \text{nonnegative integer} \qquad l \in L; k \in K \tag{1.12}$$

The set of pilots that need to be trained is specified in advance and serves as input to the model (MP). Each pilot has a current and a future position defined by the triplet (Base, Fleet, Seat), a course type that specifies the training required to move the pilot to the new position, and a set of known absences during which no training can be assigned. Each course has an associated capacity, cost, frequency, and duration. The number of weeks in the time horizon and the set of positions, each with their associated demand, serve as additional inputs to the model. Initial preprocessing of the demand values is required in order to take into account retirees that are leaving on their sixtieth birthday and pilot planned absences. New hires may also be created and assigned to training in this model. A new hire cost for each position is also included.

The model's objective function minimizes the total cost of classes scheduled and new hires employed. Constraints (1.1) and (1.2) ensure that every pilot will be trained and the training will occur only during his/her available weeks. Constraints (1.3) and (1.4) maintain that pilots and new hires will only be assigned to start training in week k when a corresponding course is scheduled to start in that week. Course frequencies are addressed in constraint (1.5). For each course type j, a parameter m_j specifies the minimum interval allowed between scheduled classes. The flow conservation constraint specifying the current staffing levels at each position in each week is enforced in constraint (1.6). Constraint (1.7) ensures that the business plan (demand) is met for each position during every week in the planning horizon. The course capacity constraint is enforced in constraint (1.8). Finally, constraints (1.9) through (1.12) are required to enforce the binary and integral restrictions on pilots, classes, and staffing levels.

The training assignment problem under consideration is accurately described using the mathematical model presented. In practice however, a feasible solution with respect to the business plan demand and course scheduling constraints often does not exist. There are two reasons for this. First, the set of pilots initially provided as input to the system was manually generated using localized data and short-term planning techniques. Thus, this input tends to more accurately reflect the business plan demand in the beginning of the time horizon, becoming infeasible toward the end of the time horizon. Second, when changes to the business plan are entered in order to perform what-if analysis, they may be infeasible with respect to training capacity and/or training time constraints. Thus, it may be impossible to train enough new hires and employees in time to meet the new demand. Continental Airlines was well aware of these issues, and they requested a solution schedule that maximized, rather than required, compliance with the business plan for each week in the time horizon at every position in the schedule. Finally, accurate and reliable cost data for the objective function did not exist at the time this system was developed. For these reasons, a heuristic approach was taken. The details of the heuristic solution approach to the training assignment problem are presented in the next section.

3 HEURISTIC METHOD FOR SOLVING THE MANPOWER PLANNING PROBLEM

The heuristic presented in this section uses an iterative method to schedule pilots for training. The driving factor in this method is demand. Demand at a position is defined to be the number of additional pilots above the current capability level needed to meet the business plan on a given date. A negative value indicates surplus capability. The time horizon has been discretized into weekly segments. For each week in the time horizon, the heuristic iterates through the list of unassigned pilots and, if all relevant constraints are satisfied, assigns them to training. Updates are performed each time the entire unassigned pilot list has been traversed in order to prepare the heuristic for the next week's iteration. The following pseudocode outlines the basic structure of the heuristic.

```
Algorithm Pilot Training Scheduling
Input data and parameters
Initialization/Pre-processing
For each WEEK in the time horizon
    Assign Pilots
    Perform Updates
End For
Output Solution
```

We now present the details of the subroutines in the Pilot Training Scheduling algorithm.

Input data and parameters

In this procedure, the position data, pilot data, retiree data, and course data from the database are input, including the following:

- set of pilots that need to be trained;
- specification of current and future positions for each pilot and corresponding course type in order to move the pilot to the new position;
- capacity, cost, frequency, and duration for each course;
- length of the planning horizon;
- all positions and their associated demand;
- known absences for each pilot;
- new hire cost for each position; and
- pilots' birthdate information.

Initialization/Pre-processing

The initialization/pre-processing step initializes the parameters used in the heuristic. Additional preprocessing is performed on two lists that are maintained throughout the algorithm. The Daily Change List is used to track the number of pilots entering and leaving each position each week. It is also used in the Perform Updates routine to calculate the current number of pilots available at the start of each week. This is initialized based on the list of retirees that will be leaving on their birth dates. The Demand List is used to track the number of additional pilots needed at each position in each week of the time horizon (i.e. position-week demand). The Demand List is used as a basis for sorting the list of unassigned pilots in the Assign Pilot routine. It is initialized from the weekly business plan requirements, the current number of pilots at each position, and the list of retirees that will be leaving on their birth dates.

Finally, a feasibility check is performed at each position for each week in the planning horizon. This check is based on the training constraints, the total number of pilots moving to each position, and the business plan demand at each position for each week in the time horizon. Infeasible position-weeks indicate that the business plan cannot be met at a position due to either: (1) there is a shortage in the number of pilots specified for advancement into that position, and/or (2) the training time and capacity constraints do not allow enough pilots to be trained in time to meet the increasing demand. Infeasible position-weeks are highlighted in the solution report.

Assign Pilots

This procedure assigns pilots to classes in captain/first-officer and second-officer/second-officer pairs, in proper seniority order, and prior to their due dates whenever possible. The pairing of pilots in classes utilizes training resources most efficiently. A violation of the proper seniority order or missing a due date results in variable penalty costs to the airline. Furthermore, pilots will never be assigned to training during a planned absence or prior to their earliest training date which becomes a factor only when a sixty-year-old captain or first officer is moving to a second officer position. The pseudocode for the Assign Pilots routine is shown below.

```
For Demand Week, W = current week to MAX_TIME_HORIZON
    Sort Pilots by Demand at assigned position for week W
    For each Unassigned Pilot
        If Check if Due Date Approaching
            If Check Pre-conditions
                Assign if Sufficient Capacity
        If Check if Destination Demand
            If Check Pre-conditions
                If Find Partner
                    Assign Pair if Sufficient Capacity
    End for each Unassigned Pilot
End for Demand
Assign without Demand Pull
```

The outer loop traverses the Demand Week used in the *Sort Pilots by Demand* routine to control the order in which pilots are considered for assignment to a class. An attempt to assign a pilot is made if the pilot's due date is approaching, based on a parameterized measure called the frequency_parameter. An attempt to assign a pair of pilots is made if there is a positive demand at the first pilot's destination position.

A pilot, or pair of pilots, is assigned to a class if all relevant constraints (pre-conditions) are met. If one or more constraints are not met, the first pilot is flagged as ineligible, and further attempts for assignment are not made until the next week's iteration.

Pilots may be assigned without partners if their due dates are approaching (based on a second parameterized measure called the duration_parameter), or if they are the last of an odd number of pilots required to take the course.

Upon completion of the Demand Week loop, an attempt to assign all pilots that have not been marked as ineligible is made if their due dates are approaching (based on the duration_parameter) or if their destination position has been marked in need of excess capability.

The following gives more details for each step.

Sort Pilots by Demand

The list of unassigned pilots is sorted by each pilot's destination position's demand, then by due date, and then by seniority/move value. Destination demand is the demand at the pilot's destination position on the week corresponding to the Demand Week. The Demand List stores each position's demand in each week of the time horizon and is updated as pilots are assigned to classes. For this reason, the list of unassigned pilots must be resorted upon each iteration of the Demand Week loop. The due date is the latest date a pilot may be assigned to a course without incurring a penalty cost. Seniority/move value sorting requires that the more senior pilots be assigned first if the assignment is an upward move; otherwise, they are assigned last. A final sort criteria places all ineligible pilots at the end of the list. Pilots are marked as ineligible for the given week if assigning them to a class violates one or more course frequency constraints, or if a valid partner cannot be found for the pilot.

Check if Due Date Approaching

The frequency parameter is used to decide if a pilot's due date is approaching. The following equation is used to determine if the due date is nearing:

IF

[(pilot's due date) - (frequency_parameter) * (pilot's required course frequency)] $\leq$ (current_date)

THEN

approaching due date

Because the revised objective of the optimization model is to maximize compliance with the business plan, pilots are not assigned to training if their current positions do not have sufficient capability without them to meet the business plan. However, during the development of this heuristic, manpower planning managers indicated that all pilots must begin training by their due dates. In order to satisfy this constraint as the pilots' due dates near, they are assigned to training without regard to their current positions' capabilities. Thus, the Check if Due Date Approaching routine flags pilots with approaching due dates in order to skip the excess capability check in the Check Pre-conditions routine performed next.

Check Pre-conditions

The pilot, course, and position constraints are checked here. The pilot constraints consist of earliest training dates and the set of planned absences. Pilots may not be assigned to class prior to their earliest training date or during a known absence. The legal start date of the pilot's required course is also checked here, based on the course frequency constraints.

Excess capability at a pilot's origin position is also checked here. A look_ahead parameter is used to determine how many weeks into the future to check for excess capability at a position. In most instances, an excess of one pilot is sufficient to allow a pilot to leave his/her position and enter training. New hires, retirees, and pilots marked with approaching due dates are not required to have excess capability at their origin position before leaving their current position.

Finally, for all weeks on or past the stop_date parameter, a check on excess capability is no longer performed. The stop_date parameter was created in order to guarantee the assignment of all pilots to training by the end of the time horizon. The input data received from the airline most accurately reflects the business plan demand in the beginning of the time horizon, but becomes infeasible toward the end of the time horizon. Capability deficiencies due to pilot movement in the last quarter of the planning horizon are normally addressed with future pilot system bids. Due to the infeasibility in this data set, the stop_date was essential to the completion of the training assignment schedule.

Assign (Pair) if Sufficient Capacity

The pair of pilots or singleton pilot will be officially assigned to a course if the class enrollment has not yet reached its maximum level. Upon official assignment, the pilot's destination position's Daily Change list is updated to reflect the pilot's expected arrival into the position, determined by the course duration. The origin and destination positions' Demand Lists are also updated to reflect the pilot's leaving the originating position on the current date (if not a retiree or new hire) and entering his destination position sometime in the future depending on the course duration. Course enrollment is also updated in this routine.

Check if Destination Demand

If there is positive demand at the pilot's destination position on the week indicated by the Demand Week, an attempt will be made to assign the pilot to training. Demand is stored in the Demand List and updated with each pilot assignment made. Furthermore, a position's demand may change throughout each week in the time horizon due to fluctuations in the business plan requirements. If there is a positive demand at the pilot's destination position at any time from the current week to the end of the time horizon, an attempt will be made in the current week to assign the pilot to training.

Find Partner

The Find Partner routine attempts to assign pilots into classes in pairs. The pairing of pilots into classes is a valuable attribute of a training assignment schedule, as it directly corresponds to the greater, more efficient utilization of training resources.

Given an initial pilot from the list of unassigned pilots with a positive destination demand, the Find Partner routine traverses the list of unassigned pilots beginning from the point ahead of the initial pilot. If a pilot match is found, Check Pre-conditions is called on the second pilot. If the partner is valid, an attempt is made to assign the two together in Assign Pair if Sufficient Capacity, as described in the next section.

If no partner (eligible or otherwise) is found in the previous search, the beginning of the list will be traversed. If no pilot match is found throughout the entire list, the pilot is marked as a singleton and will be assigned alone. If a pilot match is found in the beginning of the list, the pilot will not be assigned alone, but will wait for his/her partner to become eligible to be assigned with him/her. An exception to this rule exists, allowing the pilot to be assigned alone, if the pilot's due date is approaching. In this routine a second parameter, the duration_parameter, is used to determine if the due date is approaching.

The value of this parameter may be set differently based on the duration of the pilot's required course. This value is added to the current date and compared to the pilot's due date. If it is greater than or equal to the due date, then the latest training date is approaching.

IF
  [(current date) + (duration_parameter)] ≥ (pilot's due date)
THEN
  approaching due date

Assign without Demand Pull

An attempt is made to assign pairs of pilots with no demand at their destination position if their due date, based on the duration_parameter described in the Find Partner routine, is nearing or if moving them will help to create a needed excess at their destination position. A list of the positions in need of excess is created and stored on each iteration of the weekly loop. A position is in need of excess if a pilot tries to leave due to demand at his/her destination position, but is prevented due to insufficient excess capability at his/her current position.

Perform Updates

Various bookkeeping and updates are performed at the end of each weekly iteration in order to prepare the heuristic for the start of the next week. The legal start dates for all courses created in the current week are updated, based on the course frequency constraints. All pilots scheduled to finish training at the start of the next week will have the destination position's Daily Change List decremented from the week after next to the end of the time horizon to reflect that they are now included in the capability count at that position. For all retirees whose birthday falls in the next week, their Daily Change List at their originating position is incremented from the week after next to the end of the time horizon to reflect they have now left that position (no longer allowed to fly). The Daily Change List entry for the next week remains unaltered, as it will be used to calculate the position's pilot capability level at the start of that week. The Daily Change List simplifies this calculation as shown below:

(Capability at start of next week) = (Capability at end of current week + Daily Change List entry for the next week)

Once used in the above calculation, the Daily Change List entry is set to zero to reflect that all changes have been accounted for. In the No Limits version of this heuristic, new hire needs are also identified for all but captain positions and only in weeks far enough into the time horizon to take into account the time required to train the new hires. The number of new hires required at each position in the current week is equal to the position's positive demand. All new hires identified in the No Limits version are immediately included in the appropriate position's capability count.

Output Solution

The heuristic algorithm outputs all relevant pilot, new hire, position and course data. This includes pilot course assignments, infeasible position-weeks, class enrollment, class start dates and, when run in No Limits mode, the number of additional new hires identified at each position in each week. An exception report is also generated that identifies all pilots that have either missed their due dates or could not be assigned to a class.

4 COMPUTATIONAL RESULTS

The data set provided by Continental Airlines and used to validate the prototype spans a 52-week time horizon. Fifty-one sub-base/fleet/seat positions, made up of a combination of 8 fleets, 5 bases, and 3 seats (captain, first officer, and second officer), were considered. A total of 1,239 pilots required training, of which 1,015 were regular pilots. The remaining 224 were the new hires identified in advance by Continental. Fifty-two retiring pilots were also included in the data set. Each pilot required assignment to one out of a total of sixteen different course types for their required training. Of the sixteen course types, eight were long, eight-week courses with a minimum frequency of three weeks, and eight were four-week short courses with a minimum frequency of four weeks.

During the prototype's development, various combinations of parameters were evaluated. The best results had the look_ahead parameter set at ten weeks, the frequency_parameter set at two weeks, and the stop_week parameter set at week forty. The duration_parameter was set at 52 weeks which meant that the due date was always considered to be approaching in the *Find Partner* and *Assign Without Demand* routines.

The results of running the solver in Limits mode showed that the total number of position-weeks for which the capability did not meet demand was 36 prior to week nine, and 319 from week nine on. The total number of position-weeks for which the business plan was infeasible due to training and time constraints was 210. The total number of classes scheduled was 125. All pilots were assigned to training, and all pilots met their due dates.

The results of running the solver in No Limits mode showed that the total number of position-weeks for which the capability did not meet demand dropped from 319 to 159 from week nine on. Because the solver does not identify new hires in the first eight weeks, the count remained the same prior to week nine. New hires require an eight-week training cycle, and thus it would be infeasible to consider them for use until the ninth week. The number of scheduled courses was 116. As this only took into account regular pilots, no new hires were assigned to courses. Finally, all pilots were assigned to training, and all due dates were met.

A significant reduction was found in the total number of new hires identified by the No Limits version of the solver. The number of new hires dropped to 196, over a 12% reduction from the 224 identified by the manpower planners for input in the Limits version. Table 1 displays the monthly aggregates of the yearly total of new hires. It can be seen that the new hires are required later in the time horizon when the capability is less likely to handle the business plan. In many cases, the business plan changes before these weeks are encountered in the schedule. Because of this, the No Limits version of the solver puts off hiring new hires until as late in the time horizon as possible, reducing the costs associated with surplus capability. By comparing the Monthly Aggregates Tables of the Limits and No Limits solutions of Tables 1 and 2, the No Limits solution reduces the surplus capability by an average of 6,500 hours per month. Average block hours assigned to a pilot is approximately sixty per month. Conservatively estimating an average annual salary of $60,000 for a pilot, the total yearly savings exceeds $6 million.

In addition to reducing the number of new hires required to meet the business plan, the No Limits version is more flexible to meet the needs of a changing business plan and allows superior what-if analysis. Because the Limits version of the optimization model solver incorporates the static list of new hires provided by Continental Airlines and does not allow the assignment of any other new hires, it may not be appropriate for all what-if scenarios. In some cases the solver reports back an increased number of weeks for which the business plan cannot be accommodated, but provides few suggestions as to what is needed to comply with the business plan. For these cases, running the No Limits solver would offer more insight by identifying when, where, and how many new hires are required to meet the business plan.

Table 1: Monthly Aggregates Table – Original Information, No Limits Version

Month	Business Plan (block hours)	Capability (block hours)	New Hires
APR-97	222,693	254,339	0
MAY-97	229,706	250,131	22
JUN-97	236,241	254,516	4
JUL-97	233,643	258,201	0
AUG-97	234,282	260,731	3
SEP-97	236,911	263,005	0
OCT-97	237,828	265,145	1
NOV-97	239,251	266,181	0
DEC-97	233,706	265,444	0
JAN-98	234,837	270,310	126
FEB-98	237,450	271,917	28
MAR-98	243,669	275,064	12
TOTAL	2,820,220	3,154,986	196

Table 2: Monthly Aggregates Table – Original Information, Limits Version

Month	Business Plan (block hours)	Capability (block hours)	New Hires
APR-97	222,693	254,471	0
MAY-97	229,706	250,322	0
JUN-97	236,241	255,552	0
JUL-97	233,643	262,370	0
AUG-97	234,282	268,135	0
SEP-97	236,911	273,210	0
OCT-97	237,828	277,625	0
NOV-97	239,251	279,101	0
DEC-97	233,706	278,444	0
JAN-98	234,837	276,985	0
FEB-98	237,450	276,968	0
MAR-98	243,669	279,232	0
TOTAL	2,820,220	3,232,420	0

5 CONCLUSIONS AND FUTURE WORK

Due to the dynamic and highly competitive nature of the airline industry, effective manpower planning tools for scheduling pilots to training and efficiently managing training resources are becoming more and more critical for success. Currently, most airlines tackle manpower planning issues manually. Thus, they are unable to utilize a long-term, global perspective of the data when creating the pilot training schedules. Consequently, staffing levels are padded with surplus capability in order to handle fluctuations in the business plan, pilot unavailability due to training is extensive, training resources are underutilized, and pilot pay protection penalty costs are often inevitable. In addition, missed opportunity costs arise as potential revenue-enhancing modifications to the business plan cannot be evaluated quickly enough and with sufficient confidence to exploit. An optimal training assignment schedule assigns pilots to training, while meeting the business plan requirements and incorporating staffing capability levels, and efficiently manages training resources. Moreover, these solutions must be attained quickly in order to properly evaluate the impact of changes to the original plan.

This paper describes a heuristic solution method designed to solve the pilot training assignment problem. The training assignment problem assigns pilots to training in such a manner that the business plan requirements at each position are met throughout the planning horizon. The heuristic described in this paper was deployed at Continental Airlines as a PC-based application with a PowerBuilder interface. Two versions of this heuristic were evaluated, the Limits and No Limits versions. The former version is more flexible for meeting the needs of a changing business plan allowing for superior what-if analysis. In addition, this version of the solver was shown to significantly reduce the surplus capability levels and new hire requirements.

Based on the results of this heuristic, a full-scale, production level system is expected to drastically reduce administrative, training, and staffing costs. Moreover, the what-if capability of the system will serve to increase the airline's revenue by allowing the user to quickly and confidently evaluate proposed changes to the business plan. In order to build the full-scale production system model, detailed cost data must be added to allow for dollar-based comparisons of multiple scenarios. Additionally, the system will require live feeds from real-time systems that will act as the source for most of the data required to conduct scheduling and business plan adherence checking.

The pilot training assignment problem is only a piece of the entire manpower planning puzzle in the airline industry. To build an integrated decision support system for manpower planning, other modules are needed for monthly planning optimization, training resource scheduling and management, and vacation and seniority management, as well as various administration modules that provide and maintain data and relationships to the optimization modules.

Pilot Fleet Optimization, which contains the pilot training assignment discussed in this paper, produces base/fleet capability deployment plans which maximize compliance with the stated business plan. Pilot Training Optimization generates pilot training schedules that maximize training resource utilization while minimizing the time that pilots are unavailable due to training. These two modules are integrated to provide base/fleet capability deployment plans, pilot training schedules, and training resource schedules that collectively minimize the cost of assigning, training, and redeploying pilots to meet the business plans.

Flight Attendant Fleet Optimization and Flight Attendant Training Optimization are integrated to identify optimal base and language deployment requirements while minimizing the dependency on new hires. New hires are scheduled for initial training and assigned for duty as needed and with respect to the capacity of the training resources available. System users are able to conduct what-if analysis by modifying the base/fleet block hours and/or flight attendant complements as well as the training resource availability. The fleet optimization and training optimization modules for both pilots and flight attendants are most effective for intermediate- and long-term staffing and training planning. They determine the optimal staffing requirements for meeting the business plan beyond the next few months. In contrast, Flight Attendant Monthly Planning Optimization facilitates the management of flight attendants for compliance with next month's business plan. It processes transfer, leave, and jobshare requests as well as new hire deployments to generate the most cost-effective flight attendant capability at each base. It also constructs optimal vacation relief lines for which line holders may bid.

Administration modules are critical for providing the data visibility required to make sound staffing and training decisions as well as the functionality for keeping data valid and consistent. The staffing administration modules provide access to pilot and flight attendant personnel data. Updates to this data are made through these modules. These modules also present the known capability and rosters for all bases, fleets, and language types for any upcoming month requested. Additionally, all access to the optimization modules is through these administrative modules. All reports, graphs, charts, and maps corresponding to capabilities and block hour requirements are produced in these modules. It is through these modules that the user is able to manage personnel data, view assigned capabilities, determine how to effectively deploy staffing resources to meet next month's business plan, and evaluate the benefits of modifying the long-term business plan. Similarly, Pilot Training Administration provides access to pilot training resource data, allows updates to training resource availability, and facilitates the day-to-day management of the training resources. The seniority administration and vacation administration modules are responsible for the management of seniority number recalculation and vacation modification request processing, respectively. The seniority administration modules are important for maintaining the correct system and base seniority values for all pilots and flight attendants.

Such an integrated manpower planning system is currently under evaluation at Continental Airlines. Based on a conservative estimate, the annual savings that can be generated from such a system is over $10 million.

Acknowledgements

The authors would like to thank Anna White, Bryan Green, Robert Benton, Gary Wagner, and Al Wilson of Continental Airlines for offering us the opportunity to work on this project and for supporting us with insights, data, and information.

REFERENCES

[1] T. Aykin, 1996, Optimal Shift Scheduling with Multiple Break Windows, *Management Science*, 42, 591-602.

[2] K. Baker, 1974, Scheduling a Full-Time Workforce to Meet Cyclic Staffing Requirements, *Management Science*, 20, 1561-1568.

[3] W.S. Brownell and J.M. Lowerre, 1976, Scheduling of Work Forces Required in Continuous Operations under Alternative Labor Policies, *Management Science*, 22, 597-605.

[4] M.J. Brusco and L.W. Jacobs, 1993, A Simulated Annealing Approach to the Cyclic Staff-Scheduling Problem, *Naval Research Logistics Quarterly*, 40, 69-84.

[5] R.N. Burns and M.W. Carter, 1985, Workforce Size and Single Shift Schedules with Variable Demands, *Management Science*, 31, 599-607.

[6] R.N. Burns and G.J. Koop, 1987, A Modular Approach to Optimal Multiple-Shift Manpower Scheduling, *Operations Research*, 35, 100-110.

[7] B. Faaland and T. Schmitt, 1993, Cost-based Scheduling of Workers and Equipment in a Fabrication and Assembly Shop, *Operations Research*, 41, 253-268.

[8] F. Glover, C. McMillan, and R. Grover, 1984, A Heuristic Programming Approach to the Employee Scheduling Problem and Some Thoughts on 'Managerial Robots,' *Journal of Operations Management*, 4, 113-128.

[9] R.C. Grinold and K.T. Marshall, 1977, *Manpower Planning Models*, Elsevier North-Holland, Inc., New York, NY.

[10] E.G., Keith, 1979, Operator Scheduling, *AIIE Transactions*, 11, 37-41.

[11] P.S. Park and P.M. Bobrowski, 1989, Job Release and Labor Flexibility in a Dual Resource Constrained Job Shop, *Journal of Operations Management*, 8, 230-249.

[12] M. Segal, 1974, The Operator-Scheduling Problem: A Network Flow Approach, *Operations Research*, 22, 808-823.

[13] P.E. Taylor and J. Huxley, 1989, A Break from Tradition for the San Francisco Police: Patrol Officer Scheduling Using an Optimization based Decision Support System, *Interfaces* 19, 4-24.

[14] M.D. Treleven and D.A. Elvers, 1985, An Investigation of Labor Assignment Rules in Dual Constrained Job Shop, *Journal of Operations Management*, 6, 51-68.

[15] G. Yu, L. Zhou, P. Brockett, and S. Nielsen, 1996, A Network Model for the Navy's Awaiting Instruction Problem, Research Report, Center for Management of Operations and Logistics, The University of Texas at Austin.

2

SHORTCOMINGS OF MRP II AND A NEW PLANNING META–METHOD

Alf Kimms and Andreas Drexl

Lehrstuhl für Produktion und Logistik
Institut für Betriebswirtschaftslehre
Christian–Albrechts–Universität zu Kiel
Olshausenstr. 40
24118 Kiel
GERMANY

ABSTRACT

Lot sizing when done for the short–term heavily interacts with the sequencing decisions for the operations to be performed. Especially for real–world situations where capacities are scarce, demand is dynamic, and precedence relations among the operations have to be taken into account the MRP II logic which is implemented in most production planning systems does not satisfy. In this paper, we will reveal the shortcomings of MRP II by means of an example. A mixed–integer programming model is then defined to specify the problem of capacitated, dynamic, multi–level lot sizing and scheduling. Also, we present a generic solution method (a so–called meta–method) which may be used as a basis of more advanced implementations that may replace the traditional MRP II systems.

1 PROBLEM CONTEXT

Consider the organization of an in–house production system. Typically, the architecture of such a system is build up from several production cells, so–called segments, which may be implemented in different fashions (flow lines or work centers for instance). This macro–structure further refines into a micro–structure as each segment provides the capability to perform a bunch of operations.

G. Yu (ed.), Industrial Applications of Combinatorial Optimization, 25–43.

Raw materials and component parts are floating concurrently through this complex system in order to be processed and assembled until a final product comes out being ready for deliverance.

Production planning is one of the most challenging subjects for the management there. It appears to be a hierarchical process ranging from long– to medium– to short–term decisions (see for instance [6, 11, 31, 32, 33, 35]). Our focus will be the short–term scope which links to medium–term decisions via the master production schedule (MPS). The MPS defines the external (or independent) demand, i.e. due dates and order sizes for final products. The goal now is to find a feasible production plan which meets the requests and provides release dates and amounts for all products including component parts. For economical reasons, finding a feasible plan is not sufficient. In the usual case, production plans can be evaluated by means of an objective function (e.g. a function which measures the setup and the holding costs). Then, the aim is to find a feasible production plan with optimum (or close to optimum) objective function value.

2 PROBLEM OUTLINE

Let the manufacturing process be triggered by orders which originate from customers or from other facilities. Suppose now, that the output of the *make-to-order* system under concern is or at least includes a set of *non-customized products*. Certainly, this is a valid assumption for many firms no matter what industry they belong to and no matter of what size they are.

To motivate a planning activity, we first need to identify a subject of concern that is worth (in terms of economical rationale) to be considered. A first clue are large inventories. Due to the opportunity costs of capital and the direct costs of storing goods, holding items in inventory and thus causing holding costs should be avoided. On the other hand, if different parts are making use of common resources, say machines, and a setup action must take place to prepare proper operation, then opportunity costs (i.e. setup costs) are incurred since production is delayed. Another aspect of sharing resources is that the production of such parts cannot coincide if different setup states are required. Hence, orders must be sequenced. If production planning is about the timing of production and not about what to produce (e.g. make–or–buy decisions), then production costs need not to be considered as long as they are time invariant. In summary, we have a trade–off between low setup costs (favoring large production lots) and low holding costs (favoring a lot–for–lot–like production where sequence decisions have to be made due to sharing common resources). Essentially, the problem of short-term production planning turns out to be a *lot sizing and scheduling problem* then.

If we ask about how to solve this production planning problem, we first need a deeper understanding of its basic attributes. The first key element we have to remember is the stream of component parts floating through a complex production system. Operations may be executed only if parts being subject of these particular operations are indeed available. In other words, a production plan must respect the precedence relations of operations. Hence, *multi-level* structures must be taken into account. For the sake of convenience, we do not further distinguish between operations and items (also called products or parts). Each operation produces an item, and each item is the output of an operation. Both terms are used as synonyms. Apparently, we face a *multi-item* problem here.

The second key element of our problem is the presence of scarce capacity. As usual in in-house production systems, producing an item requires a certain amount of one or more resources (e.g. manpower, machine time, energy, ...) with limited capacity per time unit. Thus, production planning must take *scarce capacity* into account. Following the terminology in the literature, we also say machine instead of resource.

Furthermore, we have the following situation: The time interval which is the focus of the planning process is *finite* and subdivided into several *discrete time periods*. To refer to these periods we number them consecutively beginning with period 1. The length of the overall time interval is called the planning horizon, or horizon for short, and is counted in number of time periods. This properly reflects the real-world situation where we face a planning horizon of say four weeks (or 20 days or 40 shifts) and discrete time periods are naturally given. The (known or estimated) external demand (given by the MPS) is given in units per item per period. It is to be met promptly at the end of each period. Backlogging and shortages are not allowed here which enforces a high service level. The demand may vary from period to period. This is called *dynamic* demand. All relevant data for the planning process are assumed to be *deterministic* which is justified by having a short-term planning problem on hand.

3 CURRENT PRACTICE

The basic working principle of today's decision support systems for Manufacturing Resource Planning (MRP II) is more or less the same in all current implementations. It should be reviewed here by means of a small example (see also [8]). For an overview of more than a hundred modern MRP II software packages we refer to [1, 2] where a detailed list of features is enclosed.

Assume the following data: Three items $j = 1, \ldots, 3$ are to be produced sharing a single machine. The gozinto– structure[1] of these items is given in Figure 1. The planning horizon is 10 periods ($t = 1, \ldots, 10$) long and inventory is empty. Table 1[2] provides the MPS (with external demands d_{jt}), the capacity limit per period C_t, the need of capacity per item p_j, and the item–specific setup costs s_j and holding costs per period h_j. We assume that the minimum lead time is zero.

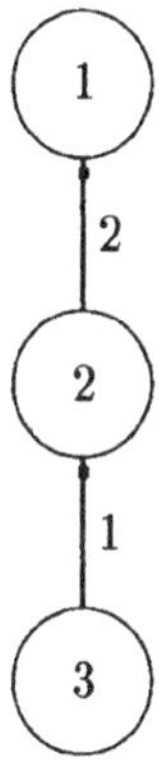

Figure 1 A Gozinto–Structure with Three Items

Starting with the MPS, lot sizes for all items are determined by a level–by–level approach disregarding capacity constraints. In our example, we start to

[1] The term gozinto–structure was coined by Vazsonyi [34] who gave a reference to an italian mathematician named Zepartzat Gozinto. This person is pure fiction. But, if you read his name in English pronunciation, it turns out what is meant: The product–structure. Nodes represent items, and arcs depict precedence relations between the items. Arc weights are production coefficients.

[2] Missing entries in tables are assumed to be zero throughout the text.

Table 1 Data of the Example

d_{jt}	$t=1$	...	6	7	8	9	10	h_j	s_j	p_j
$j=1$			20		20		20	25	900	1
$j=2$								10	850	1
$j=3$								10	800	1
C_t	100	...	100	100	100	100	100			

compute lot sizes for item 1 using some lot sizing rule. Let us say, we decide to use a lot–for–lot policy. Then, we derive internal demands and due dates for the next level which is item 2. As a result we have an internal demand for 40 units of item 2 in periods 6, 8 and 10. Again, we employ some lot sizing rule, but, this time considering item 2. Suppose, we decide to produce the demands of periods 8 and 10 in one lot in period 8. Following the same lines, we then compute internal demands for item 3 and again employ some lot sizing rule. As a result we might get what is shown in Table 2 where q_{jt} denotes the production quantity of item j in period t. Note, this is not a valid production plan since capacity restrictions are violated in period 8.

Table 2 A Production Plan with Capacity Constraint Violations

q_{jt}	$t=1$	2	3	4	5	6	7	8	9	10
$j=1$						20		20		20
$j=2$						40		80		
$j=3$						40		80		

In a next step, the intermediate result is modified to find a plan without requiring an excess of capacity. This is done by shifting lots to the left or to the right until the capacity profile is met. In our example for instance, this can be achieved by shifting the lot for item 2 in period 8 to the left (see Table 3).

But, because precedence constraints are not taken into account when lots are shifted in order to find a plan that does not violate the capacity restrictions, the conventional approach obviously fails. Looking at the example reveals, it would never be possible to produce item 2 in period 7 since we lack a sufficient amount of item 3 there.

Table 3 A Production Plan with Precedence Relation Violations

q_{jt}	$t=1$	2	3	4	5	6	7	8	9	10
$j=1$						20		20		20
$j=2$						40	80			
$j=3$						40		80		

The traditional way to overcome this is to introduce lead times. To make this idea clear, let us start with the MPS again (see Table 1). Going through the level–by–level approach as described above, we now use an offset of, say, two periods — the so–called lead time — when we compute the due dates of the internal demand. The outcome of the procedure is given in Table 4.

Table 4 A Production Plan with Positive Lead Times

q_{jt}	$t=1$	2	3	4	5	6	7	8	9	10
$j=1$						20		20		20
$j=2$				40		80				
$j=3$		40		80						

Note, this plan is infeasible again. Once more, we are off capacity limits. Shifting the lot of item 2 in period 4 to the left, leads us to the solution in Table 5.

Table 5 A Feasible Production Plan

q_{jt}	$t=1$	2	3	4	5	6	7	8	9	10
$j=1$						20		20		20
$j=2$			40			80				
$j=3$		40		80						

Now we are done. The result represents a feasible production plan. The sum of setup and holding costs is 9,800. Due to the introduction of positive lead times we had more flexibility when shifting lots which helped. And, indeed this is what can be observed in practice. Known as the lead–time–syndrome [36], planners tend to increase lead times arbitrarily whenever they detect backorders or high work-in-process inventories. But, as a result most firms suffer from long makespans and large total holding costs. To be convinced, compare the feasible plan in Table 5 with the optimum plan in Table 6. Figure 2 shows the schedule of the optimum solution by means of a Gantt-chart. The sum of setup and holding costs is 6,700.

Table 6 An Optimum Production Plan

q_{jt}	$t=1$	2	3	4	5	6	7	8	9	10
$j=1$						20		20		20
$j=2$						40		80		
$j=3$						40	80			

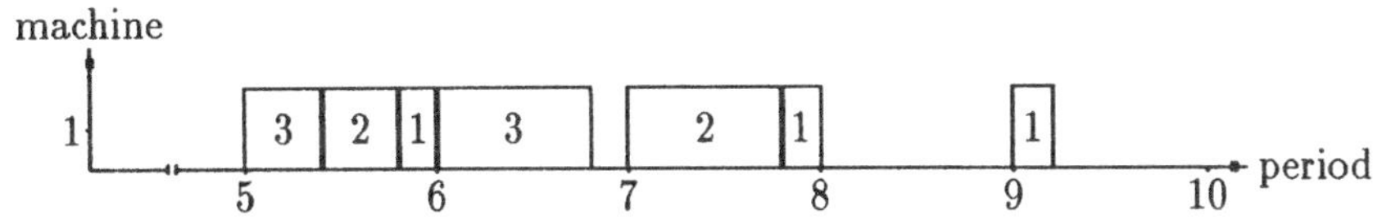

Figure 2 An Optimum Schedule

The reason for this dilemma in production planning apparently is the traditional level-by-level approach. This widely used method does not take capacity constraints and precedence relations simultaneously into account. Unfortunately, good alternatives are not available yet. This gives the motivation to develop such.

4 BASIC ASSUMPTIONS

Several items are to be produced in order to meet some known (or estimated) dynamic demand without backlogs and stockouts. Precedence relations among these items define an acyclic gozinto-structure of the general type. In contrast to many authors who allow demand for end items only, now, demand may occur for all items including component parts. The finite planning horizon is subdivided into a number of discrete time periods. Positive lead times are given due to technological restrictions such as cooling or transportation for instance.

Furthermore, items share common resources. Some (maybe all) of them are scarce. The capacities may vary over time. Producing one item requires an item–specific amount of the available capacity. All data are assumed to be deterministic.

Items which are produced in a period to meet some future demand must be stored in inventory and thus cause item–specific holding costs. Most authors assume that the holding costs for an item must be greater than or equal to the sum of the holding costs for all immediate predecessors. They argue that holding costs are mainly opportunity costs for capital which occurs no matter a component part is assembled or not. Two reasons persuade us to make no particular assumptions for holding costs. First, as it is usual in the chemical industry for instance, keeping some component parts in storage may require ongoing additional effort such as cooling, heating, or shaking. While these parts need no special treatment when processed, storing component parts might be more expensive than storing assembled items. Second, operations such as cutting tin mats for instance make parts smaller and often easier to handle. The remaining "waste" can often be sold as raw material for other manufacturing processes. Hence, opportunity costs may decrease when component parts are assembled. However, it should be made clear that the assumption of general holding costs is the most unrestrictive one. All models and methods developed under this assumption work for more restrictive cases as well.

Each item requires at least one resource for which a setup state has to be taken into account. Production can only take place if a proper state is set up. Setting a resource up for producing a particular item incurs item–specific setup costs which are assumed to be sequence independent. Setup times are not considered. Once a certain setup action is performed, the setup state is kept up until another setup changes the current state. Hence, same items which are produced having some idle time in–between do not enforce more than one setup action. To get things straight, note that some authors use the word changeover instead of setup in this context.

The most fundamental assumption here is that for each resource at most one setup may occur within one period. Hence, at most two items sharing a common resource for which a setup state exists may be produced per period. Due to this assumption, the problem is known as the proportional lot sizing and scheduling problem (PLSP) [7, 14, 25]. By choosing the length of each time period appropriately small, the PLSP is a good approximation to a continuous time axis. It refines the well–known discrete lot sizing and scheduling problem (DLSP) [5, 12, 17, 27, 30] as well as the continuous setup lot sizing problem (CSLP) [3, 19, 18]. Both assume that at most one item may be produced per period. All three models could be classified as small bucket models since only a few (one or two) items are produced per period. In contrast to this, the well–known capacitated lot sizing problem (CLSP) [4, 10, 16, 26, 28, 29] represents a large bucket model since many items can be produced per period. Remember, the CLSP does not include sequence decisions and is thus a much "easier" problem. An extension of the single–level CLSP with partial sequence decisions can be found in [13]. In [15] a large bucket single–level lot sizing and scheduling model is discussed. A review of different lot sizing and scheduling models can be found in [9].

5 A MIXED–INTEGER PROGRAMMING MODEL

An important variant of the PLSP is the one with multiple machines (PLSP–MM). Several resources (machines) are available and each item is produced on an item–specific machine. This is to say that there is an unambiguous mapping from items to machines. Of course, some items may share a common machine. Special cases are the single–machine problem for which models and methods are given in [23, 20, 24, 22], and the problem with dedicated machines where items do not share a common machine. For the latter optimal solutions can be easily computed with a lot–for–lot policy [21].

Let us first introduce some notation. In Table 7 the decision variables are defined. Likewise, the parameters are explained in Table 8. Note, from these data, we can easily derive the net requirement nr_j for an item j and the set $\mathcal{P}_j$ of all immediate predecessors of an item j. Furthermore, $\bar{\mathcal{P}}_j$ denotes the set of all predecessors of item j and id_{ji} is the internal demand for item i that is directly or indirectly caused by having a demand for one unit of item j. All these will later on be needed for describing a solution meta–method. Using this notation, we are now able to present a MIP–model formulation.

Table 7 Decision Variables for the PLSP–MM

Symbol	Definition
I_{jt}	Inventory for item j at the end of period t.
q_{jt}	Production quantity for item j in period t.
x_{jt}	Binary variable which indicates whether a setup for item j occurs in period t ($x_{jt} = 1$) or not ($x_{jt} = 0$).
y_{jt}	Binary variable which indicates whether machine m_j is set up for item j at the end of period t ($y_{jt} = 1$) or not ($y_{jt} = 0$).

$$\min \sum_{j=1}^{J} \sum_{t=1}^{T} (s_j x_{jt} + h_j I_{jt}) \tag{2.1}$$

subject to

$$I_{jt} = I_{j(t-1)} + q_{jt} - d_{jt} - \sum_{i \in \mathcal{S}_j} a_{ji} q_{it} \qquad \begin{array}{l} j = 1, \ldots, J \\ t = 1, \ldots, T \end{array} \tag{2.2}$$

$$I_{jt} \geq \sum_{i \in \mathcal{S}_j} \sum_{\tau = t+1}^{\min\{t+v_j, T\}} a_{ji} q_{i\tau} \qquad \begin{array}{l} j = 1, \ldots, J \\ t = 0, \ldots, T-1 \end{array} \tag{2.3}$$

$$\sum_{j \in \mathcal{J}_m} y_{jt} \leq 1 \qquad \begin{array}{l} m = 1, \ldots, M \\ t = 1, \ldots, T \end{array} \tag{2.4}$$

$$x_{jt} \geq y_{jt} - y_{j(t-1)} \qquad \begin{array}{l} j = 1, \ldots, J \\ t = 1, \ldots, T \end{array} \tag{2.5}$$

$$p_j q_{jt} \leq C_{m_j t}(y_{j(t-1)} + y_{jt}) \qquad \begin{array}{l} j = 1, \ldots, J \\ t = 1, \ldots, T \end{array} \tag{2.6}$$

$$\sum_{j \in \mathcal{J}_m} p_j q_{jt} \leq C_{mt} \qquad \begin{array}{l} m = 1, \ldots, M \\ t = 1, \ldots, T \end{array} \tag{2.7}$$

$$y_{jt} \in \{0, 1\} \qquad \begin{array}{l} j = 1, \ldots, J \\ t = 1, \ldots, T \end{array} \tag{2.8}$$

$$I_{jt}, q_{jt}, x_{jt} \geq 0 \qquad \begin{array}{l} j = 1, \ldots, J \\ t = 1, \ldots, T \end{array} \tag{2.9}$$

Table 8 Parameters for the PLSP–MM

Symbol	Definition
a_{ji}	"Gozinto"–factor. Its value is zero if item i is not an immediate successor of item j. Otherwise, it is the quantity of item j that is directly needed to produce one item i.
C_{mt}	Available capacity of machine m in period t.
d_{jt}	External demand for item j in period t.
h_j	Non–negative holding cost for having one unit of item j one period in inventory.
I_{j0}	Initial inventory for item j.
$\mathcal{J}_m$	Set of all items that share the machine m, i.e. $\mathcal{J}_m \stackrel{def}{=} \{j \in \{1, \ldots, J\} \mid m_j = m\}$.
J	Number of items.
M	Number of machines.
m_j	Machine on which item j is produced.
p_j	Capacity needs for producing one unit of item j.
s_j	Non–negative setup cost for item j.
$\mathcal{S}_j$	Set of immediate successors of item j, i.e. $\mathcal{S}_j \stackrel{def}{=} \{i \in \{1, \ldots, J\} \mid a_{ji} > 0\}$.
T	Number of periods.
v_j	Positive and integral lead time of item j.
y_{j0}	Unique initial setup state.

The objective (2.1) is to minimize the sum of setup and holding costs. Equations (2.2) are the inventory balances. At the end of a period t we have in inventory what was in there at the end of period $t-1$ plus what is produced minus external and internal demand. To fulfill internal demand we must respect positive lead times. Restrictions (2.3) guarantee so. Note, v_j represents a minimum lead time that is needed for transportation or cooling, for instance. If no minimum lead time is to be considered, one may choose $v_j = 1$. Note, $v_j = 0$ does not guarantee a feasible schedule. But, since we consider small time buckets, $v_j = 1$ does not harm. Larger lead times than the minimum lead time result from the planning process only which is in contrast the the practice of MRP II where large lead times are introduced arbitrarily. Constraints (2.4) make sure that the setup state of each machine is uniquely defined at the end of each period. Those periods in which a setup happens are spotted by

(2.5). Note that idle periods may occur in order to save setup costs. Due to (2.6) production can only take place if there is a proper setup state either at the beginning or at the end of a particular period. Hence, at most two items can be manufactured on each machine per period. Capacity constraints are formulated in (2.7). Since the right hand side is a constant, overtime is not available. (2.8) define the binary-valued setup state variables, while (2.9) are simple non-negativity conditions. The reader may convince himself that due to (2.5) in combination with (2.1) setup variables x_{jt} are indeed zero-one valued. Hence, non-negativity conditions are sufficient for these. For letting inventory variables I_{jt} be non-negative backlogging cannot occur.

6 A META-METHOD

There is a generic construction scheme that forms the basis of potential methods. It is a backward oriented procedure which schedules items period by period starting with period T and ending with period one. We choose here a recurrent representation which enables us to develop the underlying ideas in a stepwise fashion. Now, let us assume that *construct*$(t, \Delta t, m)$ is the procedure to be defined. Keep in mind that $t + \Delta t$ is the period and m is the machine under concern. We use $\Delta t \in \{0, 1\}$ where $\Delta t = 1$ indicates that the setup state for machine m at the beginning of period $t + 1$ is to be fixed next and $\Delta t = 0$ indicates that we already have chosen a setup state at the end of period t. In the specification below, j_{mt} will denote the setup state for machine m at the end of period t. Assume $j_{mt} = 0$ for $m = 1, \ldots, M$ and $t = 1, \ldots, T$ initially.

Before the construction mechanism starts, the decision variables y_{jt} and q_{jt} are assigned zero for $j = 1, \ldots, J$, $m = 1, \ldots, M$, and $t = 1, \ldots, T$. Remember, given the values for y_{jt} and q_{jt} the values for x_{jt} and I_{jt} are implicitly defined. Furthermore, assume auxiliary variables $\tilde{d}_{jt}$ and CD_{jt} for $j = 1, \ldots, J$ and $t = 1, \ldots, T$. The former ones represent the entries in the demand matrix and thus are initialized with $\tilde{d}_{jt} = d_{jt}$. The latter ones stand for the cumulative future demand for item j which is not been met yet. As we will see, the cumulative demand can be efficiently computed while moving on from period to period. For the sake of convenience we introduce $CD_{j(T+1)} = 0$ for $j = 1, \ldots, J$. The remaining capacity of machine m in period t is denoted as RC_{mt}. Initially, $RC_{mt} = C_{mt}$ for $m = 1, \ldots, M$ and $t = 1, \ldots, T$.

The initial call is *construct*$(T, 1, 1)$ and initiates the fixing of setup states at the end of period T. Table 9 gives all the details.

Table 9 Evaluating $construct(T, 1, \cdot)$

---

choose $j_{mT} \in \mathcal{I}_{mT}$.
if ($j_{mT} \neq 0$)
    $y_{j_{mT}T} := 1$.
if ($m = M$)
    $construct(T, 0, 1)$.
else
    $construct(T, 1, m + 1)$.

The choice of j_{mT} needs to be refined, but at this point we do not need any further insight and suppose that the selection is done somehow. All we need to know is that $\mathcal{I}_{mt} \subseteq \mathcal{J}_m \cup \{0\}$ for $m = 1, \ldots, M$ and $t = 1, \ldots, T$ is the set of items among which items are chosen. Item 0 is a dummy item. As one can see, once a setup state is chosen for all machines at the end of period T, a call of $construct(T, 0, 1)$ is made. Table 10 provides a recipe of how to evaluate such calls.

Table 10 Evaluating $construct(t, 0, \cdot)$ where $1 \leq t \leq T$

---

for $j \in \mathcal{J}_m$
    $CD_{jt} := \min\left\{CD_{j(t+1)} + \tilde{d}_{jt}, \max\{0, nr_j - \sum_{\tau=t+1}^{T} q_{j\tau}\}\right\}$.
if ($j_{mt} \neq 0$)
    $q_{j_{mt}t} := \min\left\{CD_{j_{mt}t}, \frac{RC_{mt}}{p_{j_{mt}}}\right\}$.
    $CD_{j_{mt}t} := CD_{j_{mt}t} - q_{j_{mt}t}$.
    $RC_{mt} := RC_{mt} - p_{j_{mt}} q_{j_{mt}t}$.
    for $i \in \mathcal{P}_{j_{mt}}$
        if ($t - v_i > 0$ and $q_{j_{mt}t} > 0$)
            $\tilde{d}_{i(t-v_i)} := \tilde{d}_{i(t-v_i)} + a_{ij_{mt}} q_{j_{mt}t}$.
if ($m = M$)
    $construct(t - 1, 1, 1)$.
else
    $construct(t, 0, m + 1)$.

The situation when calling *construct*$(t, 0, m)$ is that the setup state j_{mt} has already been chosen. Remarkable to note, how easy it is to take initial inventory into account. This is due to the backward oriented scheme. Evaluating

$$\min\left\{CD_{j(t+1)} + \tilde{d}_{jt}, \max\{0, nr_j - \sum_{\tau=t+1}^{T} q_{j\tau}\}\right\} \tag{2.10}$$

makes sure that for an item j no more than the net requirement nr_j is produced. Note, cumulating the production quantities is an easy task which can be done very efficiently. Given the cumulative demand $CD_{j_{mt}t}$, production quantities $q_{j_{mt}t}$ can be determined with respect to capacity constraints. Afterwards, we simply update the $\tilde{d}_{jt}$–matrix to take internal demand into account and proceed. Table 11 describes how to evaluate *construct*$(t, 1, \cdot)$–calls.

Table 11 Evaluating *construct*$(t, 1, \cdot)$ where $1 \leq t < T$

---

choose $j_{mt} \in \mathcal{I}_{mt}$.
if $(j_{mt} \neq 0)$
    $y_{j_{mt}t} := 1$.
    if $(j_{mt} \neq j_{m(t+1)})$
        $q_{j_{mt}(t+1)} := \min\left\{CD_{j_{mt}(t+1)}, \frac{RC_{m(t+1)}}{p_{j_{mt}}}\right\}$.
        $CD_{j_{mt}(t+1)} := CD_{j_{mt}(t+1)} - q_{j_{mt}(t+1)}$.
        $RC_{m(t+1)} := RC_{m(t+1)} - p_{j_{mt}} q_{j_{mt}(t+1)}$.
        for $i \in \mathcal{P}_{j_{mt}}$
            if $(t + 1 - v_i > 0$ and $q_{j_{mt}(t+1)} > 0)$
                $\tilde{d}_{i(t+1-v_i)} := \tilde{d}_{i(t+1-v_i)} + a_{ij_{mt}} q_{j_{mt}(t+1)}$.
if $(m = M)$
    construct$(t, 0, 1)$.
else
    construct$(t, 1, m + 1)$.

These lines closely relate to what is defined in Table 10. Differences lie in the fact that a setup state is chosen for the end of period t but items are scheduled in period $t + 1$. For computing production quantities we must therefore take into account that item $j_{m(t+1)}$ may already be scheduled in period $t + 1$.

Note, the combination of what is given in Tables 10 and 11 enforces that every item j_{mt} that is produced at the beginning of a period $t + 1$ is also produced at the end of period t if there is any positive cumulative demand left.

Turning back to the specification of the *construct*–procedure, it remains to explain what shall happen when the first period is reached. Table 12 describes how to schedule those items in period 1 for which the machines are initially set up for. In contrast to what is given in Table 11 the initial setup state is known and thus needs not to be chosen.

Table 12 Evaluating $construct(0, 1, \cdot)$

---

if $(j_{m0} \neq j_{m1})$

$$q_{j_{m0}1} := \min\left\{CD_{j_{m0}1}, \frac{RC_{m1}}{p_{j_{m0}}}\right\}.$$

$$CD_{j_{m0}1} := CD_{j_{m0}1} - q_{j_{m0}1}.$$

if $(m = M)$

$construct(0, 0, 1).$

else

$construct(0, 1, m + 1).$

A call to $construct(0, 0, \cdot)$ terminates the construction phase. What is left is a final feasibility test where

$$nr_j = \sum_{t=1}^{T} q_{jt} \tag{2.11}$$

must hold for $j = 1, \ldots, J$ for being a feasible solution. Eventually, the objective function value of a feasible solution can be determined.

If we have no initial inventory, we can also perform a capacity check testing

$$\sum_{j \in \mathcal{J}_m} \sum_{i \in \{\bar{\mathcal{P}}_j \cup \{j\}\} \cap \mathcal{J}_m} p_i id_{ji} CD_{j(t+\Delta t)} > \sum_{\tau=1}^{t+\Delta t} C_{m\tau} \tag{2.12}$$

which must be false for $m = 1, \ldots, M$ if period $t + \Delta t$ is under concern and thus, when true, indicates an infeasible solution.

The procedure is a meta–method only, since some details are left unspecified. The rule to select an item is, for instance, not discussed any further. For concrete implementations which are based on this meta–method, these details need to be defined. In [25] we find some meta–method instances which work well. All of them are sampling methods which run the construction phase several times and come up with the best solution that was found. Good suboptimal results, say within a 10% range from the optimum objective function value, are reported to be available after a few seconds when running on a modern personal computer.

7 CONCLUSION

We have pointed out that the underlying concept of modern MRP II systems is insufficient. Many things, such as long lead times and high work–in–process, practitioners complain about are not as fate would have it, but are inherent in modern production planning systems. A mathematical programming model is defined to specify the multi–level lot sizing and scheduling problem which is to be solved in order to avoid these shortcomings. Moreover, we have given a generic method that outlines efficient solution procedures to attack the planning problem. Future work should integrate concrete implementations of the meta–method into computer based planning tools and compare the results with those of existing MRP II systems.

Acknowledgements

This work was done with partial support from the DFG–project Dr 170/4–1.

REFERENCES

[1] American Production and Inventory Control Society Inc., 1995, MRP II Software/Vendor Directory, APICS — The Performance Advantage, 9, 38-48.

[2] American Production and Inventory Control Society Inc., 1995, MRP II Software/Vendor Directory Addendum, APICS — The Performance Advantage, 11, 54-56.

[3] G.R. Bitran and H. Matsuo, 1986, Approximation Formulations for the Single–Product Capacitated Lot Size Problem, *Operations Research*, 34, 63-74.

[4] M. Diaby, H.C. Bahl, M.H. Karwan, and S. Zionts, 1992, A Lagrangean Relaxation Approach for Very–Large–Scale Capacitated Lot–Sizing, *Management Science*, 38, 1329-1340.

[5] W. Dinkelbach, 1964, Zum Problem der Produktionsplanung in Ein – und Mehrproduktunternehmen, Würzburg, Physica, 2nd edition.

[6] A. Drexl, B. Fleischmann, H.O. Günther, H. Stadtler, and H. Tempelmeier, 1994, Konzeptionelle Grundlagen kapazitätsorientierter PPS–Systeme, *Zeitschrift für betriebswirtschaftliche Forschung*, 46, 1022-1045.

[7] A. Drexl and K. Haase, 1995, Proportional Lotsizing and Scheduling, *International Journal of Production Economics*, 40, 73-87.

[8] A. Drexl, K. Haase, and A. Kimms, 1995, Losgrössen– und Ablaufplanung in PPS–Systemen auf der Basis rando misierter Opportunitätskosten, *Zeitschrift für Betriebswirtschaft*, 65, 267-285.

[9] A. Drexl and A. Kimms, 1996, Lot Sizing and Scheduling: Survey and Extensions, *European Journal of Operational Research*, to appear.

[10] G.D. Eppen and R.K. Martin, 1987, Solving Multi–Item Capacitated Lot–Sizing Problems Using Variable Redefinition, *Operations Research*, 35, 832-848.

[11] B. Fleischmann, 1988, Operations–Research– Modelle und –Verfahren in der Produktionsplanung, *Zeitschrift für betriebswirtschaftliche Forschung*, 58, 347-372.

[12] B. Fleischmann, 1990, The Discrete Lot–Sizing and Scheduling Problem, *European Journal of Operational Research*, 44, 337-348.

[13] K. Haase, 1993, Capacitated Lot–Sizing with Linked Production Quantities of Adjacent Periods, Working Paper 334, University of Kiel.

[14] K. Haase, 1994, Lotsizing and Scheduling for Production Planning, Lecture Notes in Economics and Mathematical Systems, 408, Berlin, Springer.

[15] K. Haase and A. Kimms, 1996, Lot Sizing and Scheduling with Sequence Dependent Setup Costs and Times and Efficient Rescheduling Opportunities, Working Paper 393, University of Kiel.

[16] K.S. Hindi, 1996, Solving the CLSP by a Tabu Search Heuristic, *Journal of the Operational Research Society*, 47, 151-161.

[17] S. van Hoesel and A. Kolen, 1994, A Linear Description of the Discrete Lot–Sizing and Scheduling Problem, *European Journal of Operational Research*, 75, 342-353.

[18] U.S. Karmarkar, S. Kekre, and S. Kekre, 1987, The Deterministic Lotsizing Problem with Startup and Reservation Costs, *Operations Research*, 35, 389-398.

[19] U.S. Karmarkar and L. Schrage, 1985, The Deterministic Dynamic Product Cycling Problem, *Operations Research*, 33, 326-345.

[20] A. Kimms, 1993, A Cellular Automaton Based Heuristic for Multi–Level Lot Sizing and Scheduling, Working Paper 331, University of Kiel.

[21] A. Kimms, 1994, Optimal Multi–Level Lot Sizing and Scheduling with Dedicated Machines, Working Paper 351, University of Kiel.

[22] A. Kimms, 1994, Demand Shuffle — A Method for Multi–Level Proportional Lot Sizing and Scheduling, *Naval Research Logistics*, to appear.

[23] A. Kimms, 1996, Multi–Level, Single–Machine Lot Sizing and Scheduling (with Initial Inventory), *European Journal of Operational Research*, 89, 86–99.

[24] A. Kimms, 1996, Competitive Methods for Multi–Level Lot Sizing and Scheduling: Tabu Search and Randomized Regrets, *International Journal of Production Research*, 34, 2279–2298.

[25] A. Kimms, 1997, Multi–Level Lot Sizing and Scheduling — Methods for Capacitated, Dynamic, and Deterministic Models, Heidelberg, Physica.

[26] Ö. Kirca and M. Kökten, 1994, A New Heuristic Approach for the Multi–Item Dynamic Lot Sizing Problem, *European Journal of Operational Research*, 75, 332-341.

[27] L.S. Lasdon and R.C. Terjung, 1971, An Efficient Algorithm for Multi–Item Scheduling, *Operations Research*, 19, 946-969.

[28] V. Lotfi and W.H. Chen, 1991, An Optimal Algorithm for the Multi–Item Capacitated Production Planning Problem, *European Journal of Operational Research*, 52, 179-193.

[29] J. Maes and L.N. van Wassenhove, 1988, Multi–Item Single–Level Capacitated Dynamic Lot–Sizing Heuristics: A General Review, *Journal of the Operational Research Society*, 39, 991-1004.

[30] M. Salomon, L.G. Kroon, R. Kuik, and L.N. van Wassenhove, 1991, Some Extensions of the Discrete Lotsizing and Scheduling Problem, *Management Science*, 37, 801-812.

[31] V. Söhner and C. Schneeweiss, 1995, Hierarchically Integrated Lot Size Optimization, *European Journal of Operational Research*, 86, 73-90.

[32] H. Stadtler, 1988, Hierarchische Produktionsplanung bei losweiser Fertigung, Heidelberg, Physica.

[33] M. Switalski, 1989, Hierarchische Produktionsplanung - – Konzeption und Einsatzbereich, Heidelberg, Physica.

[34] A. Vazsonyi, 1958, Scientific Programming in Business and Industry, New York, Wiley.

[35] A. Villa, 1989, Decision Architectures for Production Planning in Multi–Stage Multi–Product Manufacturing Systems, *Annals of Operations Research*, 17, 51–68.

[36] G. Zäpfel and H. Missbauer, 1993, New Concepts for Production Planning and Control, *European Journal of Operational Research*, 67, 297–320.

3

THE DEVELOPMENT, IMPLEMENTATION AND USE OF A PACKAGE TO FACILITATE PLANNING AND PRODUCTION SCHEDULING IN A TOBACCO PROCESSING PLANT

Miles G. Nicholls

School of Information Systems
Swinburne University of Technology
P.O. Box 218, Hawthorn, Victoria, 3122
AUSTRALIA

ABSTRACT

In this paper the operations associated with making cigarettes (from prepared leaf to packing and warehousing) are modeled and a scheduling "package" developed to provide an addition to the MIS. This package interfaces with the relevant on-site databases. The model allows management to determine whether the forecast demands for the complete range of products for the next year are able to be produced, facilitates production scheduling on a monthly basis for up to a year ahead and allows a wide range of "what if" questions to be answered. The model encompasses all operations of the plant including the assignment of forecast product demand to appropriate making machines, production of the appropriate quantity of filters, assignment of the cigarettes produced to specific packers and then the determination as to whether the final product produced in the current period will be used to satisfy the current month's demand or demand in the future. It should be noted that the making and packing machines cannot make or pack all products, only specific subsets. The model is developed on a monthly basis and is then replicated for up to a year ahead, with the individual months joined by the closing stock constraints. The development of this model has allowed management for the first time to test the feasibility of annual forecast demand and then on a monthly basis gain an idea of how it might be scheduled given the likely available resources.

G. Yu (ed.), Industrial Applications of Combinatorial Optimization, 44–71.

1 INTRODUCTION

Management of the Philip Morris Ltd. tobacco processing plant at Moorabbin (in Victoria, Australia) had for some time expressed concern regarding their lack of an objective, relatively simple and user friendly way of obtaining "best practice" answers to questions such as:

(a) whether the forecast demand (on a monthly basis) as supplied from the forecasting package currently in use was able to be produced by the factory;

(b) how on an **"average" monthly basis** should production be scheduled (i.e., what product on what machine and in what quantity together with determining whether it would then be stored for future demand or used to satisfy current demand) so that total machine use was minimised;

(c) how should **daily** production be scheduled having already established the feasibility of meeting demand and ascertaining a macro (or target) monthly production schedule.

Additionally, an objective (nonperson dependent) mechanism was required that allowed the evaluation of various scenarios with respect to new equipment, products and other technological changes. Hitherto, these requirements were unmet and analysis was very approximate and unable to be done other than on a very rough one to two month basis. This exercise also required a considerable amount of time and was resource intensive. A mechanism was also required to assist in the determination of where bottle-necks (if any) were in the production process and where expansion (if any) should take place.

In the first instance, management were concerned only with (a), (b) and (c) above in the context of the Secondary Floor where the making and packing of cigarettes takes place. The Primary area where leaf preparation and blending is performed experienced no pressing difficulties compared to the Secondary Floor. Additionally, with respect to (c) above, this was considered a second phase to the project and would be attended to in the future.

The requirement to concentrate on (a) and (b) and leave (c) until the future was helpful from the technical point of view, since the modeling of the plant on a **daily** basis for a year ahead is not feasible. This is due to the size of the resultant model (around 90,000 constraints and 360,000 variables with 7,200 binary variables).

Thus, by initially determining annual demand feasibility and developing monthly production schedules using the monthly based model, one has then established "target" feasible production quantities which can then be inputs into a subsequent monthly or two week "rolling" model operating on a **daily** basis. These are much smaller and able to be modeled realistically. While this approach may not produce the global optimum it will produce feasible and at least "near optimal" results, a vast improvement over the process of "guesstimation" used in the past (which, as already indicated, was at best limited to dealing with a couple of months at a time).

Subsequent **daily scheduling/sequencing** as required in (c) can be approached in a similar way to that adopted by Claassen and van Beek [1]. The pressing need however, was to find a mechanism that would provide answers to (a) and (b).

Management also had some special requirements, namely:

(i) SAS/OR (1989) Version 6.10 was to be used as the "optimizer" and **basis** for data manipulation and report writing, since it was a known and reliable product that the company had already had experience with;

(ii) the computing platform was to be a PC running on a Pentium Chip (with a one gigabyte hard disk and 32 megabytes of RAM) as the mainframe was not available for use with this project;

(iii) the "package" finally developed must reflect the user's needs, be user-friendly and not require "external expertise" to set up and use (i.e., the selection of data, preparation of the "model" and the running and report generation must all be simply menu driven with full flexibility); and

(iv) the "package" must interface with existing data bases and the output from various proprietary packages already used by the company.

The model is required to be able to schedule the production of product in such a way as to minimize the total number of machines (making, filter and packing) required over the planning period (normally one year). It must also determine the quantity of each product to be made and packed on specific machines, taking into account the individual machines' abilities to produce and pack a limited range of products. The quantity destined for immediate demand (sales) and future demand (stock) for each of the products also needs to be ascertained. Its frequency of use was expected to be a maximum of several times weekly in order to incorporate ongoing changes in, for example, forecast demand and machine configurations.

By using a model to determine the best way to produce the forecast demand over the coming year on a monthly basis, one is also **implicitly** evaluating the feasibility of producing the demanded quantity, i.e., whether the forecast annual demand **can** be produced with the given plant and other resources. If producing the forecast demand is declared infeasible, then the model (through a simple infeasibility analyzer) will assist in determining the necessary plant capacity increase (capacity **per se** or numbers of machines) required (assuming that overtime can be made available). Alternatively, the model could be instructed to determine that part of the demand that must be left unsatisfied.

Management's planning requirements, prior to the development of a model, were passed on to production schedulers for their consideration, using nothing more than experience and some simple spreadsheets. This approach was ill suited to a complex production scheduling problem as represented by the Plant's operations and also a very **time-consuming** one that did not necessarily produce a correct (feasible) result. Additionally, the process also failed to produce robust results in terms of monthly production scheduling beyond a couple of months planning horizon and did not provide a structured solution that could be subjected to "what if" analysis and other sensitivity questions. This is not surprising given the relatively large nature of the problem with thousands of combinations of machines available to produce the products.

The development of the package described in this paper has revolutionized production scheduling activities and allowed management for the first time to achieve their list of requirements.

2 THE PRODUCTION PROCESS

The model developed encompasses each of the phases of production, with the Primary area being treated essentially as an **input** into the model with some specific capacity limitations (i.e., like a raw material).

Figure 1 summarizes the production operations involved. The first phase is the provision of the required leaf blend from the Primary area. This is determined by the production scheduled for each of the products. The next phase is the

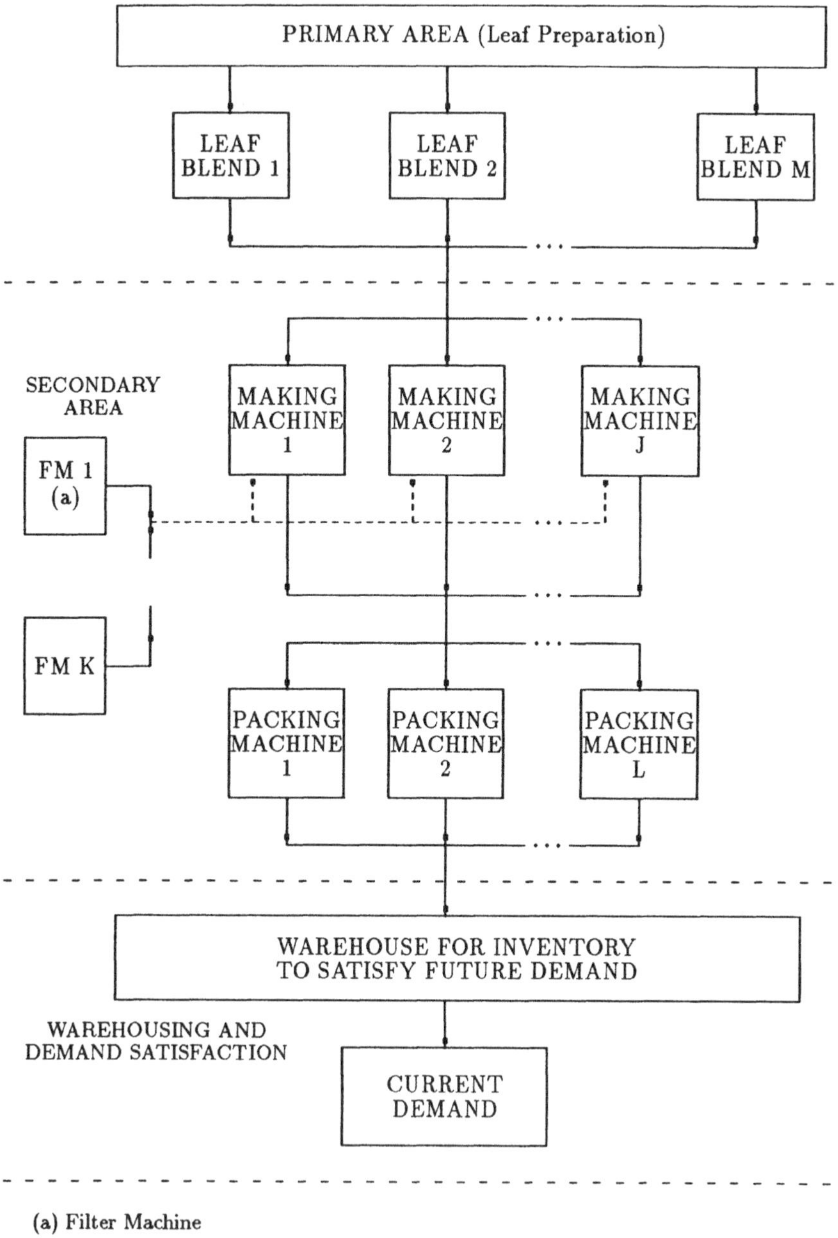

Figure 1 The Phases of Production - Secondary Floor

allocation of the production of products to the making machines, taking into account that each machine can only produce certain products.

The third phase of production is the provision of the appropriate quantities of filters (by type) to the various making machines. This is done on an aggregate basis, since each of the filter machines can produce each of the fourteen filter types required and supply them to all the making machines on the Secondary Floor.

The fourth phase of production is the packing of the product. Again, some of the products can only be packed by specific machines and this limited machine allocation is reflected in the model. The unit of production throughout all these phases is thousands of "sticks" (i.e., thousands of cigarettes).

The last phase of production is the delivery of the product to inventory for the satisfaction of future and/or immediate demand. All production is bonded, and therefore must go the warehouse prior to ultimate distribution.

3 THE CORE MATHEMATICAL MODEL

At the heart of the "package" that will meet all management's requirements, is the **core mathematical model**. It is developed here for the "t^{th}" month in the planning period and is a mixed integer linear programming one incorporating an intermediate product modeling approach. Having developed this "t^{th} month" model, it can then replicated for all the months in the planning period (up to one or more years ahead) and joined together by the opening and closing stocks of the products. The full model is then said to be a "staircase" model. See [10], [11] for examples of such models and the replication process. A discussion of the core model is also to be found in [12].

As will be seen later, this core model will be incorporated into a sophisticated "package" that is oriented towards management use, requiring no "expert" input. The package facilitates the attainment of all the objectives specified by management.

The model is generated using SAS (discussed in detail later in the paper) and solved using the branch and bound technique (incorporated in SAS/OR) to establish an optimal (or near optimal) solution to the mixed integer programming problem. The model is made mixed integer since it requires the use of a relatively small number of binary variables.

In summary, the core model is required to allocate the production of cigarettes to the various making machines, ensuring that the quantity of leaf blend is not exceeded and taking into account the capacity of each making machine together with their ability to make only certain brands. Additionally, the appropriate number of filters is required to be produced (again with the capacity of the filter machines in mind). The made cigarettes are then allocated to the allowable packing machines (of which only certain brands can be packed by a specific group of packers). The production process is "driven" by the forecast product demand.

It should also be borne in mind that the safety stocks of each brand must be maintained, and that no more than a specified number of months demand for each product may be held in stock at any given time. All this must occur such that the demand for each brand is met now and in the future, with the total number of machines used to undertake the production being a minimum for the entire planning period.

The **explicit** objective of the model, the minimization of the number of machines used throughout each of the months (i.e., makers, filter machines and packers) in the planning period, has the effect of also minimizing the number of operators require d for the same period. This is due to the fact that even if a machine is required to be operated for only a small part of the month in question, an operator will be required on a **full time basis** (i.e., for all months) due to industrial relations considerations. Additionally, management places equal importance on the current month versus the last month of the planning period, thus negating any "time discount" considerations associated with machine use.

This application involves what the literature would term "master scheduling" where the production and associated activities are organized bearing in mind the product demands (see for example [2] and [14]). It is however, **the objective function** and the formulation of the core model that makes this application different from the above examples.

The model developed in this paper does not share much in common with the multi-level production scheduling models where the objective is for example minimization of makespan [3], minimization of variation of production rates [9] or minimization of mean tardiness in flowshops [8]. Additionally, the fact that this application has evolved into a package that can be readily used by the people who need it together with the observation that the package has been integrated into the plant MIS and has at its disposal all the relevant information required, adds some additional interest also.

It is not the intention to minimize inventory **per se**, but an **implicit** objective to keep inventory within strict limits. The upper limit is determined by the fact that stock cannot be kept for more than a maximum period of time and the lower limit by safety stock considerations. The main expenses associated with the plant are directly associated with the operation of the various machines.

Note that in the core mathematical model, the variables associated with the model are defined in the way normally associated with intermediate product modeling formulations, i.e., names are longer than in simplified production flow models (i.e., more meaningful names are used such as PM for production from the making process rather than just x_1). Note also that all variables are in multiple Roman upper case except for the binary variables which are single Greek letters and that all right hand sides are in single upper case Roman letters. Coefficients (including net rates of production and equivalence values) are single lower case Roman letters.

In the mathematical model below, the **variables** are defined as follows:

$\mathbf{PA_{i,j,t}}$ = the quantity of the i th product assigned for manufacture to the j th making machine in the **t** th month;

$\mathbf{PP_{i,t}}$ = the quantity of product i finally produced from the J making machines and available for packing in month t;

$\mathbf{PM_{l,i,t}}$ = the quantity of product i assigned to the l th packing machine in month t;

$\mathbf{PPP_{i,t}}$ = the quantity of product i produced and packed and despatched to the warehouse in the current month t;

$\mathbf{DP_{i,t}}$ = the demand satisfied from stock of the i th product in the current month t;

$\mathbf{QLB_{m,t}}$ = the total quantity of leaf of blend m required in **t**;

$\mathbf{CSP_{i,t}}$ = the closing stock level of the i th product for the current month t;

$\mathbf{CSP_{i,t-1}}$ = the closing stock level of the i th product for the previous month **t-1**;

$\delta\mathbf{MAK_{j,t}}$ = a binary variable whose value is one if the j th making machine is used, and zero otherwise during month t;

$\delta\mathbf{FIL_{n,t}}$ = a binary variable that is one if the n th filter machine is used and zero otherwise in month t;

$\delta\mathbf{PAK_{l,t}}$ = a binary variable equal to one if the l th packing machine is being used in the current month t and zero otherwise.

The **right hand sides and bounds** are defined thus:

$\mathbf{C_{MAK_{j,t}}}$ = the capacity of the j th making machine in month t;

$\mathbf{C_{FIL_{n,t}}}$ = the capacity of the n th filter machine in month t where the filter machines are ranked in descending order of capacity;

$\mathbf{C_{PAK_{l,t}}}$ = the capacity of the l th packing machine in month t;

$\mathbf{B_{m,t}}$ = the maximum quantity of blend m leaf available in the current month t;

$\mathbf{F_{i,t}}$ = the demand for product i in month t

$\mathbf{D_{i,t}}, \mathbf{G_{i,t}}$ = are the limits on stocks of product i in month t;

$\mathbf{H_{i,0}}$ = the opening stock of product i in period 1.

$\mathbf{L_t}, \mathbf{U_t}$ = are the upper and lower limits to the quantity of filters produced in month t;

The **coefficients** used in the model are as follows:

$\mathbf{a_{i,t}}$ = is the proportion of i th product that when passing through the making machines is not lost in production in month t (note the loss here is product not machine specific);

$\mathbf{b_{i,t}}$ = is the standardizing coefficient that ensures that the capacity used by each product is expressed in terms of the product that the making machine capacity (right hand side) is expressed in for month t;

$\mathbf{c_{i,t}}$ = is the reciprocal of $a_{i,t}$ averaged across the products using blend m tobacco in month t;

$\mathbf{d_{l,t}}$ = the proportion of product successfully packed in month t on the l th packing machine (note the loss here is machine related);

$\mathbf{e_{l,t}}$ = the same as $b_{i,t}$ but for the packing machines in month t;

$\mathbf{f_{i,t}}$ = the same as $b_{i,t}$ but for the filter machines in month t (e.g., if no filter, than $f_{i,t} = 0$);

The **sets and subsets** used in the model are as follows:

$\mathbf{J_1}$ = is the integer variable subset;

$\mathbf{J^o(i)}$ = is that subset of permissible making machines that are used to make product i;

$\mathbf{I^o(j)}$ = is that subset of permissible products that are made on the j th making machine;

$\mathbf{L^o(i)}$ = is that subset of permissible packing machines that are used to pack product i;

$\mathbf{I^1(l)}$ = is that subset of permissible products packed on packing machine **l**.

While **T** = the planning period (in months).

The mathematical model in **general** terms is as follows:

$$\min Z = \sum_{j=1}^{J} \delta MAK_{j,t} + \sum_{n=1}^{K} \delta FIL_{n,t} + \sum_{l=1}^{L} \delta PAK_{l,t}$$

subject to

$$-\sum_{j \in J^o(i)} PA_{i,j,t} a_{i,t} + PP_{i,t} = 0 \qquad (i=1,\ldots,I) \qquad (1)$$

$$\sum_{i \in I^o(j)} PA_{i,j,t} b_{i,t} - \delta MAK_{j,t} C_{MAK_{j,t}} \leq 0 \qquad (j=1,\ldots,J) \qquad (2)$$

$$-\sum_{i \in I^1(m)} PP_{i,t} c_{i,t} + QLB_{m,t} = 0 \qquad (m=1,\ldots,M) \qquad (3)$$

$$QLB_{m,t} \leq B_{m,t} \qquad (m=1,\ldots,M) \qquad (4)\text{B}$$

$$\sum_{i=1}^{I} PP_{i,t} f_{i,t} + \sum_{k=n}^{K} C_{FIL_{k,t}} (1 - \delta FIL_{n,t}) - \sum_{k=1}^{K} C_{FIL_{k,t}} \leq 0 \qquad (n=1,\ldots,K) \qquad (5)$$

$$\sum_{i=1}^{I} PP_{i,t} \leq U_t \qquad (6)B$$

$$\sum_{i=1}^{I} PP_{i,t} \geq L_t \qquad (7)B$$

$$-PP_{i,t} + \sum_{l \in L^o(i)} PM_{l,i,t} = 0 \qquad (i=1,\ldots,I) \qquad (8)$$

$$-\sum_{l \in L^o(i)} PM_{l,i,t} d_{l,t} + PPP_{i,t} = 0 \qquad (i=1,\ldots,I) \qquad (9)$$

$$\sum_{i \in I^1(l)} PM_{l,i,t} e_{l,t} - \delta PAK_{l,t} C_{PAK_{l,t}} \leq 0 \qquad (l=1,\ldots,L) \qquad (10)$$

$$CSP_{i,t-1} + PPP_{i,t} - DP_{i,t} - CSP_{i,t} = 0 \qquad (i=1,\ldots,I) \qquad (11)$$

$$DP_{i,t} = F_{i,t} \qquad (i=1,\ldots,I) \qquad (12)B$$

$$CSP_{i,t} \geq G_{i,t} \qquad (i=1,\ldots,I) \qquad (13)B$$

$$CSP_{i,t} \leq D_{i,t} \qquad (i=1,\ldots,I) \qquad (14)B$$

$$CSP_{i,0} = H_{i,0} \qquad (i=1,\ldots,I) \qquad (15)B$$

$$(t = 1, .., T)$$

$$\delta MAK_{j,t},\ \delta FIL_{n,t},\ \delta PAK_{l,t} \in \{0,1\}$$

$$\delta MAK_{j,t},\ \delta FIL_{n,t},\ \delta PAK_{l,t} \in J_1$$

$$\text{all variables} \geq 0 \qquad (3.1)$$

Note also that a **B** beside the constraint number indicates that it will not be a constraint in the SAS model, but rather it will be treated as a bound thus reducing the number of constraints and thus the solution time.

3.1 A Discussion of the Mathematical Model

In the above model the objective function

$$\min Z = \sum_{j=1}^{J} \delta MAK_{j,t} + \sum_{n=1}^{K} \delta FIL_{n,t} + \sum_{l=1}^{L} \delta PAK_{l,t}$$

represents the total number of machines used during month t. When any of the **J** making machines are used $\boldsymbol{\delta MAK_{j,t}}$ is by definition one. Similarly, when any of the **K** filter and **L** packing machines are operating during the month, then the $\boldsymbol{\delta FIL_{n,t}}$ and $\boldsymbol{\delta PAK_{l,t}}$ variables are one respectively. The value of the objective function will thus represent the number of machines and therefore operatives required for the month to produce the necessary product for immediate and future demand satisfaction.

By minimizing the number of machines that are required to be operated in a given month, the number of personnel are thereby minimized as well (given the existence of a one for one relationship). This then implicitly minimizes the major cost factors encountered in production, namely labour and machines. Embodied in the above logic is the fact that if a machine is required to be operated, even for a small part of the month, then an operative is required but for the **entire month** since fractional employment is not possible here.

This formulation of the objective function has an important side benefit. The formulation facilitates the development of a simple Infeasibility Resolution Heuristic (IRH) that is able (because of its nature) to track down infeasibilities and either automatically rectify the problem by making additional capacity available (through pre-specified quantities of overtime) or make recommendations for management to consider. This will be discussed later in the paper.

Constraint (**1**),

$$-\sum_{j \in J^o(i)} PA_{i,j,t} a_{i,t} + PP_{i,t} = 0 \quad (i = 1, \ldots, I)$$

is a material balance equation (in intermediate product modeling terms) whose purpose is to allocate the total production of product **i** [$\mathbf{PP_{i,t}}$] (of which there are a total of **I** products) to making machine **j** (of which there are a total of **J**) taking into account the loss that will be incurred in the making process [$\mathbf{a_{i,t}}$]. However, only some products can be made on each of the making machines. This allowable subset of **J** for each **i** is $\mathbf{J^o(i)}$ (i.e., $\mathbf{j \in J^o(i)}$).

Constraint (**2**),

$$\sum_{i \in I^o(j)} PA_{i,j,t} b_{i,t} - \delta MAK_{j,t} C_{MAK_{j,t}} \leq 0 \quad (j = 1, \ldots, J)$$

determines the total standardized [$\mathbf{b_{i,t}}$] usage by each of the allowable subset of products on each machine ($\mathbf{I \in I^o(j)}$, i.e., the equivalent of $\mathbf{j \in J^o(i)}$ from the making machines perspective) and ensures that if production on that **j** th machine is greater than zero, then $\delta\mathbf{MAK_{j,t}}$ is one and the usage of the machine does not exceed its capacity (to produce the base product), $\mathbf{C_{MAK_{j,t}}}$. The standardization coefficient [$\mathbf{b_{i,t}}$] is based on the product which is used when expressing the machine's capacity.

By adopting the use of the binary variable to impose the capacity limitation on the making machine, one has an **indicator** of when a making machine is used ($\delta\mathbf{MAK_{j,t}} = 1$) and thus an element for the objective function (which aims to minimize the number of machines in use). Hence the constraint performs two functions.

Constraint (**3**),

$$-\sum_{i\in I^o(j)} PP_{i,t}c_{i,t} + QLB_{m,t} = 0 \quad (m = 1,\ldots,M)$$

Constraint (**4**),

$$QLB_{m,t} \leq B_{m,t} \quad (m = 1,\ldots,M)$$

Constraint (**3**) seeks to determine the quantity of each blend of tobacco required by the production schedule. Each of the products are made up of tobacco of a certain blend (of which there are **M**), and for each blend the number of products (**i**) that contain it are represented by the subset $\mathbf{I^1(m)}$. The total quantity of blend **m** over the subset of allowable products is $\mathbf{QLB_{m,t}}$ and constraint (**4**) ensures that this required quantity (derived from production) does not exceed the throughput capacity of the tobacco blend silos ($\mathbf{B_{m,t}}$), and is treated as an upper bound rather than a constraint. The final production of each product made is used here to reduce the density of the model, and therefore this must be loss adjusted to represent the quantity of the **i** th product **input** into makers. The quantity of blend **m** per one thousand sticks of product **i** (within the allowable subset), loss adjusted, is $c_{\mathbf{i,t}}$.

Constraint (**5**),

$$\sum_{i=1}^{I} PP_{i,t}f_{i,t} + \sum_{k=n}^{K} C_{FIL_{k,t}}(1 - \delta FIL_{n,t}) - \sum_{k=1}^{K} C_{FIL_{k,t}} \leq 0 \quad (n = 1,\ldots,K)$$

ensures that the capacity of each of the filter machines (measured in terms of a base product) is not exceeded, but in such a way as to require the **minimum** number of constraints. The problem size in reality is of critical importance in terms of computer data handling and CPU solution time. The model in its current replicated form is approximately 3,000 constraints with some 12,000 variables. Constraint (**5**), in its **modus operandi** is very similar to that of the capacity constraint on the making machines. The ordering of the filter machines in (**1**) is important, with the list of machines ($\mathbf{n = 1,\ldots,K}$) being in **descending order of capacity**. The coefficient $\mathbf{f_{i,t}}$ represents the standardization factor. If the product being produced requires a filter needing the same effort to produce as the base product used in determining machine capacity, then $\mathbf{f_{i,t}} = 1$. If there is no filter required, then $\mathbf{f_{i,t}} = 0$ and generally, $0 \leq \mathbf{f_{i,t}} \leq 1.5$.

When the first filter machine is used (which if there is any production of product it will be by definition) then $\delta\mathbf{FIL_{1,t}} = 1$. If the extent of the production is such that the total capacity of filter machine one is exceeded, then $\delta\mathbf{FIL_{1,t}} = 1$ and $\delta\mathbf{FIL_{2,t}} = 1$ etc. The "universal" nature of the filter machines (i.e., the fact that all filter machines can produce all filters and supply them manually or in automatic mode with or without storage to any making machine) facilitates the way in which they are treated in terms of capacity determination, and the savings obtained due to this fact should not be squandered.

If it is assumed that there are three filter machines, then the first constraint in Constraint type (**5**) will require the total number of filters demanded to be less than or equal to the first filter machine's capacity. If this is not sufficient capacity then the constraint has the total **remaining machines** capacity added to it (via the binary variable becoming one) and the next constraint considered. The second constraint in (**5**) represents the capacity of the largest machine (the first) **plus** the second largest (the second machine). If this constraint is not violated then the capacity provided by the two largest machines is adequate and only one binary variable was required to be set to one. The third (and last constraint) in (**5**) will under these circumstances be slack and the binary variable left at zero due to the objective function. Note that in the specification of (**5**), if a constraint (other than the first) is violated then the preceding constraints are (by definition) violated also. Thus, due to the structure of (**5**) and the objective function, the filter machines are used in descending order of capacity.

Constraints (**6**) and (**7**),

$$\sum_{i=1}^{I} PP_{i,t} \leq U_t$$

$$\sum_{i=1}^{I} PP_{i,t} \geq L_t$$

are upper and lower bounds respectively on the number of filters that can be handled (not necessarily produced) in month t. The number of filters produced in a month has been found by management to be a rough indication of the level of labour activity, and since on a month-to-month basis the level of activity should be relatively equal across all the months of the time period. One can't in reality effectively run a plant with virtually no activity one month and maximum production the next. Consequently, management has specified that the "level of activity," bearing in mind the number of employees at the beginning of the planning period, should be within the specified levels of U_t and L_t.

In constraint (**8**),

$$-PP_{i,t} + \sum_{l \in L^o(i)} PM_{l,i,t} = 0 \quad (i = 1, \ldots, I)$$

the made product **i** ($\mathbf{PP_i}$) is assigned to an allowable subset of packing machines ($\mathbf{l} \in \mathbf{L^o(i)}$) since not all products can be packed by all packing machines.

Constraint (**9**),

$$- \sum_{l \in L^o(i)} PM_{l,i,t} d_{l,t} + PPP_{i,t} = 0 \quad (i = 1, \ldots, I)$$

aggregates the amounts of product i packed across all permissible machines (loss adjusted using [$\mathbf{d_{l,t}}$]) at the end of the process ($\mathbf{PPP_i}$) that will then go into the warehouse for either the satisfaction of immediate or future demand.

Constraint (**10**),

$$\sum_{i \in I^1(l)} PM_{l,i,t} e_{l,t} - \delta PAK_{l,t} C_{PAK_{l,t}} \leq 0 \quad (l = 1, \ldots, L)$$

ensures that the standardized capacity of the **L** packing machines is not exceeded. Note that the $\mathbf{i} \in \mathbf{I^1(l)}$ is the equivalent of $\mathbf{l} \in \mathbf{I^o(i)}$ from the products' perspective. Thus the total amount of product **i** from the allowable subset, packed on the $\mathbf{l}^{\text{th}}$ packer, must not exceed $\mathbf{C_{PAK_{l,t}}}$ the capacity of the machine. Note that when a packer is used, then the binary variable is set at one ($\delta\mathbf{PAK_{l,t}} = 1$) i.e., using the same mechanism as embodied in constraint (**2**).

Constraint (**11**),

$$CSP_{i,t-1} + PPP_{i,t} - DP_{i,t} - CSP_{i,t} = 0 \quad (i = 1, \ldots, I)$$

represents the inventory of the various products. It operates on the basis that the opening stock (for the first month $\mathbf{OSP_i}$ and then $\mathbf{CSP_{i,t-1}}$ for the remaining months in the planing period) has added to it the monthly production amount $\mathbf{PPP_i}$ less the amount that has been withdrawn to satisfy demand for the current month ($\mathbf{DP_{i,t}}$), leaving the closing stock for the current month ($\mathbf{CSP_{i,t}}$). This is the only constraint that SAS/OR will indicate has an infeasibility should one occur and is one of the keys to the Infeasibility Resolution Heuristic. This is discussed later in the paper, as it is an important part of the "package" that allows management to be self sufficient in terms of its use.

The remainder of the constraints are essentially bounds, with Constraint (**12**) specifying what the forecast demand is for the $\mathbf{i}^{th}$ product for the current month (obtained from the forecasting package output file) and the remainder of the Constraints [(**13**) and (**14**)] establishing the safety stock and, based on FIFO usage of stock, an upper limit ensuring that stock cannot be held for more than six months for any of the **i** products. For a twelve month planning period, the upper limit is approximated by half the annual demand.

Constraint (**15**) is only in the model for the first month in order to establish the initial opening stocks, and is then replaced in successive months as already described above.

4 THE DEVELOPMENT OF A MANAGEMENT-ORIENTED PACKAGE FOR PRODUCTION SCHEDULING BASED ON THE CORE MATHEMATICAL MODEL

Before the actual logic of the package and its operation can be described, it is necessary to cover two main areas, that of the **replication of the core mathematical** model into a multi-period "staircase" model and the operation of the all important **Infeasibility Resolution Heuristic**. Both these aspects are critical to the operation of the package and its ability to meet management's requirements.

4.1 Replication of the Core Mathematical Model into the Planning Period Model

The core mathematical model, as indicated previously, contains around 1,000 variables (47 of which were binary) and 300 constraints. When the model was replicated into the "staircase" yearly model (the most common planning horizon), the eventual size expands to approximately 12,000 variables (564 of which were binary) with 3,000 constraints.

In order to obtain the model representing this entire planning period, the core model is designated the first month and then replicated by SAS with the opening stock variables deleted and replaced by the closing stock variables of the previous period for all months after the first. This is done easily in SAS by inserting a series of 1's under the closing stock variables from the previous period in the current periods closing stock constraints. Figure 2 below demonstrates the process.

After this simple replication, other data must be inserted into the model in order to complete it. The information required includes demand forecasts, working days per month, "name plate" capacities of all machines and the construction of capacity dependent values in the filter constraints (see (5)) as well as values for upper and lower limits for closing stocks, etc.

Hence, the planning period over which the core model can be replicated depends on the extent to which other plant data bases are able to supply the necessary data. This aspect of the package and the **modus operandi** will be covered later.

4.2 The Infeasibility Resolution Heuristic

Considerable detail relating to the Infeasibility Resolution Heuristic (IRH) can be found in [13], however, it is important to specify at the beginning, that the heuristic is simple and does not rely on any aspect of the simplex algorithm itself (other than the values of the binary variables in the infeasible continuous solution vector), as do implicitly or explicitly the majority of infeasibility analyzers (see for example the excellent summary of these in Greenberg's [4], [5], [6] and [7]).

The IRH is an important part of the package, since the resolution of infeasibilities in Phase I of the simplex algorithm (where a feasible solution is sought

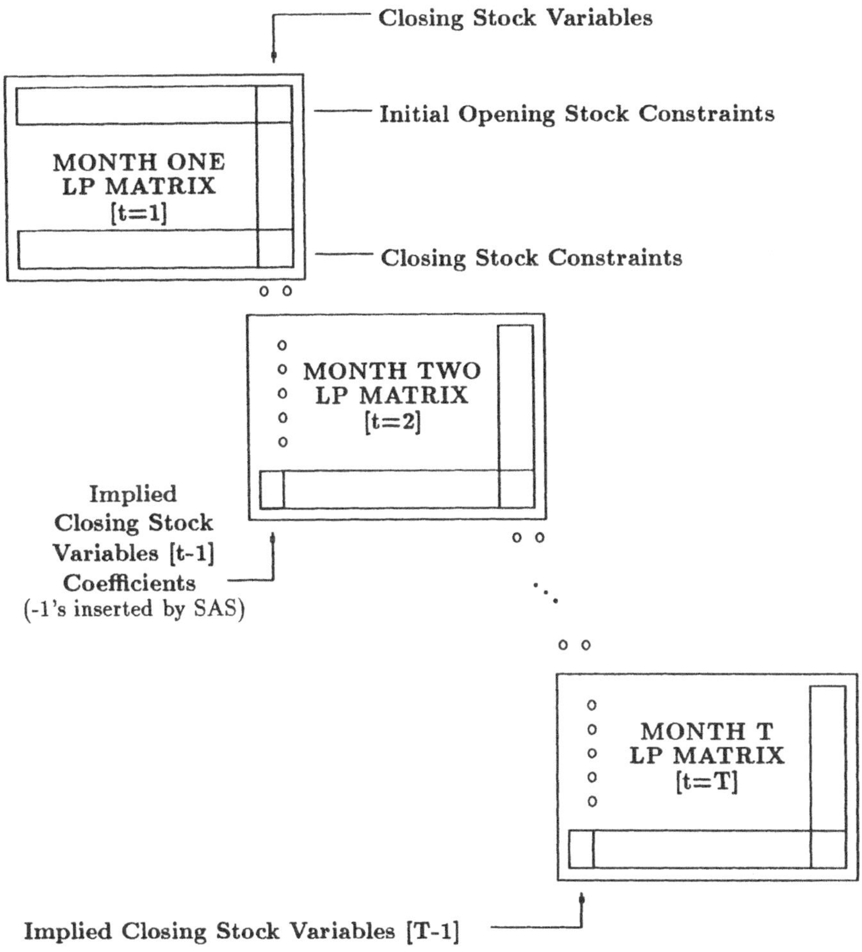

Figure 2 The Simplified Structure of the Replicated Linear Programming Constraints - Omitting the Objective Function and Right Hand Sides

prior to seeking optimal continuous and integer solutions) is not necessarily able to be achieved by other than "experts." An understanding of the constraints, the model formulation and the solution process is normally required. But in this model, because of the nature of the objective function and the model formulation itself, an IRH can be invoked that requires none of the detailed understanding listed above.

The fact that: the objective function seeks to **minimize the sum of the binary variables**; the only place where infeasibilities occur within the model (and flagged by the SAS/OR package) is the closing stock constraints; and **that the binaries will only be 1 if full capacity is reached** on a machine during Phase I (as distinct from the binaries being 1 by definition in the optimal integer solution even if only a small part of the capacity has been used), facilitates the IRH.

When an infeasibility is detected, the product involved and the period in which it occurs are noted. Then the subsets of each of the permissible machines (for each category of machine) are accessed and the associated binary variables examined. If **all** the binary variables of at least one of the relevant permissible subsets are equal to one, then a capacity limitation has been found and the necessary increase in capacity can be determined (and provided by overtime). The capacity shortage will be associated with the period in which it was detected if the lower limit of stock has been violated (13) (i.e., there is not enough stock available to make up the shortfall in production for meeting demand in the current period) whereas, if the upper limit is violated (14) (i.e., more stock is required to be built up now for the future than is permitted) then the infeasibility will occur in a future period and the IRH algorithm will then search for this and make the necessary capacity alteration.

Then in summary:

when an infeasibility of detected in Phase I of the SAS/OR solution procedure, either;

$$CSP_{i,t} \leq G_{i,t} \quad \text{or} \quad CSP_{i,t} \geq D_{i,t} \tag{3.2}$$

i.e., the infeasibility, the shortage/surplus required of stock will be negative or positive i.e.,

the infeasibility amount:

$$\Gamma_{i,t} = |CSP_{i,t-1} + PPP_{i,t} - DP_{i,t} - CSP_{i,t}| \tag{3.3}$$

will be positive (nonzero) or negative according to whether:

$$\Gamma_{i,t} \leq 0 \quad \begin{cases} PPP_{i,t} & \leq \quad DP_{i,t}; \qquad D_{i,t-1} = D_{i,t} \\ PPP_{i,t} & \leq \quad DP_{i,t} - (D_{i,t-1} - D_{i,t}) \\ & \qquad\qquad \text{where} \quad D_{i,t-1} \geq D_{i,t} \\ PPP_{i,t} & \leq \quad DP_{i,t} - (D_{i,t-1} - D_{i,t}) \\ & \qquad\qquad \text{where} \quad D_{i,t-1} \leq D_{i,t} \end{cases} \tag{3.4}$$

infeasibility in the current period, and

$$\Gamma_{i,t} \geq 0 \quad \begin{cases} PPP_{i,t} \geq DP_{i,t}; \quad G_{i,t-1} = G_{i,t} \\ PPP_{i,t} \geq DP_{i,t} - (G_{i,t-1} - G_{i,t}) \\ \qquad \text{where} \quad G_{i,t-1} \geq G_{i,t} \\ PPP_{i,t} \geq DP_{i,t} - (G_{i,t-1} - G_{i,t}) \\ \qquad \text{where} \quad G_{i,t-1} \leq G_{i,t} \end{cases} \tag{3.5}$$

infeasibility in a future period.

The exact IRH algorithm used will depend on whether (3.4) or (3.5) has occurred since (3.4) concerns itself with a violation in period t, while (3.5) is associated with a violation in future period(s).

Having determined the appropriate product(s) then all that remains is to **determine if**:

$$\sum_{n=1}^{K} \delta FIL_{n,t} = K \tag{3.6}$$

(where **K** is the total number of filter machines)

$$\sum_{j \in J^{o}(i)} \delta MAK_{j,t} = j^{**} \tag{3.7}$$

(where $\mathbf{j^{**}}$ is the total number of making machines in the permissible subset $\mathbf{J^{o}(i)}$)

$$\sum_{l \in L^{o}(i)} \delta PAK_{l,t} = l^{**} \tag{3.8}$$

(where $\mathbf{l^{**}}$ is the total number of machines in the permissible subset $\mathbf{L^{o}(i)}$).

If at least one of (3.6), (3.7) or (3.8) are met, then the infeasibility has been detected (either in the current or a future period).

The SAS/OR infeasible solution file (infeasible solution output summary) together with the SAS/OR **Infeasible Information Summary** is downloaded into an appropriate file that can be accessed by a simple spreadsheet routine which constitutes the IRH and feeds back information via amended capacity values for the appropriate machine(s) in the relevant time period. Full details are available in [13] where several examples are offered.

This IRH makes the package truly independent of outside experts and facilitates the normally time consuming task of infeasibility detection and resolution. Due to high product demands and often very restrictive permissible subsets, infeasibilities in Phase I occur quite often.

4.3 The Production Scheduling Package

The package is currently resident on a PC linked to the relevant plant data bases and is based on SAS/OR and written in SAS **per se**. The interfaces with the user via a series of menus and screens and depending on what options are requested at each stage, additional menus are called. Additionally, data input occurs either through the keyboard (for small amounts of data) or through interfacing with other plant data bases and downloading the required information. The package operates totally within a user-friendly Windows environment.

The user can construct a completely new model for scenario analysis, modify and existing model or, if it is required, simply "roll" the model forward a month. A roll-over is where the planning period can be extended (by say a month through new demand forecasts becoming available) and the first month is removed and the latest (new) month added at the point of the old planning period (the old last month). There is always an "official" base model (unreplicated) that can be accessed, but not altered, to avoid destruction of the core model. Figure 3 below summarizes the operation of the package.

As can be seen from 3 the package allows the user to either:

(i) roll the model over and resolve (for use when the demand forecasts are updated by a month [plus]);

(ii) select a previous model that has been replicated and make selected changes to it, e.g., a new set of forecast demands (and subsequent safety stock limits and opening stock values) or changes to individual items (capacity, etc);

(iii) generate a completely new model allowing for changes to be made prior to replication in items such as the machine assignment to products, additions/subtractions to the product list, and so on, as well as individual changes.

In Figure 3, a number of macro operations have been referred to. These are discussed below.

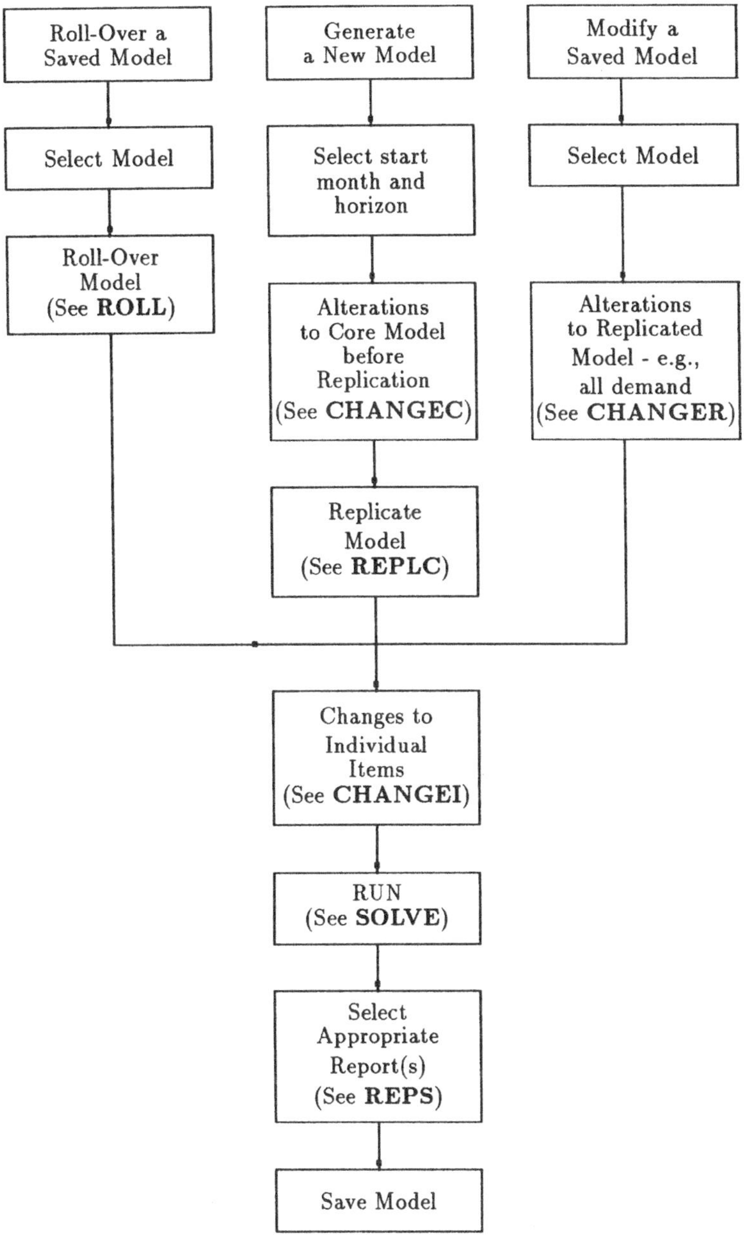

Figure 3 Summary Schematic of the Operation of the Production Scheduling Package

ROLL

This is a macro routine in SAS that allows for an existing model to be rolled over for up to five months ahead, providing sufficient forecasts are available from the forecasting package to sustain this. The "roll-over" is (for a one month operation) the removal of the first month and its replacement with a new "last" month. The first month opening stock configuration is automatically inserted into the new first month.

In order to operate this procedure, access is gained to the forecasting package output file. For each new month rolled onto, the safety stock limits are automatically recalculated and a new opening stock inserted in the "new" first month (which is the old "second" month). This is achieved through access to the stock-on-hand file. Additionally, for each new month added to the "old" model, the capacities are recalculated automatically by the procedure using its access to the working day calendar. The new "last" month added to the model is a carbon copy of the previous month with respect to all structural model details. This ROLL procedure is all performed automatically.

Note here also that there is the facility to make some individual changes to the model as these might be needed to alter some of the model aspects in the new months.

CHANGEC

This macro procedure facilitates changes being made to the core mathematical model (where, irrespective of whether it is the "base" model or merely a previously saved core model, it is down loaded from a secure location upon entry to this procedure) prior to replication into the full model. **CHANGEC** allows products to be added or deleted from the model, reassignments to be made between machines and products, addition or deletion of machines (and with the addition of machines the description of the new product assignments possible), changes to base capacities of machines and alterations to losses and equivalence and standardization coefficients. With this procedure, a completely new core model can be constructed.

REPLC

The **REPLC** procedure takes the core model existing (and named) at this point (perhaps after **CHANGEC** has been used) and using the mechanism discussed in **Section 4.1** above, produces the full model for the selected planning period starting at the chosen month. When the model has been replicated, the forecast demands are accessed and put in place and the safety stocks (upper and lower limits) automatically calculated. The capacities of all machines are then calculated using the working day calendar accessed by the procedure. The opening stock for the first month is then accessed from the stock-on-hand file. The full replicated model is now ready for any individual changes (through **CHANGEI**) or to be solved.

CHANGEI

This procedure facilitates individual changes rather than wholesale changes such as replacing all demands or replacing a specific coefficient throughout the model. With **CHANGEI**, alterations can be made to single or multiple occurrences of capacities, demands (and thus automatically to safety stocks), all coefficients, and all other bounds and right hand sides. This facility allows the model to be "fine tuned" prior to solution. It also allows the setting up of "what if" scenarios very easily.

CHANGER

CHANGER allows extensive changes such as selecting a new starting month and requiring the forecast demand figures to be completely changed. With a change to the starting month made, the working days calendar would then alter and the capacities of machines would be automatically recalculated for the full replicated model. Again the extent of change here is limited by the availability of information from the other plant files that the package has access to.

SOLVE

As the name suggests, this macro procedure is concerned with the solution of the model. The macro uses SAS/OR PROCLP procedure (branch and bound). The Infeasibility Resolution Heuristic is located here and invoked as discussed earlier.

REPS

This macro procedure has access to the SAS output files containing the solution. Depending on the reports required (there are a number of reports that can be selected prior to or after the run) various spreadsheet routines are accessed and the output rearranged into the appropriate form, with all the variable names in the SAS output files replaced by the standard product codes and names used by the plant. Reports can be requested on machine/product allocation, product quantity made, packed, and placed into stock and also grouped according to products' "families."

The package is a very powerful tool assisting in forecast demand feasibility determination and its subsequent "average monthly" production scheduling. All the programming for the package including the use of the SAS **package** and the interface with the plant data bases was carried out by a specialist consultant applications programmer.

5 IMPLEMENTATION AND OPERATION OF THE PRODUCTION SCHEDULING PACKAGE

The development of the package saw two distinct phases. Initially, the core mathematical model was developed, coded and solved with very little data manipulation facilities being available. This was to ensure that the data and model was correct and to determine the usefulness of the resultant output. Additionally, this stage of development saw considerable interest being directed to solution times and numerical stability of the model. This latter aspect became necessary to investigate due to the large difference in magnitude of coefficients (10^7) both between and across constraints. This coupled with an average density of the LP matrix of around 0.06% – 1%, necessitated adjustments to some of the SAS package operating tolerances. However, numerical instability was not a problem in the end, with only occasionally the full inverse being required to be obtained by SAS/OR.

With respect to solution times, initially with the original plant configuration a twelve month replicated model required typically seventeen to twenty five hours to have performed approximately 1000 **integer iterations** (of the branch and bound integer solution algorithm using SAS/OR) with around three to four integer solutions being generated. The difference between the solutions (with respect to the objective function) was minimal and the first solution was deemed by management to be acceptable. This meant that the initial solution time was around the seventeen hours (on a PC). However, when some plant reconfiguration occurred (and was incorporated into the model) solutions were able to be obtained within three hours. This is quite acceptable as a solution time, since the model is not something required to be produced for day-to-day operation of the plant.

During the first phase of implementation, the solutions were examined in detail to determine whether they made sense to the production schedulers and whether they could be easily implemented. The answers to these questions were indeed favorable. The resultant solutions contained a number of new ways of setting up the monthly production targets and overall machine allocations in order to meet the forecast annual demands that had not been tried before.

The second phase of implementation saw the development of the infrastructure of the package as illustrated in Figure 3. This development was extremely important as it gave management the power to build their own models quickly and efficiently. The building of a new model requires on average (allowing for interaction and discussion) around fifteen to twenty five minutes for a twelve month planning period in a form ready for submission for a solution.

Considerable time was also spent on the development of the reports required after solution had been affected. Reports are obtained (without having to leave the package) by simply clicking a mouse on the appropriate icon and selecting the report desired. A spreadsheet package was used to access the SAS output files and then order and massage the data into the appropriate forms

6 CONCLUDING REMARKS

The model (and subsequent package) developed in this paper has seen, for the first time at Philip Morris in Australia, the application of an objective mechanism for determining the feasibility of forecast demand (on a monthly basis) for up to a year ahead and the determination of a master production schedule (for the same planning time frame) that meets this demand, minimizes the use of machines and meets all the stock on hand requirements, i.e., an optimal solution in Philip Morris' terms.

The method of handling the problem to date was to employ skilled but substantially overworked production schedulers. The resultant solutions while usable and feasible for up to a month or two ahead, ran into problems thereafter due to the sheer complexity of the production situation. The necessity to have stocks available in say six months time but produced now and in the next month together with all the other production requirements was just too much to reasonably cope with.

The necessity to **repeat** the year ahead "manual" production scheduling also often arose, due to forecast updates, causing an enormous amount of time to be devoted, to arriving at "solutions." The changes to machinery (i.e., the purchase of new machines and reconfiguration of old ones) also necessitated these reruns. The implementation of the master production scheduling package has revolutionized production planning within the plant.

Acknowledgements

The author gratefully acknowledges the help and encouragement received from the management of Philip Morris Limited (Moorabbin), especially that of David Busse. Additionally, appreciation is expressed for the permission to publish and for the valuable comments of an anonymous referee.

REFERENCES

[1] D. G. H. Claassen and P. van Beek, 1993, Planning and scheduling packaging lines in the food industry, *European Journal of Operational Research*, 70, 150-158.

[2] S. Das and K. Sarin, 1994, An integrated approach to solving master aggregate scheduling problems, *International Journal of Production Economics*, 34, 167-77.

[3] B. T. Eck and M. Pinedo, 1993, On the minimization of the makespan subject to flowtime optimality, *Operations Research*, 4, 797-801.

[4] H. Greenberg, 1993a, How to analyze results of linear programs. Part 1: preliminaries, *Interfaces*, 23, 56-67, .

[5] H. Greenberg, 1993b, How to analyse results of linear programs. Part 2: price interpretation, *Interfaces*, 23, 97-114.

[6] H. Greenberg, 1993c, How to analyse results of linear programs. Part 3: infeasibility diagnosis, *Interfaces*, 23, 120-139.

[7] H. Greenberg, 1994, How to analyse results of linear programs. Part 4: forcing substructures, *Interfaces*, 24, 121-130.

[8] Y-D. Kim, 1993, A new branch and bound algorithm for minimizing mean tardiness in two-machine flowshops, *Computers and Operations Research*, 20, 391-401.

[9] W. Kubiak, 1993, Minimizing variation of production rates in just-in-time systems: A survey, *European Journal of Operational Research*, 66, 259-271.

[10] M. G. Nicholls, 1993, The development of a mathematical model of the operations of a national glass manufacturer, *Journal of the Operational Research Society*, 44, 935-943.

[11] M. G. Nicholls, 1995a, Scheduling production in a heavily constrained plant - anode manufacture in an aluminium smelter, *Journal of the Operational Research Society*, 46, 579-591.

[12] M. G. Nicholls, 1995b, The development of a mathematical model to assess the feasibility of forecast product demand and assist in production scheduling in a tobacco processing plant, Proceedings of the 24 th Annual Meeting of the Western Decision Sciences Institute, Ed. Marcoulides, G., San Fransisco, USA, 483-487.

[13] M. G. Nicholls, 1996, Heuristic interpretation of infeasible solutions to a production scheduling problem in cigarette manufacturing, *Journal of the Operational Research Society*, 47, 963-979.

[14] R. A. Reid and D. W. Clements, 1994, Facility shutdown: a model-based approach to production planning, *Production and Inventory Management Journal*, 35, 43-49.

[15] SAS Institute Inc.,1989, SAS/OR Users Guide, Version 6, First Edition, Cary, NC:SAS Institute Inc.

4

STABILITY RADIUS OF AN OPTIMAL SCHEDULE: A SURVEY AND RECENT DEVELOPMENTS

Yuri N. Sotskov, Vyacheslav S. Tanaev, and Frank Werner*

Institute of Engineering Cybernetics
Surganov Str. 6, 220012 Minsk
BELARUS

**Otto-von-Guericke-Universität Magdeburg*
Fakultät für Mathematik
PSF 4120, 39016 Magdeburg
GERMANY

ABSTRACT

The usual assumption that the processing times of the operations are known in advance is the strictest one in deterministic scheduling theory and it essentially restricts its practical aspects. Indeed, this assumption is not valid for the most real-world processes. This survey is devoted to a stability analysis of an optimal schedule which may help to extend the significance of scheduling theory for some production scheduling problems. The terms 'stability', 'sensitivity' or 'postoptimal analysis' are generally used for the phase of an algorithm at which a solution (or solutions) of an optimization problem has already been found, and additional calculations are performed in order to investigate how this solution depends on the problem data. We survey some recent results in the calculation of the stability radius of an optimal schedule for a general shop scheduling problem which denotes the largest quantity of independent variations of the processing times of the operations such that this schedule remains optimal. We present formulas for the calculation of the stability radius, when the objective is to minimize mean or maximum flow time. The extreme values of the stability radius are of particular importance, and these cases are considered more in detail. Moreover, computational results on the calculation of the stability radius for randomly generated job shop scheduling problems are briefly discussed. We also show that the well-known test problem with 6 jobs and 6 machines has both stable and unstable optimal makespan schedules.

G. Yu (ed.), Industrial Applications of Combinatorial Optimization, 72–108.

1 INTRODUCTION

In *deterministic* scheduling theory [11, 29] the processing times of the operations, which have to be processed, are supposed to be given in advance, i.e. before applying a scheduling procedure. Unfortunately, such problems do not often arise in practice. Even if the processing times are known beforehand, OR workers are forced to take into account possible changes and errors within the practical realization of a schedule, the precision of equipment for calculating the processing times, round-off errors in the calculation of a schedule on the computer, machine breakdowns, additionally arriving jobs with high priorities and so on.

More general scheduling settings have been considered in *stochastic* scheduling [2, pp. 33-59],[20], where the processing times are random variables with a known distribution function of probabilities. However, in practice such functions also may be unknown. The results presented in this survey may be considered as an attempt to initialize investigations of scheduling problems *under conditions of uncertainty*. We study the influence of round-off errors of the processing times on the property of a schedule to be optimal.

Different scheduling problems may be represented as extremal problems on disjunctive graphs (see e.g. [11, 28, 29]). The only requirement for such a representation is the prohibition of preemptions of operations. As it follows from recent results [2, pp. 277-293] this approach is the most suitable one for traditionally difficult scheduling problems. In this survey, we use the disjunctive graph model to represent the input data of the following *general shop* scheduling problem denoted by $G//F$.

There are a set $Q = \{1, 2, \ldots, q\}$ of operations and a set $M = \{M_1, M_2, \ldots, M_m\}$ of machines. Let Q_k be the set of operations that have to be processed on machine M_k:

$$Q = \cup_{k=1}^{m} Q_k;\; Q_k \neq \emptyset;\; Q_k \cap Q_l = \emptyset;\; k = 1, 2, \ldots, m;\; l = 1, 2, \ldots, m;\; k \neq l.$$

At any time each machine can process at most one operation. If $i \in Q_k$, then the non-negative real value p_i denotes the processing time of operation i on machine $M_k \in M$. Preemptions of operations are not allowed and a schedule of the operations Q on the machines M may be defined by the completion times c_i or by the starting times $c_i - p_i$ of all operations $i \in Q$.

The set of operations Q is supposed to be partially ordered by the given precedence constraints $\rightarrow$. We write $i \rightarrow j$ if operation $i \in Q$ is the immediate predecessor of operation $j \in Q$. So, $i \rightarrow j$ implies

$$c_i \leq c_j - p_j. \tag{4.1}$$

In particular, for the *job shop* problem, denoted by $J//F$, the set of operations Q is partitioned into n chains

$$Q = \cup_{k=1}^{n} Q^{(k)},$$

where each chain includes the set $Q^{(k)}$ of all operations of a job J_k, $1 \leq k \leq n$, and it represents the technological route of this job.

Since at any time a machine can process at most one operation, the conditions $i \in Q_k$ and $j \in Q_k$ imply one of the following inequalities:

$$c_i \leq c_j - p_j \quad \text{or} \quad c_j \leq c_i - p_i. \tag{4.2}$$

The general shop scheduling problem $G//F$ is to find a feasible schedule $(c_1, c_2, \ldots, c_q)$ in order to minimize the value of a given non-decreasing objective function $F(C_1, C_2, \ldots, C_n)$, where C_i is the completion time of job J_i. Such a criterion is called *regular*. We concentrate mainly on two frequently used regular criteria: the maximum flow time (makespan) $C_{max} = \max_{i=1}^{n} C_i$ and the mean flow time $\sum C_i = \sum_{i=1}^{n} C_i$.

The input data of such a problem can be represented by means of a disjunctive graph $G = (Q, A, D)$, where
– the set Q of operations is the set of vertices, a non-negative weight p_i being assigned to each vertex $i \in Q$,
– A is the set of directed (conjunctive) arcs, representing conditions (4.1):

$$A = \{(i,j) : i \rightarrow j,\ i \in Q,\ j \in Q\},$$

– D is the set of pairs of directed (disjunctive) arcs, representing conditions (4.2):

$$D = \{\{(i,j),(j,i)\} : i \in Q_k;\ j \in Q_k;\ i \not\rightarrow j;\ j \not\rightarrow i;\ k = 1, 2, \ldots, m\}.$$

While solving problem $G//F$, each pair of disjunctive arcs $\{(i,j),(j,i)\}$ must be settled. It means that one of these arcs must be added to a subset $D_s \subset D$ of chosen arcs and the other one must be rejected from the disjunctive graph:

(*) $(i,j) \in D_s$ if and only if $(j,i) \in D \backslash D_s$.

The choice of arc (i, j) (arc (j, i), respectively) defines a precedence of operation i (operation j) over operation j (operation i) on their common machine $M_k \in M$. Obviously, not each of such subsets D_s may be feasible. Every time an arc (i, j) is chosen, all disjunctive arcs that are made redundant due to the transitive and antireflexive nature of the precedence relations, should be excluded. Therefore, a feasible schedule s is defined by a subset $D_s \subset D$ such that along with the above condition (*) the following condition (**) is satisfied:

(**) digraph $G_s = (Q, A \cup D_s, \emptyset)$ has no circuits.

Since the objective function is non-decreasing, we may consider only *semiactive* schedules: a schedule is called semiactive if no operation can start earlier without delaying the processing of some other operation or/and without violating the sequence of operations on some machine. Let $P(G) = \{G_1, G_2, \ldots, G_\lambda\}$ be the set of all digraphs G_s that satisfy both conditions (*) and (**). On the one hand, each digraph $G_s \in P(G)$ defines a unique semiactive schedule

$$s = (c_1(s), c_2(s), \ldots, c_q(s)),$$

where $c_i(s)$ is the earliest completion time of operation $i \in Q$ with respect to the digraph G_s. On the other hand, each semiactive schedule defines a unique digraph $G_s \in P(G)$. In the following we call the digraph $G_s \in P(G)$ *optimal* if and only if s is an optimal schedule.

Although problem $G//F$ is NP-hard in the strong sense for any regular criterion considered in scheduling theory [11, 29], we can find an optimal schedule $s = (c_1(s), c_2(s), \ldots, c_q(s))$ in $O(q^2)$ time after having constructed an optimal digraph G_s. In other words, the 'main difficulty' of problem $G//F$ is to construct an optimal digraph $G_s = (Q, A \cup D_s, \emptyset)$, i.e. to find the best set D_s of chosen arcs. Due to the particular importance of the set D_s, it is called the *signature* of a schedule s [1, 23, 26, 27, 28].

This survey is devoted to the *stability ball* of an optimal digraph G_s, i.e. a ball in the space of the numerical input data such that within this ball schedule s remains optimal. Section 2 contains a formal definition of the *stability radius*, i.e. the maximal radius of such a stability ball.

As already mentioned, the main reason for performing a stability analysis is that in most practical cases the processing times of the operations, which have to be processed in the considered processing system, are inexact or uncertain before applying a scheduling procedure. In such cases a stability analysis is necessary to investigate the credibility of an optimal schedule at hand. If possible errors of the processing times are larger than the stability radius of an optimal schedule, this schedule may not be the best in a practical realization and, consequently, there is not much sense in large efforts to construct an 'optimal schedule'. In such a case it may be more advisable to restrict the scheduling procedure to the construction of an approximate or heuristic solution. On the other hand, this is not the case when all real changes of the processing times are less than or equal to the stability radius of an optimal schedule: an a priori constructed optimal schedule will remain optimal (the best) in the practical realization.

Another reason for investigating the stability radius may be connected with the frequent need to solve a set of similar scheduling problems. Since in reality the characteristics of the processing systems (the number of machines, the type of technological routes, the range of variations of the processing times and so on) do not change quickly, it is sometimes possible to use previous experience for solving a new similar scheduling problem. It is particularly important for NP-hard problems to have information on the stability because for these problems it is undesirable to resolve the problem for each new data set.

Moreover, since the majority of scheduling problems is NP-hard, enumeration schemes such as branch and bound are often used for finding an optimal schedule. To this end, it is necessary to construct a large branching tree and most of the information, contained in this tree, is lost after having solved the problem. In such situations the stability radius of the constructed schedule gives the possibility to use at least a part of this information for further similar scheduling problems.

In Section 3 the calculation of the stability radius of an optimal schedule for problem $G//F$ is reduced to a non-linear mathematical programming problem if the criterion is regular. Formulas for calculating the stability radius along with characterizations of its extreme values for problems $G//C_{max}$ and $G//\sum C_i$ are presented in Sections 4 and 5, respectively. A small example is given in Section 6 to illustrate the notations and some presented results. In Section 7 we briefly discuss computational results on the calculation of the stability radii or at least of their upper bounds for randomly generated job shop scheduling problems with the mean flow time and the makespan criteria. Section 8 surveys some related approaches to stability analysis in combinatorial optimization.

2 THE DEFINITION OF THE STABILITY RADIUS

The main question under consideration is: How can one vary the processing times $p_i \geq 0, i \in Q$, in the problem $G//F$ such that an optimal schedule of the problem remains optimal and how can one calculate the largest quantity of such variations of the processing times?

Note that any small variation $p_i \pm \epsilon$, $\epsilon > 0$, of a processing time p_i changes at least the completion time $c_i(s)$ of operation i in an optimal schedule $s = (c_1(s), \dots, c_i(s), \dots, c_q(s))$ and, as a result, we obtain another semiactive schedule: $(\dots, c_i(s) + \epsilon, \dots)$ or $(\dots, c_i(s) - \epsilon, \dots)$. However, an optimal digraph $G_s = (Q, A \cup D_s, \emptyset)$ for the new problem obtained due to such a variation of p_i usually remains the same if ϵ is sufficiently small. Thus, a signature D_s of an optimal schedule s is essentially more stable. Moreover, in practice it is often more important to keep in mind not the calendar times when the operations have to be started or have to be completed, but only m optimal sequences in which the operations Q_k have to be processed on machine M_k, $k = 1, 2, \dots, m$.

Therefore, the research in [1, 10, 22, 23, 25, 26] was mainly devoted to the stability of an optimal digraph G_s, which represents an optimal solution of problem $G//F$ in a rather compact form. Due to these arguments, we can concretize the above question: Under which largest independent changes in the components of the vector of the processing times $p = (p_1, p_2, \dots, p_q)$, digraph G_s remains optimal? Next, we introduce these notions in a formal way.

Let R^q be the set of all non-negative q-dimensional real vectors p with the maximum metric, i.e. the *distance* $r(p, p')$ between the vectors $p \in R^q$ and $p' = (p'_1, p'_2, \dots, p'_q) \in R^q$ is defined as follows:

$$r(p, p') = \max_{i \in Q} |p_i - p'_i|,$$

where $|p_i - p'_i|$ denotes the absolute value of the difference $p_i - p'_i$.

Definition 1 *The closed ball $O_\varrho(p)$ with the radius ϱ and the centre p in the space of the q-dimensional real vectors is called a stability ball of G_s, if for any vector $p' \in O_\varrho(p) \cap R^q$, schedule s remains optimal. The maximum value of the radius ϱ of a stability ball $O_\varrho(p)$ of G_s is called the stability radius of G_s and it is denoted by $\varrho_s^{max}(p)$ for the makespan, by $\varrho_s^{\Sigma}(p)$ for the mean flow time and by $\varrho_s^{F}(p)$ for an arbitrarily given regular criterion.*

This definition implies a general approach for calculating $\varrho_s^F(p)$, which is discussed in Section 3 and concretized for $F = C_{\max}$ and $F = \sum C_i$ in Sections 4 and 5, respectively.

3 REGULAR CRITERION

Let operation $j_i \in Q^{(i)} \subseteq Q$ be the last operation of job J_i, $1 \leq i \leq n$. We denote by $\{\mu\}$ the set of vertices which form a path μ in the digraph G_s and by $l^p(\mu)$ the weight of this path:

$$l^p(\mu) = \sum_{i \in \{\mu\}} p_i.$$

Let $\tilde{H}_s^i$ be the set of all paths in the digraph $G_s = (Q, A \cup D_s, \emptyset)$ ending in vertex $j_i \in Q^{(i)}$. Obviously, the value of $c_{j_i}(s)$ for a schedule s is equal to the largest weight of a path in $\tilde{H}_s^i$. The path $\mu \in \tilde{H}_s^i$ is called *dominant* if there is no other path $\nu \in \tilde{H}_s^i$ such that $\{\mu\} \subset \{\nu\}$. Otherwise, we shall say that path μ is dominated by path ν. Let H_s^i denote the set of all dominant paths in $\tilde{H}_s^i$. Since $p_k \geq 0$ for all $k \in Q$, we obtain

$$c_{j_i}(s) = \max_{\mu \in H_s^i} l^p(\mu).$$

Thus, the schedule $s = (c_1(s), c_2(s), \ldots, c_q(s))$ is optimal if and only if

$$\begin{aligned} &F(\max_{\mu \in H_s^1} l^p(\mu), \max_{\mu \in H_s^2} l^p(\mu), \ldots, \max_{\mu \in H_s^n} l^p(\mu)) = \\ &\min_{k=1,2,\ldots,\lambda} F(\max_{\nu \in H_k^1} l^p(\nu), \max_{\nu \in H_k^2} l^p(\nu), \ldots, \max_{\nu \in H_k^n} l^p(\nu)). \end{aligned} \tag{4.3}$$

If $\phi^F(p)$ denotes the set of all optimal semiactive schedules with respect to criterion F and $s \in \phi^F(p)$, then by Definition 1 we have

$$\varrho_s^F(p) = \inf\{r(p, x) : x \in R^q, s \notin \phi^F(x)\}. \tag{4.4}$$

So, it follows from (4.3) and (4.4) that, in order to calculate $\varrho_s^F(p)$, it is sufficient to know the optimal value of the objective function $f(x_1, x_2, \ldots, x_q)$ of the following non-linear mathematical programming problem:

minimize

$$f(x_1, x_2, \ldots, x_q) = \max_{i=1,2,\ldots,q} |x_i - p_i| \tag{4.5}$$

subject to

$$\begin{aligned} &F(\max_{\mu \in H_s^1} l^x(\mu), \max_{\mu \in H_s^2} l^x(\mu), \ldots, \max_{\mu \in H_s^n} l^x(\mu)) > \\ &\quad \min_{k=1,2,\ldots,\lambda; k \neq s} F(\max_{\nu \in H_k^1} l^x(\nu), \max_{\nu \in H_k^2} l^x(\nu), \ldots, \max_{\nu \in H_k^n} l^x(\nu)) \end{aligned} \tag{4.6}$$

$$x_i \geq 0, \quad i = 1, 2, \ldots, q. \tag{4.7}$$

If condition (4.6) is not satisfied for any $x \in R^q$, then digraph G_s is optimal for any vector $x \in R^q$ of processing times: $s \in \phi^F(x)$ for any $x \in R^q$. In this case we shall say that the stability radius is *infinitely large* and we shall write

$$\varrho_s^F(p) = \infty.$$

In all other cases there exists an optimal value f^* of the objective function of problem (4.5)-(4.7):

$$f^* = \inf \max_{i=1,2,\ldots,q} |x_i - p_i|,$$

where the infimum is taken over all vectors x satisfying conditions (4.6) and (4.7). To find the value f^*, it is sufficient to know a solution $x^0 = (x_1^0, x_2^0, \ldots, x_q^0)$ of problem (4.5)-(4.7), where the sign $>$ in inequality (4.6) is replaced by the sign $\geq$. It is clear that

$$f^* = \max_{i=1,2,\ldots,q} |x_i^0 - p_i| = r(x^0, p) = \varrho_s^F(p)$$

and for any small $\epsilon > 0$, there exists a vector $x^\epsilon = (x_1^\epsilon, x_2^\epsilon, \ldots, x_q^\epsilon)$ such that $r(x^\epsilon, p) = \varrho_s^F(p) + \epsilon$ and $s \notin \phi^F(x^\epsilon)$. It may occur that solution x^0 is equal to p. In the latter case we have

$$\varrho_s^F(p) = r(p, p) = 0$$

and the optimal digraph G_s is *unstable*. If x^0 is not equal to p, we have $\varrho_s^F(p) > 0$ and the optimal digraph G_s is *stable*.

4 MAKESPAN CRITERION

The best studied case of problem $G//F$ is the one with $F = C_{\max}$. Let H and H_s denote the sets of all dominant paths in digraph $(Q, A, \emptyset)$ and in digraph

$G_s \in P(G)$, respectively. The value of $\max_{i=1}^{n} C_i$ of a schedule s is given by the weight of a maximum-weight (also called *critical*) path in the digraph $G_s \in P(G)$. Obviously, at least one critical path in G_s is dominant and for any $\mu \in H$, there exists a path $\nu \in H_s$ that dominates path μ or $\mu \in H_s$. Thus, equality (4.3) for problem $G//C_{\max}$ is converted to

$$\max_{\mu \in H_s} l^p(\mu) = \min_{k=1,2,\ldots,\lambda} \max_{\nu \in H_k} l^p(\nu) \tag{4.8}$$

and an optimal makespan schedule s is defined by the digraph $G_s \in P(G)$ satisfying equality (4.8).

Let $H_k(p)$ denote the set of all critical dominant paths in the digraph $G_k \in P(G)$ (with respect to vector p). Obviously, we have $H_k(p) \subseteq H_k$. To present necessary and sufficient conditions for $\varrho_s^{max}(p) = 0$ proven in [22, 23], we need the following auxiliary claim from [17].

Lemma 1 *There exists a real $\epsilon > 0$ such that the set $H_k \backslash H_k(p)$ contains no critical path of digraph $G_k \in P(G)$ for any vector of processing times $p^\epsilon = (p_1^\epsilon, p_2^\epsilon, \ldots, p_q^\epsilon) \in O_\epsilon(p) \cap R^q$, i.e.*

$$H_k(p^\epsilon) \subseteq H_k(p).$$

Proof. We calculate the positive real number

$$\epsilon_k = \min_{\nu \in H_k \backslash H_k(p)} \frac{l_k^p - l^p(\nu)}{2q}. \tag{4.9}$$

Hereafter $l_k^p = \max_{\mu \in H_k} l^p(\mu)$ is the value of the objective function C_{max} for schedule k with the vector p of processing times. For any real ϵ, which satisfies the inequalities $0 < \epsilon < \epsilon_k$, the difference in the right side of equality (4.9) remains positive when vector p is replaced by any vector $p^\epsilon \in O_\epsilon(p) \cap R^q$. Indeed, the number of vertices in any path ν in digraph G_k is at most equal to q and, therefore, the difference $l_k^p - l^p(\nu)$ may not be 'overcome' by a vector p^ϵ if $r(p, p^\epsilon) < \epsilon_k$. ∎

Theorem 1 *For an optimal schedule $s \in \phi^{max}(p)$, equality $\varrho_s^{max}(p) = 0$ holds if and only if there exists another optimal schedule $k \in \phi^{max}(p), k \neq s$, and there exists a path $\mu^* \in H_s(p)$ such that there does not exist any path $\nu^* \in H_k(p)$ with $\{\mu^*\} \subseteq \{\nu^*\}$.*

Proof. *Sufficiency.* We shall show that if the conditions of the theorem are satisfied, then we have $\varrho_s^{max}(p) < \epsilon$ for any given $\epsilon > 0$.

To this end, we construct a vector $p^* = (p_1^*, p_2^*, \dots, p_q^*)$ with the components

$$p_i^* = \begin{cases} p_i + \epsilon^*, & \text{if } i \in \{\mu^*\} \\ p_i & \text{otherwise,} \end{cases}$$

where $\epsilon^* = \min\{\epsilon_k, \epsilon_s, \epsilon\}$ with ϵ_s and ϵ_k defined as in (4.9). If $\max_{\nu \in H_k} l^{p^*}(\nu) = l^{p^*}(\nu^0)$, then due to Lemma 1 and the inequalities $p_i^* \geq p_i$, $i \in Q$, we have

$$\begin{aligned} l^{p^*}(\nu^0) = \max_{\nu \in H_k(p)} l^{p^*}(\nu) = \\ l^p(\nu^*) + \epsilon^* |\{\mu^*\} \cap \{\nu^*\}| = l^p(\mu^*) + \epsilon^* |\{\mu^*\} \cap \{\nu^*\}|. \end{aligned} \quad (4.10)$$

Since $\{\mu^*\} \subseteq \{\nu^*\}$ does not hold for any $\nu^* \in H_k(p)$, the inequality

$$|\{\mu^*\} \cap \{\nu^*\}| < |\{\mu^*\}|$$

holds (hereafter $|a|$ denotes the number of elements in the set a) and we can continue (4.10) in the following way:

$$l^p(\mu^*) + \epsilon^* |\{\mu^*\} \cap \{\nu^*\}| < l^p(\mu^*) + \epsilon^* |\{\mu^*\}| = l^{p^*}(\mu^*) = \max_{\mu \in H_s} l^{p^*}(\mu).$$

Thus, we obtain $l_k^{p^*} < l_s^{p^*}$ and $s \notin \phi^{max}(p^*)$, which imply $\varrho_s^{max}(p) < \epsilon$ because of $r(p, p^*) = \epsilon^* \leq \epsilon$.

Next, we prove *necessity* by contradiction. Suppose that $\varrho_s^{max}(p) = 0$ but the conditions of the theorem are not satisfied. We consider two cases i) and ii) of violating these conditions.

i) Assume that there does not exist another optimal makespan schedule: $\phi^{max}(p) = \{s\}$. Then we calculate the real number

$$\epsilon^0 = \frac{1}{2q} \min\{l_t^p - l_s^p : t = 1, 2, \dots, \lambda;\ t \neq s\}.$$

Since s is the only optimal makespan schedule, we have $\epsilon^0 > 0$. For any positive real $\epsilon < \epsilon^0$, the difference $l_t^p - l_s^p$ remains (strictly) positive when vector p is replaced by an arbitrary vector $p^0 \in O_\epsilon(p) \cap R^q$. So we can conclude that digraph

G_s remains optimal for any such vector p^0 of the processing times. Therefore, we have $\varrho_s^{max}(p) \geq \epsilon > 0$ which contradicts the assumption $\varrho_s^{max}(p) = 0$.

ii) Assume that $|\phi^{max}(p)| > 1$ and for any schedule $k \in \phi^{max}(p)$, $k \neq s$, and for any path $\mu^* \in H_s(p)$, there exists a path $\nu_k^* \in H_k(p)$ such that $\{\mu^*\} \subseteq \{\nu_k^*\}$. In this case we can take any ϵ that satisfies the inequalities

$$0 < \epsilon < \min\Big\{ \min_{k \in \phi^{max}(p)} \epsilon_k, \frac{1}{2q} \min\{l_t^p - l_s^p : t = 1, 2, \ldots, \lambda;\ t \notin \phi^{max}(p)\}\Big\}.$$

From Lemma 1, due to inequality $\epsilon < \epsilon_s$, we get equality

$$l_s^{p^0} = \max_{\mu \in H_s(p^0)} l^{p^0}(\mu) = \max_{\mu \in H_s(p)} l^{p^0}(\mu) \tag{4.11}$$

for any vector $p^0 \in O_\epsilon(p) \cap R^q$. Since $\epsilon < \epsilon_s$ and $\epsilon < \epsilon_k$, and there exists a path $\nu_k^* \in H_k(p)$, $k \in \phi^{max}(p)$, $k \neq s$, for any path $\mu^* \in H_s(p)$ such that $\{\mu^*\} \subseteq \{\nu_k^*\}$, we obtain the inequality

$$\max_{\mu \in H_s(p)} l^{p^0}(\mu) \leq \max_{\nu \in H_k(p)} l^{p^0}(\nu). \tag{4.12}$$

Thus, due to (4.11) and (4.12), we have

$$l_s^{p^0} \leq \max_{\nu \in H_k(p)} l^{p^0}(\nu) \tag{4.13}$$

for any optimal schedule $k \in \phi^{max}(p)$, $k \neq s$. Since

$$\epsilon < \frac{1}{2q} \min\{l_t^p - l_s^p : t = 1, 2, \ldots, \lambda; t \notin \phi^{max}(p)\},$$

condition $t \notin \phi^{max}(p)$ implies $t \notin \phi^{max}(p^0)$. Taking into account (4.13) and the latter implication, we can conclude that $s \in \phi^{max}(p^0)$ for any vector $p^0 \in R^q$ with $r(p, p^0) \leq \epsilon$. Consequently, we have $\varrho_s^{max}(p) \geq \epsilon > 0$, which contradicts the assumption $\varrho_s^{max}(p) = 0$. ■

Obviously, the conditions of Theorem 1 are violated if $H_s(p) \subseteq H$ and so we get

Corollary 1 *If $s \in \phi^{max}(p)$ and $H_s(p) \subseteq H$, then $\varrho_s^{max}(p)00$.*

For the following corollary it is not necessary to know the set $H_s(p)$.

Corollary 2 *If $\phi^{max}(p) = \{s\}$, then $\varrho_s^{max}(p) > 0$.*

In [22, 23] the following characterization of an infinitely large stability radius was given.

Theorem 2 *For an optimal schedule $s \in \phi^{max}(p)$, we have $\varrho_s^{max}(p) = \infty$ if and only if for any path $\mu \in H_s \backslash H$ and any digraph $G_t \in P(G)$, there exists a path $\nu \in H_t$ such that $\{\mu\} \subseteq \{\nu\}$.*

Proof: *Necessity.* Following the contradiction method, we suppose that $\varrho_s^{max}(p) = \infty$ but there exist a path $\mu \in H_s \backslash H$ and a digraph $G_t \in P(G)$ such that for any $\nu \in H_t$ the relation $\{\mu\} \subseteq \{\nu\}$ does not hold. We set $\epsilon' = \max_{i=1}^{q} p_i$ and consider the vector $p' = (p'_1, p'_2, \ldots, p'_q) \in R^q$, where

$$p'_i = \begin{cases} \epsilon', & \text{if } i \in \{\mu\} \\ 0 & \text{otherwise.} \end{cases}$$

For any path $\nu \in H_t$, we have

$$l^{p'}(\nu) = \epsilon' |\{\mu\} \cap \{\nu\}|.$$

Since relation $\{\mu\} \subseteq \{\nu\}$ does not hold, we have $l^{p'}(\nu) < l^{p'}(\mu)$. Therefore, we obtain $l_t^{p'} < l^{p'}(\mu) = l_s^{p'}$ and hence $s \notin \phi^{max}(p')$. We get a contradiction:

$$\varrho_s^{max}(p) < r(p', p) \leq \max_{i=1}^{q} p_i < \infty.$$

Sufficiency. Let ϵ be a positive number as large as desired. We take any vector $p^0 \in O_\epsilon(p) \cap R^q$ and suppose that

$$l_s^{p^0} = l^{p^0}(\mu).$$

If $\mu \in H$, then $l^{p^0}(\mu) \leq l_t^{p^0}$ holds for every $t = 1, 2, \ldots, \lambda$. If $\mu \in H_s \backslash H$, then, due to the conditions of the theorem, for any feasible schedule t, there exists a path $\nu \in H_t$ such that $\{\mu\} \subseteq \{\nu\}$. Therefore, we get $l^{p^0}(\mu) \leq l^{p^0}(\nu) \leq l_t^{p^0}$. Thus in both cases we have $s \in \phi^{max}(p^0)$. ∎

Directly from the above proof of the necessity we obtain a simple upper bound of the stability radius.

Corollary 3 *If $\varrho_s^{max}(p) < \infty$, then we have $\varrho_s^{max}(p) < \max_{i=1}^{q} p_i$.*

Due to Theorem 2 one can identify a problem whose optimal makespan schedule is implied only by the given precedence constraints and the distribution of the operations to the machines, but independent from the processing times of the operations. However, because of the generality of problem $G//C_{max}$, it is difficult to check the above conditions. In [10] it has been shown that for problem $J//C_{max}$, there are necessary and sufficient conditions for $\varrho_s^{max}(p) = \infty$ which can be verified in $O(q^2)$ time. To present the latter conditions, we need the following notations.

Let B_k (C_k, respectively) be the set of all operations $i \in Q$ such that $i \rightarrow j$ ($j \rightarrow i$) and $j \in Q_k, i \notin Q_k$. For a set B of operations, let $n(B)$ be the number of jobs having at least one operation in B.

Theorem 3 *For problem $J//C_{max}$, there exists an optimal digraph G_s with an infinitely large stability radius if and only if we have $\max\{|B_k|, |C_k|\} \leq 1$ for any machine M_k with $n(Q_k) > 1$ and, if there exist two operations $g \in B_k$ and $f \in C_k$ of job J_l, then there exists a path from f to g in the digraph $(Q, A, \emptyset)$ (possibly $f = g$).*

From Theorem 3 it follows that there are problems $J//C_{max}$ with an optimal schedule with an infinitely large stability radius for any given number of jobs n and number of machines m. On the other hand, for flow shop and open shop problems such a schedule can exist only for n and m not greater than 2. Moreover, in [10] the analogies to Theorems 2 and 3 for the job shop problem $J//L_{max}$ with minimizing maximum lateness have been proven and it has been shown that there does not exist an optimal schedule s with $\varrho_s^F(p) = \infty$ for all other regular criteria (see [11]), which are considered in scheduling theory.

Formulas for calculating $\varrho_s^{max}(p)$ have been derived in [22, 23, 26]. Next we prove a formula given in [26].

Assume that $\varrho_s^{max}(p) < \infty$ holds for a given optimal schedule $s \in \phi^{max}(p)$ of problem $G//C_{max}$. Using equality (4.8), we can conclude that equality (4.4) for $F = C_{max}$ is converted to the following:

$$\varrho_s^{max}(p) = \inf\left\{r(p,x) : x \in R^q,\ \max_{\mu \in H_s} l^x(\mu) > \min_{k=1,2,\ldots,\lambda;\ k \neq s} \ \max_{\nu \in H_k} l^x(\nu)\right\}.$$

Therefore, to find the stability radius $\varrho_s^{max}(p)$ it is sufficient to construct a vector $x \in R^q$ that satisfies the following three conditions:

1) there exists a digraph $G_k \in P(G)$, $k \neq s$, such that $l_s^x = l_k^x$, i.e.

$$\max_{\mu \in H_s} l^x(\mu) = \max_{\nu \in H_k} l^x(\nu); \tag{4.14}$$

2) for any given real $\epsilon > 0$, which may be as small as desired, there exists a vector $p^\epsilon \in R^q$ such that $r(x, p^\epsilon) = \epsilon$ and inequality

$$\max_{\mu \in H_s} l^{p^\epsilon}(\mu) > \max_{\nu \in H_k} l^{p^\epsilon}(\nu) \tag{4.15}$$

is satisfied for at least one digraph $G_k \in P(G)$;
3) the distance $r(p, x)$ achieves its minimal value among the distances between vector p and the other vectors in the space R^q which satisfy both above conditions 1 and 2.

After having constructed such a vector $x \in R^q$ one can define the stability radius of digraph G_s:

$$\varrho_s^{max}(p) = r(p, x),$$

since the critical path in digraph G_s becomes larger than that of digraph G_k for any $p^\epsilon \in R^q$ with positive real ϵ, which may be as small as possible (see condition 2), and so digraph G_s is no longer optimal, while in the ball $O_{r(p,x)}(p)$ digraph G_s remains optimal (see condition 3). Digraph G_k, which satisfies conditions 1, 2 and 3 will be called a *competitive* digraph for the optimal digraph G_s.

Thus, the calculation of the stability radius may be reduced to a rather sophisticated extremal problem on the given set of weighted digraphs $P(G) = \{G_1, G_2, \dots, G_\lambda\}$ with a variable vector of weights assigned to the vertices of each digraph $G_k \in P(G)$. As it follows from (4.14) and (4.15), the main objects for such a calculation are the sets of dominant paths H_k, $k = 1, 2, \dots, \lambda$.

To satisfy conditions 1, 2 and 3, we look next for a vector $x = p(r) = (p_1(r), p_2(r), \dots, p_q(r)) \in R^q$ with the components $p_i(r) \in \{p_i, p_i + r, p_i - r\}$ on the basis of a direct comparison of the paths from the set H_s and the paths from the sets H_k, where $k = 1, 2, \dots, \lambda$ and $k \neq s$.

Let the value $l^p(\nu)$ be greater than the length of a critical path in an optimal digraph G_s. To satisfy equality (4.14), the length of a path $\nu \in H_k$ has to be not greater than that of at least one path $\mu \in H_s$ and there must be a path $\nu \in H_k$ with a length equal to the length of a critical path in G_s. Thus, if we have calculated

$$r_\nu = \min_{\mu \in H_s} \frac{l^p(\nu) - l^p(\mu)}{|\{\mu\} \cup \{\nu\}| - |\{\mu\} \cap \{\nu\}|}, \tag{4.16}$$

we obtain equality

$$\max_{\mu \in H_s} l^{p(r)}(\mu) = l^{p(r)}(\nu) \tag{4.17}$$

for the vector $p(r) = p(r_\nu)$ with the components

$$p_i(r) = p_i(r_\nu) = \begin{cases} p_i + r_\nu, & \text{if } i \in \{\mu\}, \\ p_i - r_\nu, & \text{if } i \in \{\nu\} \backslash \{\mu\}, \\ p_i, & \text{if } i \notin \{\mu\} \cup \{\nu\}. \end{cases} \tag{4.18}$$

Here the following remark can be given: due to (4.16) vector $p(r)$ calculated in (4.18) is the closest one to the given vector p among all vectors for which equality (4.17) holds.

To reach equality (4.14) for the whole digraph G_k, we have to repeat calculation (4.16) for each path $\nu \in H_k$ with $l^p(\nu) > l^p_s$. Thus, instead of vector $p(r_\nu)$ we have to consider the vector $p(r) = p(r_{G_k})$ calculated according to formula (4.18), where

$$r_{G_k} = \min_{\mu \in H_s} \max_{\nu \in H_k;\, l^p(\nu) > l^p_s} \frac{l^p(\nu) - l^p(\mu)}{|\{\mu\} \cup \{\nu\}| - |\{\mu\} \cap \{\nu\}|}. \tag{4.19}$$

Let us now consider inequality (4.15). Since the processing times are non-negative, this inequality may not be valid for a vector $p^\epsilon \in R^q$ if path μ is dominated by path ν. Thus we can restrict our consideration to the subset H_{sk} of the set H_s of all paths, which are not dominated by paths from the set H_k:

$$H_{sk} = \Big\{ \mu \in H_s : \text{ there is no path } \nu \in H_k \text{ such that } \{\mu\} \subseteq \{\nu\} \Big\}.$$

Hence, we can replace H_s in equality (4.19) by H_{sk}.

To obtain the desired vector $x \in R^q$, we have to use equality (4.19) for each digraph $G_k \in P(G), k \neq s$. Let r denote the minimum of such a value r_{G_k}:

$$r = r_{G_{k^*}} = \min\{r_{G_k} : G_k \in P(G), k \neq s\}$$

and let $\nu^* \in H_{k^*}$ and $\mu^* \in H_{sk^*}$ be paths at which value $r_{G_{k^*}}$ has been reached:

$$r_{G_{k^*}} = r_{\nu^*} = \frac{l^p(\nu^*) - l^p(\mu^*)}{|\{\mu^*\} \cup \{\nu^*\}| - |\{\mu^*\} \cap \{\nu^*\}|}.$$

Taking into account (4.18), we note that, if $r_{\nu^*} \leq p_i$ for each $i \in \{\nu^*\} \backslash \{\mu^*\}$, vector $p(r) = p(r_{\nu^*})$ does not contain negative components, i.e. $p(r) \in R^q$.

Due to the remark given after formula (4.18), we have obtained a lower bound of the stability radius:

$$\varrho_s^{max}(p) \geq r = \min_{k=1,2,\ldots,\lambda;\ k\neq s} \ \min_{\mu\in H_{sk}} \ \max_{\nu\in H_k;\ l^p(\nu)>l_s^p} \frac{l^p(\nu)-l^p(\mu)}{|\{\mu\}\cup\{\nu\}|-|\{\mu\}\cap\{\nu\}|}. \tag{4.20}$$

The bound (4.20) is tight: if $\varrho_s^{max}(p) \leq p_i$ for each $i \in \{\nu^*\}\backslash\{\mu^*\}$, then $\varrho_s^{max}(p) = r$. For example, we have $\varrho_s^{max}(p) = r$ in (4.20) if $\varrho_s^{max}(p) \leq \min\{p_i : i \in Q\}$.

To obtain the exact value of $\varrho_s^{max}(p)$ in the general case, we construct vector $x = p^*(r) = (p_1^*(r), p_2^*(r), \ldots, p_q^*(r))$ with the components

$$p_i^*(r) = \begin{cases} p_i + r, & \text{if } i \in \{\mu\}, \\ \max\{0, p_i - r\}, & \text{if } i \in \{\nu\}\backslash\{\mu\}, \\ p_i, & \text{if } i \notin \{\mu\}\cup\{\nu\} \end{cases}$$

instead of vector $p(r)$ defined in (4.18). As it follows from the remark given after formula (4.18), such a vector $p^*(r)$ is the closest one to vector p among all vectors $x \in R^q$ which satisfy both conditions 1 and 2.

For calculating the maximal value r for vector $p^*(r)$ we can consider the operations of the set $\{\nu\}\backslash\{\mu\}$ in non-decreasing order of their processing times. Let $p_{\nu\mu}^0$ be equal to zero and let $(p_{\nu\mu}^1, p_{\nu\mu}^2, \ldots, p_{\nu\mu}^{w_{\nu\mu}})$ denote a non-decreasing sequence of processing times of operations from the set $\{\nu\}\backslash\{\mu\}$, where $w_{\nu\mu} = |\{\nu\}\backslash\{\mu\}|$. We obtain the following assertion.

Theorem 4 *If $s \in \phi^{max}(p)$ holds, then*

$$\varrho_s^{max}(p) = \min_{k=1,2,\ldots,\lambda;k\neq s} r_{ks},$$

where

$$r_{ks} = \min_{\mu\in H_{sk}} \ \max_{\nu\in H_k;l^p(\nu)>l_s^p} \ \max_{\beta=0,1,\ldots,w_{\nu\mu}} \frac{l^p(\nu)-l^p(\mu)-\sum_{\alpha=0}^{\beta} p_{\nu\mu}^{\alpha}}{|\{\mu\}\cup\{\nu\}|-|\{\mu\}\cap\{\nu\}|-\beta}.$$

Note that the formulas in Theorem 4 turn into $\varrho_s^{max}(p) = \infty$ if $H_{sk} = \emptyset$ for any $k = 1, 2, \ldots, \lambda$; $k \neq s$ (see Theorem 2). Moreover, if only a subset of the processing times (say, $P \subseteq \{p_1, p_2, \ldots, p_q\}$) can be changed but the other ones cannot be changed, formula (4.20) and Theorem 4 remain valid provided that the difference $|\{\mu\}\cup\{\nu\}|-|\{\mu\}\cap\{\nu\}|$ is replaced by $|\{\{\mu\}\cup\{\nu\}\}\cap P|-|\{\{\mu\}\cap\{\nu\}\}\cap P|$.

5 MEAN FLOW TIME CRITERION

If $F=\sum C_i$, conditions (4.3) and (4.4) are converted to the following conditions (4.21) and (4.22), respectively.

$$\sum_{i=1}^{n}\max_{\mu\in H_s^i} l^p(\mu)=\min_{k=1,2,\dots,\lambda}\sum_{i=1}^{n}\max_{\nu\in H_k^i} l^p(\nu), \tag{4.21}$$

$$\varrho_s^{\Sigma}(p)=$$
$$\inf\left\{r(p,x):x\in R^q,\ \sum_{i=1}^{n}\max_{\mu\in H_s^i} l^x(\mu)>\min_{k=1,2,\dots,\lambda;k\neq s}\sum_{i=1}^{n}\max_{\nu\in H_k^i} l^x(\nu)\right\}. \tag{4.22}$$

Therefore, to find the stability radius $\varrho_s^{\Sigma}(p)$ it is sufficient to construct a vector $x\in R^q$ that satisfies the following three conditions:
1') there exists a digraph $G_k\in P(G), k\neq s$, such that

$$\sum_{i=1}^{n}\max_{\mu\in H_s^i} l^x(\mu)=\sum_{i=1}^{n}\max_{\nu\in H_k^i} l^x(\nu); \tag{4.23}$$

2') for any given real $\epsilon>0$, which may be as small as desired, there exists a vector $p^\epsilon\in R^q$ such that $r(x,p^\epsilon)=\epsilon$ and inequality

$$\sum_{i=1}^{n}\max_{\mu\in H_s^i} l^{p^\epsilon}(\mu)>\sum_{i=1}^{n}\max_{\nu\in H_k^i} l^{p^\epsilon}(\nu) \tag{4.24}$$

is satisfied for at least one digraph $G_k\in P(G)$;
3') the distance $r(p,x)$ achieves its minimal value among the distances between vector p and the other vectors in the space R^q which satisfy both above conditions 1' and 2'.

Similarly as in the previous section (see conditions 1, 2 and 3), after having constructed such a vector $x\in R^q$ one can define the stability radius of digraph

G_s: $\varrho_s^{\Sigma}(p) = r(p,x)$. Thus, due to (4.23) and (4.24) the calculation of the stability radius may again be reduced to an extremal problem on the set of weighted digraphs $P(G)$. However, in this case we are forced to consider numerous sets of representatives of the family of sets H_k^i, $1 \le i \le n$, which may be denoted as follows.

Let Ω_k^u be a set of representatives of the family of sets $(H_k^i)_{1 \le i \le n}$. More precisely, the set Ω_k^u includes exactly one path from each set $H_k^i, 1 \le i \le n$. Since $H_k^i \cap H_k^j = \emptyset$ for each pair of different jobs J_i and J_j, we have $|\Omega_k^u| = n$ and there exist $\omega_k = \prod_{i=1}^{n} |H_k^i|$ different sets of representatives for each digraph G_k, namely: $\Omega_k^1, \Omega_k^2, \ldots, \Omega_k^{\omega_k}$. For each set Ω_k^u we can calculate the integer vector $n(\Omega_k^u) = (n_1(\Omega_k^u), n_2(\Omega_k^u), \ldots, n_q(\Omega_k^u))$, where $n_j(\Omega_k^u), j \in \{1, 2, ..., q\}$, is equal to the number of paths in Ω_k^u which includes vertex j. Since a path $\nu \in H_k^i$ includes vertex $j \in Q$ at most once, the value $n_j(\Omega_k^u)$ is equal to the number how often vertex j is contained in the multiset $\{\{\nu\} : \nu \in \Omega_k^u\}$.

In [1], vector x satisfying conditions 1' and 2' was found in the form $x(r) = (x_1(r), x_2(r), \ldots, x_q(r))$ with the components $x_i(r)$ from the set $\{p_i, p_i+r, p_i-r\}$ on the basis of a direct comparison of the set $\Omega_s^u \in \Omega_{s,k}$ of representatives of the family of sets $(H_s^i)_{1 \le i \le n}$, and the set Ω_k^u of representatives of the family of sets $(H_k^i)_{1 \le i \le n}$, where $k = 1, 2, \ldots, \lambda$; $k \ne s$ and

$$\Omega_{s,k} = \{\Omega_s^v : \text{ there does not exist a set } \Omega_k^u \text{ such that } n_i(\Omega_s^v) \le n_i(\Omega_k^u) \text{ for each } i = 1, 2, \ldots, q\}.$$

As a result, the following lower bound of the stability radius has been obtained:

$$\varrho_s^{\Sigma}(p) \ge r = \min_{k=1,2,\ldots,\lambda;\ k \ne s} \ \min_{\Omega_s^v \in \Omega_{s,k}} \ \max_{u \in \{1,2,\ldots,\omega_k\}} r_{\Omega_k^u, \Omega_s^v}, \tag{4.25}$$

where

$$r_{\Omega_k^u, \Omega_s^v} = \frac{\sum_{\nu \in \Omega_k^u} l^p(\nu) - \sum_{\mu \in \Omega_s^v} l^p(\mu)}{\sum_{i=1}^{q} |n_i(\Omega_k^u) - n_i(\Omega_s^v)|}.$$

This bound is tight: we have $\varrho_s^{\Sigma}(p) = r$ in (4.25) if $r \le \min\{p_i : i \in Q\}$.

To obtain the exact value of $\varrho_s^{\Sigma}(p)$, the vector $x^*(r) = (x_1^*(r), x_2^*(r), \ldots, x_q^*(r))$ with the components

$$x_i^*(r) = \begin{cases} p_i + r, & \text{if } n_i(\Omega_k^u) < n_i(\Omega_s^v), \\ \max\{0, p_i - r\}, & \text{if } n_i(\Omega_k^u) > n_i(\Omega_s^v), \\ p_i, & \text{if } n_i(\Omega_k^u) = n_i(\Omega_s^v) \end{cases}$$

was used in [1]. In contrast to vector $x(r)$, vector $x^*(r)$ may not have negative components.

Let the set of operations Q be ordered in the following way:

$$i_1, i_2, \ldots, i_m, i_{m+1}, \ldots, i_q, \tag{4.26}$$

where $n_{i_\alpha}(\Omega_k^u) \leq n_{i_\alpha}(\Omega_s^v)$ for each $\alpha = 1, 2, \ldots, m$ and $n_{i_\alpha}(\Omega_k^u) > n_{i_\alpha}(\Omega_s^v)$ for each $\alpha = m+1, m+2, \ldots, q$. Moreover, for the sequence (4.26) the inequalities

$$p_{i_{m+1}} \leq p_{i_{m+2}} \leq \cdots \leq p_{i_q}$$

have to be satisfied. Following [1], one can derive a formula for calculating $\varrho_s^{\Sigma}(p)$.

Theorem 5 *If $s \in \phi^{\Sigma}(p)$ holds, then*

$$\varrho_s^{\Sigma}(p) = \min_{k=1,2,\ldots,\lambda; k \neq s} R_{ks}, \tag{4.27}$$

where

$$R_{ks} = \min_{\Omega_s^v \in \Omega_{s,k}} \max_{u=1,2,\ldots,\omega_k} \max_{\beta=0,1,\ldots,q-m} \frac{\sum_{\alpha=1}^{m+\beta} p_{i_\alpha} |n_{i_\alpha}(\Omega_k^u) - n_{i_\alpha}(\Omega_s^v)|}{\sum_{\alpha=1}^{m+\beta} |n_{i_\alpha}(\Omega_k^u) - n_{i_\alpha}(\Omega_s^v)|}.$$

If only a subset of the processing times can be changed but the other ones cannot be changed, a formula similar to (4.27) was derived in [1], too.

Next, we consider the extreme values of $\varrho_s^{\Sigma}(p)$. Similarly to the notion of a critical path, which is important for problem $G//C_{max}$ (see Section 4), we introduce the notion of a *critical set* $\Omega_k^{u^*}$ of digraph $G_k \in P(G)$ for problem $G//\sum C_i$. The set $\Omega_k^{u^*}$, $u^* \in \{1, 2, \ldots, \omega_k\}$, is critical if the value of the objective function

$$L_k^p = \max_{u \in \{1,2,\ldots,\omega_k\}} \sum_{\nu \in \Omega_k^u} l^p(\nu)$$

for digraph G_k is reached on this set:

$$\sum_{\nu \in \Omega_k^{u^*}} l^p(\nu) = \max_{u \in \{1,2,\ldots,\omega_k\}} \sum_{\nu \in \Omega_k^u} l^p(\nu) = L_k^p.$$

Obviously, a critical set $\Omega_k^{u^*}$ may include a path $\nu \in H_k^i$, $i = 1, 2, \ldots, n$, if and only if

$$l^p(\nu) = \max_{\mu \in H_k^i} l^p(\mu)$$

and so for different vectors $p \in R^q$ of processing times, different sets Ω_k^u, $u \in \{1, 2, \ldots, \omega_k\}$, may be critical. Let $\Omega_k(p)$ denote the set of all critical sets $\Omega_k^{u^*}$ of digraph $G_k \in P(G)$ with the vector $p = (p_1, p_2, \ldots, p_q) \in R^q$ of processing times and let Ω_k denote the set $\{\Omega_k^u : u = 1, 2, \ldots, \omega_k\}$.

To present necessary and sufficient conditions for $\varrho_s^{\Sigma}(p) = 0$, we need the following auxiliary claim.

Lemma 2 *There exists a real $\epsilon > 0$ such that the set $\Omega_k \backslash \Omega_k(p)$ contains no critical set of digraph $G_k \in P(G)$ for any vector $p' \in O_\epsilon(p) \cap R^q$ of processing times, i.e.*

$$\Omega_k(p') \subseteq \Omega_k(p).$$

Proof: After having calculated the value

$$\epsilon_k = \frac{1}{2qn} \min \Big\{ L_k^p - \sum_{\nu \in \Omega_k^u} l^p(\nu) : \Omega_k^u \in \Omega_k \backslash \Omega_k(p) \Big\}, \tag{4.28}$$

one can verify that for any real ϵ, which satisfies the inequalities $0 < \epsilon < \epsilon_k$, the difference in the right side of equality (4.28) remains positive when vector p is replaced by any vector $p' \in O_\epsilon(p) \cap R^q$. Indeed, for any $u \in \{1, 2, \ldots, \omega_k\}$, the cardinality of set Ω_k^u may be at most equal to qn. Thus, the difference $L_k^p - \sum_{\nu \in \Omega_k^u} l^p(\nu)$ may not be 'overcome' by a vector p' if $r(p, p') < \epsilon_k$. ■

The following necessary and sufficient conditions for $\varrho_s^{\Sigma}(p) = 0$ have been derived in [1].

Theorem 6 *Let s be an optimal schedule of problem $G//\sum C_i$ with positive processing times $p_i > 0$ of all operations $i \in Q$. The equality $\varrho_s^{\Sigma}(p) = 0$ holds if and only if there exists another optimal schedule $k \in \phi^{\Sigma}(p), k \neq s$, and there exists a set $\Omega_s^{v^*} \in \Omega_s(p)$ such that for any set $\Omega_k^{u^*} \in \Omega_k(p)$ there exists an operation $i \in Q$ such that condition*

$$n_i(\Omega_s^{v^*}) \geq n_i(\Omega_k^u), \quad \Omega_k^u \in \Omega_k(p), \tag{4.29}$$

holds (or condition

$$n_i(\Omega_s^{v^*}) \leq n_i(\Omega_k^u), \quad \Omega_k^u \in \Omega_k(p), \tag{4.30}$$

holds) and inequality (4.29) (or inequality (4.30), respectively) is satisfied as a strict one for the set $\Omega_k^{u^*}$.

Proof: We prove *necessity* by contradiction. Assume that $\varrho_s^{\Sigma}(p) = 0$ but the conditions of the theorem are not satisfied. We consider three cases j), jj) and jjj) of violating these conditions.

j) Assume that there does not exist another optimal schedule, i.e. we have $\phi^{\Sigma}(p) = \{s\}$. Then we consider a real ϵ such that

$$0 < \epsilon < \frac{1}{2qn} \min_{t \neq s}(L_t^p - L_s^p)$$

holds. Similarly to the proof of Lemma 2, we can show that digraph G_s remains optimal for any vector $p^0 = (p_1^0, p_2^0, \ldots, p_q^0) \in R^q$ of the processing times provided that $r(p, p^0) \leq \epsilon$. Therefore, we have $\varrho_s^{\Sigma}(p) \geq \epsilon > 0$ which contradicts the assumption $\varrho_s^{\Sigma}(p) = 0$.

jj) Assume that $|\phi^{\Sigma}(p)| > 1$ and for any optimal schedule $k \in \phi^{\Sigma}(p)$ with $k \neq s$, and for any set $\Omega_s^v \in \Omega_s(p)$, there exists a set $\Omega_k^{u^*} \in \Omega_k(p)$ such that $n_i(\Omega_s^v) = n_i(\Omega_k^{u^*})$ for any operation $i \in Q$.

In this case, we can take any ϵ that satisfies the inequalities

$$0 < \epsilon < \min\left\{\epsilon_s, \epsilon_k, \frac{1}{2qn} \min_{t \notin \phi^{\Sigma}(p)}(L_t^p - L_s^p)\right\}. \tag{4.31}$$

¿From Lemma 2, due to inequality $\epsilon < \epsilon_s$, we get that equality

$$L_s^{p^0} = \max_{\Omega_s^v \in \Omega_s(p)} \sum_{\mu \in \Omega_s^v} l^{p^0}(\mu) \tag{4.32}$$

holds for any vector $p^0 \in O_\epsilon(p) \cap R^q$. Since there exists a set $\Omega_k^{u^*} \in \Omega_k(p)$ for any set $\Omega_s^v \in \Omega_s(p)$ and any $k \in \phi^{\Sigma}(p)$, $k \neq s$, such that $n_i(\Omega_s^v) = n_i(\Omega_k^{u^*})$, $i \in Q$, we obtain the inequality

$$\max_{\Omega_s^v \in \Omega_s(p)} \sum_{\mu \in \Omega_s^v} l^{p^0}(\mu) \leq \max_{\Omega_k^u \in \Omega_k(p)} \sum_{\nu \in \Omega_k^{u^*}} l^{p^0}(\nu),$$

because of $\epsilon < \epsilon_s$ and $\epsilon < \epsilon_k$. Therefore, due to (4.32) we have

$$L_s^{p^0} \leq \max_{\Omega_k^u \in \Omega_k(p)} \sum_{\nu \in \Omega_k^{u^*}} l^{p^0}(\nu) \tag{4.33}$$

for any optimal schedule $k \in \phi^{\Sigma}(p)$, $k \neq s$. Since $\epsilon < \frac{1}{2qn} \min_{t \notin \phi^{\Sigma}(p)} \{L_t^p - L_s^p\}$, condition $t \notin \phi^{\Sigma}(p)$ implies $t \notin \phi^{\Sigma}(p^0)$. So taking into account (4.31) and the latter implication, we conclude that $s \in \phi^{\Sigma}(p^0)$ for any vector $p^0 \in R^q$ provided that $r(p, p^0) \leq \epsilon$. Consequently, we have $\varrho_s^{\Sigma}(p) \geq \epsilon > 0$, which contradicts the assumption $\varrho_s^{\Sigma}(p) = 0$.

jjj) Assume that $|\phi^{\Sigma}(p)| > 1$ and for any optimal schedule $k \in \phi^{\Sigma}(p), k \neq s$, and for any set $\Omega_s^v \in \Omega_s(p)$, there exists a set $\Omega_k^{u^*} \in \Omega_k(p)$ such that for any operation $i \in Q$ with $n_i(\Omega_s^v) > n_i(\Omega_k^{u^*})$ there exists a set $\Omega_k^{u^0} \in \Omega_k(p)$ such that $n_i(\Omega_s^v) < n_i(\Omega_k^{u^0})$. Arguing in the same way as in case jj), we can show that $\varrho_s^{\Sigma}(p) \geq \epsilon > 0$, where ϵ is as in (4.31), since for any vector $p^0 \in O_\epsilon(p) \cap R^q$, the value $\sum_{\mu \in \Omega_s^{v^*}} l^{p^0}(\mu)$ is less than or equal to value $\sum_{\nu \in \Omega_k^{u^*}} l^{p^0}(\nu)$ or value $\sum_{\nu \in \Omega_k^{u^0}} l^{p^0}(\nu)$.

Sufficiency. We show that, if the conditions of Theorem 6 are satisfied, then $\varrho_s^{\Sigma}(p) < \epsilon$ for any given $\epsilon > 0$.

We construct a vector $p^* = (p_1^*, p_2^*, \ldots, p_q^*) \in R^q$ with components $p_i^* \in \{p_i, p_i + \epsilon^*, p_i - \epsilon^*\}$, where $\epsilon^* = \min\{\epsilon_k, \epsilon, \min_{i \in Q} p_i\}$, using the following rule: for each $\Omega_k^{u^*} \in \Omega_k(p)$, mentioned in Theorem 6, we set $p_i^* = p_i + \epsilon^*$, if inequalities (4.29) hold, or we set $p_i^* = p_i - \epsilon^*$, if inequalities (4.30) hold. Note that $\epsilon^* > 0$ since $p_i > 0, i \in Q$.

After changing at most $|\Omega_k(p)|$ components of vector p according to this rule, we obtain a vector p^* of processing times for which inequality

$$\sum_{\mu \in \Omega_s^{v^*}} l^{p^*}(\mu) > \sum_{\nu \in \Omega_k^{u^*}} l^{p^*}(\nu)$$

holds for each set $\Omega_k^{u^*} \in \Omega_k(p)$. Due to $\epsilon^* \leq \min_{i \in Q} p_i$, we have $p^* \in R^q$. Since $\epsilon^* \leq \epsilon_k$, we have

$$L_k^{p^*} = \max_{u \in \{1,2,\ldots,\omega_k\}} \sum_{\nu \in \Omega_k^u} l^{p^*}(\nu) =$$

$$\max_{\Omega_k^u \in \Omega_k(p)} \sum_{\nu \in \Omega_k^u} l^{p^*}(\nu) = \sum_{\nu \in \Omega_k^{u^*}} l^{p^*}(\nu) < \sum_{\mu \in \Omega_s^{v^*}} l^{p^*}(\mu) \leq L_s^{p^*}.$$

Thus, we conclude that $s \notin \phi^{\Sigma}(p^*)$ with $r(p, p^*) = \epsilon^*$ which implies $\varrho_s^{\Sigma}(p) < \epsilon^* \leq \epsilon$. ■

Theorem 6 directly implies the following assertion.

Corollary 4 *If $\phi^{\Sigma}(p) = \{s\}$, then $\varrho_s^{\Sigma}(p) > 0$.*

The following simple upper bound of the stability radius of an optimal schedule for problem $G//\sum C_i$ was presented in [1].

Theorem 7 *If $s \in \phi^{\Sigma}(p)$ holds for problem $G//\sum C_i$ with $\lambda > 1$ and $p_i > 0$ for at least one operation $i \in Q$, then we have*

$$\varrho_s^{\Sigma}(p) < \max_{i=1}^{q} p_i.$$

Proof: Let us consider vector $p^0 \in R^q$ with zero components: $p_i^0 = 0$ for each $i \in Q$. For this vector of processing times, each feasible digraph $G_t \in P(G)$ is optimal and each set of representatives Ω_t^u is critical. We can take a schedule $k \in \phi^{\Sigma}(p^0)$ which has only one arc $(j,i) \in D_k$ different from the arcs in D_s, i.e. $(i,j) \in D_s$ and $D_s \backslash \{(i,j)\} = D_k \backslash \{(j,i)\}$. It is easy to see that there exist sets $\Omega_s^v \in \Omega_s(p)$ and $\Omega_k^u \in \Omega_k(p)$ such that

$$n_i(\Omega_s^v) > n_i(\Omega_k^u).$$

Setting $p_i^{\epsilon} = \epsilon > 0$ and $p_l^{\epsilon} = 0$ for each $l \in Q \backslash \{i\}$, we obtain $s \notin \phi^{\Sigma}(p^{\epsilon})$ and $r(p^{\epsilon}, p) < \max\{p_i : i \in Q\}$. ∎

As it follows from Theorem 7, problem $J//\sum C_i$ with $\lambda > 1$ cannot have an optimal schedule with an infinitely large stability radius in contrast to problem $J//C_{max}$ and problem $J//L_{max}$ (see Section 4).

To illustrate the above notations and some results, we consider in Section 6 an example of a job shop scheduling problem with two jobs and two machines.

6 EXAMPLE

The job shop problem is specified by the disjunctive graph $G = (Q, A, D)$ given in Fig. 1. The first job consists of operations 1 and 2, and the second job consists of operations 3 and 4. So we have the precedence constraints $1 \to 2$ and $3 \to 4$. The assignment of the operations to the machines is as follows: $Q_1 = \{1,4\}$, $Q_2 = \{2,3\}$. The vector $p = (10, 30, 20, 40)$ defines the processing times of the operations $Q = \{1,2,3,4\}$.

For this problem we get $P(G) = \{G_1, G_2, G_3\}$ with the following signatures of semiactive schedules: $D_1 = \{(1,4),(3,2)\}$, $D_2 = \{(1,4),(2,3)\}$ and $D_3 = \{(4,1),(3,2)\}$. The corresponding sets of dominant paths are as follows: $H_1^1 = \{(1,2),(3,2)\}$, $H_1^2 = \{(1,4),(3,4)\}$, $H_2^1 = \{(1,2)\}$, $H_2^2 = \{(1,2,3,4)\}$, $H_3^1 = \{(3,4,1,2)\}$ and $H_3^2 = \{(3,4)\}$. The optimal digraph $G_1 = (Q, A \cup D_1, \emptyset)$ is represented in Fig. 2 and it defines the unique optimal semiactive schedule $(10, 50, 20, 60)$ for both criteria C_{max} and $\sum C_i$. Therefore, due to Corollary 2 and Corollary 4 we have $\varrho_1^{max}(p) > 0$ and $\varrho_1^{\Sigma}(p) > 0$.

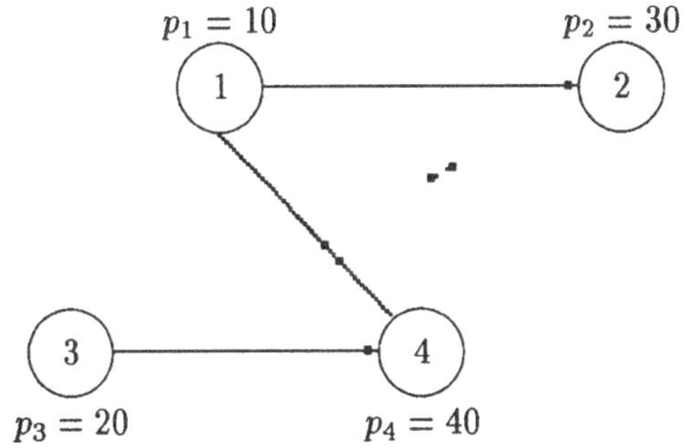

Figure 1 The disjunctive graph $G = (Q, A, D)$

We can calculate the exact value of $\varrho_1^{\Sigma}(p)$ on the basis of Theorem 5. First we compare the digraphs G_1 and G_2. We have four sets of representatives for digraph G_1, namely: $\Omega_1^1 = \{(1,2),(1,4)\}$, $\Omega_1^2 = \{(1,2),(3,4)\}$, $\Omega_1^3 = \{(3,2),(1,4)\}$ and $\Omega_1^4 = \{(3,2),(3,4)\}$. We can calculate vectors $n(\Omega_1^1) = (2,1,0,1)$, $n(\Omega_1^2) = (1,1,1,1)$, $n(\Omega_1^3) = (1,1,1,1)$ and $n(\Omega_1^4) = (0,1,2,1)$. Digraph G_2 has only one set of representatives: $\Omega_2^1 = \{(1,2),(1,2,3,4)\}$ with $n(\Omega_2^1) = (2,2,1,1)$. Obviously, we have $n_i(\Omega_1^u) \le n_i(\Omega_2^1)$ for $u \in \{1,2,3\}$ and for all $i \in \{1,2,3,4\}$. Thus we have to calculate only

$$r_{\Omega_2^1,\Omega_1^4} = \frac{(40+100)-(50+60)}{|\,2-0\,|+|\,2-1\,|+|\,1-2\,|+|\,1-1\,|} = \frac{30}{4} = 7.5$$

(cf. (4.25)).

Next we compare the digraphs G_1 and G_3. Digraph G_3 has only one set of representatives: $\Omega_3^1 = \{(3,4,1,2),(3,4)\}$ with $n(\Omega_3^1) = (1,1,2,2)$. Obviously, we have $n_i(\Omega_1^u) \leq n(\Omega_3^1)$ for $u \in \{2,3,4\}$ and for all $i \in \{1,2,3,4\}$. Thus we have to calculate only

$$r_{\Omega_3^1,\Omega_1^1} = \frac{(100+60)-(40+50)}{|\,1-2\,|+|\,1-1\,|+|\,2-0\,|+|\,2-1\,|} = \frac{70}{4} = 17.5.$$

Since $7.5 < 10 = \min\{p_1,p_2,p_3,p_4\}$, the bound (4.25) is reached and we obtain $\varrho_1^{\Sigma}(p) = \min\{7.5; 17.5\} = 7.5$ and G_2 is a competitive digraph for digraph G_1, which is optimal for $p = (10,30,20,40)$.

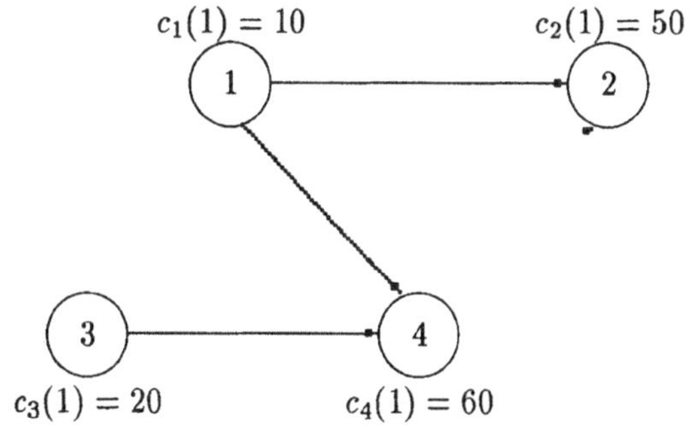

Figure 2 The optimal digraph $G_1 = (Q, A \cup D_1, \emptyset)$

Next we consider the stability radius of the optimal digraph G_1 with respect to criterion C_{max}. We have the following sets of dominant paths: $H = \{(1,2),(3,4)\}$, $H_1 = \{(1,2),(1,4),(3,2),(3,4)\}$, $H_2 = \{(1,2,3,4)\}$ and $H_3 = \{(3,4,1,2)\}$. It is clear that any path $\mu \in H_1 \backslash H$ is dominated by the path $(1,2,3,4) \in H_2$ and by the path $(3,4,1,2) \in H_3$. Thus, due to Theorem 2 we have $\varrho_1^{max}(p) = \infty$.

7 COMPUTATIONAL RESULTS

The above formulas for calculating the stability radii were coded in the language FORTRAN 77. At the worst, calculating $\varrho_s^{max}(p)$ and $\varrho_s^{\Sigma}(p)$ implies not only to have an optimal digraph G_s but to construct all feasible digraphs $G_1, G_2, \ldots, G_\lambda$. So, for small scheduling problems the program starts with generating all feasible digraphs and for each of them, which has to be compared with the optimal digraph, it finds all dominant paths. After that Theorems 4 and 5 are used for calculating $\varrho_s^{max}(p)$ and $\varrho_s^{\Sigma}(p)$, respectively.

In order to restrict the number of feasible digraphs with which a comparison of the optimal digraph G_s has to be done during the calculation of the stability radius $\varrho_s^{max}(p)$, we use the following simple bound from [25].

If there exists a digraph G_k such that

$$r_{ks} \leq \frac{l_t^p - l_s^p}{q}, \tag{4.34}$$

then digraph G_t need not to be considered during the calculation of $\varrho_s^{max}(p)$. Indeed, it is easy to show that, due to (4.34), r_{ts} cannot be smaller than r_{ks} during the calculation of $\varrho_s^{max}(p)$ (see Theorem 4).

Similarly to (4.34) for the makespan, we use the following bound (4.35) for the mean flow time criterion, and as a result, the number of comparisons of digraphs, which need to be performed for calculating $\varrho_s^{\Sigma}(p)$, were considerably reduced.

If there exists a digraph G_k such that

$$R_{ks} \leq \frac{L_t^p - L_s^p}{nq - n}, \tag{4.35}$$

then digraph G_t need not to be considered during the calculation of $\varrho_s^{\Sigma}(p)$.

So, for each considered criterion we compare the optimal digraph $G_s = G_{i_1}$ consecutively with the digraphs $G_{i_2}, G_{i_3}, \dots, G_{i_\lambda}$ from $P(G)$ in non-decreasing order of the objective function values, i.e. $F(G_{i_1}) \leq F(G_{i_2}) \leq \dots \leq F(G_{i_\lambda})$, where $F = C_{max}$ (or $F = \sum C_i$). If for the currently compared digraph $G_k = G_{i_r}$ inequality (4.34) holds for C_{max} (or inequality (4.35) holds for $\sum C_i$), we have the possibility to exclude the digraphs $G_{i_r}, G_{i_{r+1}}, \dots, G_{i_\lambda}$ from further considerations.

Note that the developed software is rather general. In principle, it allows to calculate the exact or approximate values of $\varrho_s^{max}(p)$ and $\varrho_s^{\Sigma}(p)$ for most scheduling problems (since there exists a possibility to represent them as extremal problems on a disjunctive graph, see Introduction). The only "theoretical" requirement for the considered problems is the prohibition of preemptions of operations. However, in the experiments we are forced to take into account also "practical" requirements: the running time and the memory of the used computers. It should be noted that the most critical parameter of the problem under consideration is the number $|D|$ of pairs of disjunctive arcs in G because the whole number of feasible (without a circuit) and infeasible (with circuits) digraphs generated by G is equal to $2^{|D|}$. Moreover, for each feasible digraph G_k, we have to find all dominant paths for C_{max} and (what is essentially larger) all subsets of the set of dominant paths for $\sum C_i$. Therefore, we restricted our experiments to scheduling problems with $q = 12, q = 20$ and $q = 36$ operations.

First, we present some computational results for the stability radii of optimal schedules for small randomly generated job shop problems. When generating the test problems, we distributed the operations evenly to the machines and then the operations assigned to the same machine have been evenly distributed to the jobs. The processing times of the operations are uniformly distributed real numbers from the segment [10,100]. If an instance has more than one optimal semiactive schedule, we calculate the stability radius for each of them.

In Table 1 we present the computational results for job shop problems with $q = 12$ and $q = 20$ operations. For the problems with 12 operations, we generated 50 instances for each of the considered 5 format types n x m. The stability radii $\varrho_s^{max}(p)$ and $\varrho_s^{\Sigma}(p)$ were calculated on a PC 386 usually within some seconds using only internal memory. For the problems with 20 operations we generated only 10 instances for each of the considered 5 format types n x m. Since the number of generated feasible and infeasible digraphs was equal to $2^{20} = 1048576$, and we had to use external memory on a hard disk, the running time for some of the instances with 20 operations achieved two hours on a PC 486. Table 1 gives the minimum (MIN), average (AVE) and maximum (MAX) values of the stability radius (columns 2, 3 and 4) and the minimum, average and maximum values of the stability radius divided by the average processing time of the instance (columns 5, 6 and 7). Moreover, column 8 contains the average values of the percentage of digraphs, which may be a competitive digraph (COMP) for the optimal one. In columns 9 and 10 the average and maximum numbers of optimal solutions (NOS) and in columns 11 and 12 the average and maximum numbers of feasible solutions (NFS) of the instances of each considered type are given.

It can be observed that for the small problems with 12 operations an optimal makespan schedule is more stable than an optimal mean flow time schedule. Concerning the average values of the stability radii for the problems with 20 operations, the differences between both criteria are not so large. From column 8 it can be seen that the average percentage values of competitive digraphs for the small instances with the mean flow time criterion were bounded by 7.31% and for the larger instances with 20 operations even by 2.42%. When minimizing the makespan, these values are larger (even 51.93% for instances with 6 jobs, 6 machines and 20 operations), but the latter results are mostly due to the very large numbers of optimal semiactive makespan schedules (in the following we only mean semiactive schedule when using the term feasible or optimal schedule). Moreover, it can be seen that an optimal mean flow time schedule is usually uniquely determined in contrast to the makespan criterion. Therefore, for the makespan problems it has practical sense to look for an optimal schedule that has the largest stability radius.

Table 1 Computational results for job shop problems with 12 and 20 operations

$n \times m$	RADIUS			RADIUS/p_{AVE}			COMP	NOS		NFS	
	MIN	AVE	MAX	MIN	AVE	MAX		AVE	MAX	AVE	MAX
1	2	3	4	5	6	7	8	9	10	11	12
problems with $q = 12$ operations; maximum flow time (makespan)											
3 x 3	0.01	4.33	15.71	0.02	7.85	31.51	0.96	1.98	12	443.68	876
4 x 4	0.01	3.13	20.51	0.02	5.75	33.84	1.03	2.52	11	531.86	1296
5 x 5	0.01	5.72	21.67	0.02	12.64	53.64	2.49	4.20	40	183.84	288
6 x 6	0.00	5.62	34.90	0.00	9.96	63.13	8.41	2.94	12	42.62	64
7 x 7	0.05	5.57	40.99	0.12	10.64	78.47	11.15	3.84	12	32.00	32
problems with $q = 20$ operations; maximum flow time (makespan)											
6 x 6	0.01	0.62	4.26	0.02	1.09	7.40	51.93	21.50	78	100.00	100
7 x 7	0.07	1.76	11.16	0.12	3.45	23.59	10.83	15.60	43	100.00	100
8 x 8	0.07	3.43	12.66	0.13	6.00	17.80	10.20	17.00	70	3287.30	7020
9 x 9	0.00	3.97	11.52	0.00	6.91	22.14	4.43	28.90	144	2932.80	3456
10 x 10	0.18	3.33	21.90	0.32	5.97	41.38	1.54	12.40	48	435.90	768
problems with $q = 12$ operations; mean flow time											
3 x 3	0.09	2.95	14.57	0.14	5.27	22.98	0.68	1.12	3	443.68	876
4 x 4	0.00	2.17	9.35	0.00	3.94	15.47	0.59	1.08	2	531.86	1296
5 x 5	0.08	3.59	12.33	0.16	6.43	24.88	1.55	1.02	2	183.84	288
6 x 6	0.05	5.86	21.72	0.07	10.75	36.34	5.43	1.00	1	42.62	64
7 x 7	0.03	5.08	20.88	0.07	9.75	43.53	7.31	1.00	1	32.00	32
problems with $q = 20$ operations; mean flow time											
6 x 6	0.33	1.28	5.10	0.59	2.25	8.67	2.27	1.10	2	100.00	100
7 x 7	0.23	1.33	6.57	0.40	2.32	11.19	2.42	1.20	2	100.00	100
8 x 8	0.26	1.86	6.54	0.51	3.28	11.85	0.03	1.20	2	3287.30	7020
9 x 9	0.60	2.20	4.41	1.10	3.84	8.22	0.10	1.10	2	2932.80	3456
10 x 10	0.46	3.83	8.05	0.75	6.79	13.69	0.57	1.00	1	435.90	768

More detailed computational results for further classes of randomly generated job shop problems have been given in [25], where the stability radii have been calculated for about 10000 randomly generated job shop problems with no more than 25 pairs of disjunctive arcs. Additionally we mention that in the experiments we obtained only very seldom a stability radius equal to zero for the makespan criterion (for less than 10 optimal makespan schedules) and never for the mean flow time criterion. On the other hand, for the makespan criterion an infinitely large stability radius was not seldom obtained, at least essentially more often than a zero stability radius.

From the above results it follows that we can calculate the exact value of the stability radius on the basis of a direct enumeration of the digraphs $G_1, G_2, \ldots, G_\lambda$ only for a very small number of pairs of disjunctive arcs (about 30). On the other hand, in most considered examples the competitive digraphs had objective function values that are rather close to the optimal one. So, for larger examples we decided to look for an upper bound of the stability radius, using a branch and bound algorithm for constructing k 'best' schedules (an optimal schedule or schedules and other feasible schedules with objective function values that are rather close to the optimal one).

Next, we considered the well-known job shop test problem $J//C_{max}$ from [3] with 6 jobs and 6 machines. For this test problem each job has to be processed on each machine exactly once and so we have $q = 6 \text{ x } 6 = 36$ and $|D| = 6 \text{ x } \binom{6}{2} = 90$. By the branch and bound algorithm we constructed $k = 50$ best schedules: 22 of them are optimal with $C_{max} = 55$ and the other 28 schedules have a makespan value equal to 56. We calculated an upper bound of $\varrho_s^{max}(p)$ for each optimal makespan schedule. It turned out that 18 of them have a zero stability radius and the other four optimal schedules have an upper bound of $\varrho_s^{max}(p)$ equal to 0.08333. The existence of unstable optimal schedules for this test problem is implied mainly by the fact that its processing times are integers from 1 to 10.

We also randomly generated 50 instances with 6 jobs, 6 machines and 36 operations. Again each job has to be processed on each machine exactly once as in the test problem from [3], but in contrast to the latter problem, the processing times were uniformly distributed real numbers between 1 and 10. For each generated problem with 36 operations we constructed 50 best schedules (for the makespan criterion) on the basis of the branch and bound algorithm and calculated upper bounds of $\varrho_s^{max}(p)$ for each optimal makespan schedule which was constructed. Some obtained results are given in Table 2. Note that 45 instances from Table 2 have more than one optimal makespan schedule, and among them 7 instances have 50 or even more optimal makespan schedules. We calculated also the differences between the upper bounds of $\varrho_s^{max}(p)$ for different optimal makespan schedules s of the same instance from Table 2 if this instance has two or more optimal makespan schedules. The maximum of this difference was equal to 0.84636, the average difference was 0.11709 and some optimal makespan schedules had the same stability radius. Among 50 instances presented in Table 2, there were no optimal schedule with a zero stability radius. However, in other series of instances of the same format type 6 x 6 with 36 operations we obtained two instances with zero stability radii: one with two and a second instance with four unstable optimal makespan schedules. But again unstable optimal makespan schedules arised essentially more seldom than for the test problem from [3] with integer processing times.

Unfortunately, the developed software did not allow us to find $\varrho_s^{\Sigma}(p)$ for most of the above instances with 36 operations since the calculation of the stability radius for the mean flow time criterion is essentially more time consuming than for the makespan.

Table 2 Computational results for job shop problems with 6 jobs, 6 machines, 36 operations and the makespan criterion

RADIUS (UB)			RADIUS (UB)/p_{AVE}			COMP	NOS		
MIN	AVE	MAX	MIN	AVE	MAX		MIN	AVE	MAX
0.00100	0.12939	0.87455	0.01777	2.40814	15.61688	12.30	1	$\geq$ 15.24	$\geq$ 50.00

8 RELATED APPROACHES

In spite of obvious practical importance, the literature on stability analysis in scheduling is rather small. Outside the considered approach we can mention [9, 17, 19]: in [9] the sensitivity of a heuristic algorithm with respect to the variation of the processing time of one job has been investigated, in [17] the stability of an optimal permutation schedule for the flow shop problem $F//C_{max}$ has been considered, and in [19] the results for the traveling salesman problem were used for a one machine scheduling problem with minimizing tardiness.

On the other hand, there exist a lot of papers presenting different approaches to stability analysis of optimization problems, and in this section we try to provide only a sketch of some approaches to stability analysis, which are close to the subject of this survey.

A related approach to stability analysis for so-called *linear trajectory problems* (such as the traveling salesman problem, the assignment problem, the shortest path problem and some other discrete optimization problems has been initiated in [6, 12, 13, 14, 21, 30] and developed in some other papers (see [27] for an extensive survey). Most results have been obtained for the stability radius of the whole set of solutions (optimal trajectories), i.e. for the largest radius $\varrho(p)$ of an open ball in the space of the numerical input data p such that a new optimal trajectory does not arise. A formula for calculating the stability radius $\varrho(p)$ of the set of all solutions of the traveling salesman problem has been obtained in [12, 13] and the extreme values of $\varrho(p)$ have also been determined. Analogous results for a similar problem with a bottleneck objective function have been derived in [6]. In [8] a specific transformation of a branch and bound algorithm for a traveling salesman problem for calculating $\varrho(p)$ was suggested. In [4] a polynomial algorithm has been proposed for calculating the stability radius of the whole set of solutions of some extremal problems on matroids and on the intersection of two matroids.

It should be noted that the investigation of the stability radius of one optimal trajectory of such a problem has the following drawback: the stability radius of an optimal trajectory is equal to zero if at least one further optimal trajectory exists. As it has been shown in Sections 4 and 5, this is not the case for problems $G//C_{max}$ and $G//\sum C_i$, for which the existence of two or more optimal semiactive schedules is only a necessary condition to have a zero stability radius (see Theorems 1 and 6). Such a property follows from the essentially more complicated graph-theoretical representation of a scheduling problem. This issue is confirmed by our computational experiments: although the instances of problem $J//C_{max}$ had usually a lot of optimal semiactive schedules, only a few of them had zero stability radii. For the mean flow time criterion we even did not obtain an optimal schedule with zero stability radius in our extensive experiments (although due to Theorem 6 it is not difficult to construct an unstable optimal mean flow time schedule at least for a small number of operations).

The complexity of calculating $\varrho(p)$ has been studied in [5, 7, 18, 24]. In [18] it has been shown that the problem of determining the *arc tolerance* for a combinatorial optimization problem is as hard as the problem itself (the arc tolerance is the maximum change, i.e. increase or decrease, of a single weight, which does not destroy the optimality of a solution). This means that in the case of the traveling salesman problem the arc tolerance problem is NP-hard even if an optimal tour is given. Moreover, in [5] the NP-hardness of the problem of calculating $\varrho(p)$ for the polynomially solvable shortest path problem in a digraph without negative circuits has been proven. On the other hand, in [24] it has been shown that the stability radius of an approximate solution may be calculated in polynomial time if the number of unstable components grows rather slowly, namely as $O(\log N)$, where N is the number of cities in the traveling salesman problem. In [15, 16] it has been argued that it is rather convenient from a computational point of view to use the set of k shortest tours when applying stability analysis to the symmetric traveling salesman problem.

9 CONCLUSIONS

The calculation of the stability radius based on a direct comparison of the paths for C_{max} and subsets of paths for $\sum C_i$ is very complicated and time consuming. Nevertheless, such an 'unpractical' calculation for sample problems allows to derive some properties of scheduling problems, which may be used in practically efficient methods for determining lower and upper bounds of the stability radius of an optimal schedule.

For example, our computational experiments have shown that most optimal schedules are stable, i.e. rather small errors in the determination of the processing times do not influence the property of a schedule to be optimal, or more precisely: the stability radius of such a schedule is strictly positive and so there exists a ball with the center p of processing times in the space of input data, within which the schedule remains optimal. Moreover, on the basis of the above computational results (see [25] for detail), one can make the conclusion that an optimal schedule for the makespan criterion is usually more stable than that for the mean flow time criterion.

On the other hand, the stability radius may be equal to a very small positive real number and thus with a high probability a schedule, which is (a priori) optimal, may not be the 'best' one in reality. In the latter case (i.e. if the above information is available before scheduling), a better scheduling strategy consists in restricting computations to the construction of an approximate or heuristic schedule (or schedules), which is usually not very time consuming in comparison with the construction of an optimal schedule.

Moreover, the surveyed approach gives not only the exact value or bounds of the stability radius but also 'competitive' schedules ('competitive' digraphs), which along with an optimal schedule have to be considered as possible candidates for the practical realization, when the stability radius or its upper bound is less than the possible error of the processing times known in advance. The theoretical results presented in Sections 3, 4 and 5, and the developed software may be the basis for solving scheduling problems under conditions of uncertainty when only lower and/or upper bounds of the processing times of the given operations are known in advance.

The above computational results also show that an optimal mean flow time schedule is usually uniquely determined, while two or more optimal makespan schedules are very usual (at least for the considered job shop scheduling problems with $q = 12$, $q = 20$ and $q = 36$). So, in the latter case it makes sense to look for an optimal makespan schedule with the largest value of the stability radius, since the difference of the stability radii for different optimal schedules of the same job shop problem may be very large for the makespan criterion. An optimal schedule with a larger stability radius has a better chance to be optimal in its practical realization. On the other hand, our computational experiments show that this is not valid for the mean flow time criterion, for which one can be satisfied by the first constructed optimal schedule, because even if there are two or more optimal mean flow time schedules, they usually have the same value of the stability radius.

Moreover, there exist scheduling problems for which one can look for an optimal makespan schedule with an infinitely large stability radius. In particular, if we can influence the properties of the processing system (i.e. the technological routes of the jobs, the number of used machines, the distribution of the operations to the machines, etc.), we can design a processing system with an optimal makespan schedule that has an infinitely large stability radius. In this case the variations of the processing times have no influence on such a schedule to be optimal. For some real-world scheduling problems, such a property may be desirable.

Since the stability radius of an optimal schedule may be very small, it is important to make the errors in the determination of the processing times as small as possible in order to guarantee the real optimality of a schedule at hand. Note that often only *integer* processing times are considered by scheduling theorists and since most benchmarks for scheduling problems also have only integer processing times, many scheduling algorithms use this property of the processing times essentially. On the basis of the above results, the consideration of *real* values of processing times is highly appreciated. We can illustrate this issue by the 6 x 6 test problem from [3], which has 18 unstable optimal semiactive schedules and only four stable ones (from a practical point of view only a stable optimal schedule may be considered as "really optimal" since there exists usually a positive error in the calculation of processing times). The existence of a large number of unstable optimal semiactive schedules for the 6 x 6 test problem was implied mainly by its processing times, which are integers between 1 and 10 (such an input data may be considered as a very rough estimation of the real processing times). When the processing times are supposed to be *real* numbers between 1 and 10, randomly generated job shop problems of the type 6 x 6 have mostly only stable optimal makespan schedules (see Table 2).

In conclusion we present some topics for future research. First, it would be useful to improve the bounds presented in Section 7 in order to restrict further the number of digraphs $G_k \in P(G)$, with which an optimal digraph G_s has to be compared, while calculating its stability radius. An interesting topic is also to restrict the calculations only to a part of the paths of the digraphs for the makespan criterion and only to a part of the subsets of paths for the mean flow time criterion. A more complex question is to find formulas for calculating the stability radius or at least its bounds without considering the paths of digraphs $G_k \in P(G)$.

For practical aims it is also useful to refine different branch and bound algorithms for constructing the k best schedules (instead of one, which is usually constructed) and to combine such a calculation with a stability analysis on the basis of the discussed ideas or some others (note that as it has been shown for the traveling salesman problem [15, 16] and for binary linear programming [31], the running time of such a variant of a branch and bound algorithm grows rather slowly with k). After that one can calculate upper or/and lower bounds of the stability radius of the optimal schedule. Moreover, even if we do not have the possibility of finding an optimal schedule and only approximate or heuristic schedules have been constructed, we can investigate the "stability radius" of the best of them in comparison with the other schedules that have been constructed.

Acknowledgements

This work has been supported by Deutsche Forschungsgemeinschaft (Project ScheMA) and by INTAS (Project 93-257). The authors are grateful to G.V. Andreev for his qualified contributions to the developed software and to a referee for his suggestions for improving an earlier draft of this survey.

REFERENCES

[1] H. Bräsel, Yu. N. Sotskov, and F. Werner, Stability of a schedule minimizing mean flow time, *Math. Comput. Modell.* (to appear).

[2] P. Chretienne, E.G. Coffman, J.K. Lenstra, and Z. Liu, (eds), 1995, *Scheduling Theory and its Applications*, John Wiley & Sons, New York, NY.

[3] H. Fisher, and M.L. Thompson, 1963, Probabilistic learning combinations of local job-shop scheduling rules, in: *Industrial Scheduling*; Muth, J.F. and Thompson, G.L. (eds.), Prentice-Hall, Englewood Cliffs, 225-251.

[4] E.N. Gordeev, 1987, Algorithms of polynomial complexity for computing the stability radius in two classes of trajectory problems, *U.S.S.R. Comput. Maths. Math. Phys.*, 27, 14-20.

[5] E.N. Gordeev, 1989, Solution stability of the shortest path problem, *Discrete Math.*, 1, 45-56 (in Russian).

[6] E.N. Gordeev, and V.K. Leontev, 1980, Stability in bottleneck problems, U.S.S.R. Comput. Maths. Math. Phys. 20, 275-280.

[7] E.N. Gordeev, and V.K. Leontev, 1985, The complexity of the tabulation of trajectory problems, *U.S.S.R. Comput. Maths. Math. Phys.*, 25, 199-201.

[8] E.N. Gordeev, V.K. Leontev, and I.Ch. Sigal, 1983, Computational algorithms for finding stability radius in choice problems, *U.S.S.R. Comput. Maths. Math. Phys.*, 23, 128-132.

[9] A.W.H. Kolen, A.H.G. Rinnooy Kan, C.P.M. van Hoesel, and A.P.M. Wagelmans, 1994, Sensitivity analysis of list scheduling algorithms, *Discr. Appl. Math.*, 55, 145-162.

[10] S.A. Kravchenko, Yu.N. Sotskov, and F. Werner, 1995, Optimal schedules with infinitely large stability radius, *Optimization*, 33, 271-280.

[11] E.L. Lawler, J.K. Lenstra, A.H.G. Rinnooy Kan, and D.B. Shmoys, 1993, Sequencing and Scheduling: Algorithms and Complexity, in *Logistics of Production and Inventory*, Handbook in Operations Research and Management Science 4; G.C. Graves, A.H.G. Rinnooy Kan and P.H. Zipkin (eds.), North Holland, Amsterdam, 445-522.

[12] V.K. Leontev, 1975, The stability of the traveling salesman problem, *U.S.S.R. Comput. Maths. Math. Phys.*, 15, 199-213.

[13] V.K. Leontev, 1976, Stability in combinatorial choice problems, *Soviet Mathematics Doklady*, 17, 635-638.

[14] M. Libura, 1991, Sensitivity analysis for minimum Hamiltonian path and traveling salesman problems, *Discr. Appl. Math.*, 30, 197-211.

[15] M. Libura, On accuracy of solutions for discrete optimization problems with perturbed objective functions, *Ann. Oper. Res.* (submitted).

[16] M. Libura, E.S. van der Poort, G. Sierksma, and J.A.A. van der Veen, 1996, Sensitivity analysis based on k-best solutions of the traveling salesman problem, Research report 96A14, University of Groningen.

[17] O.I. Melnikov, 1978, Optimal schedule stability for the Bellman-Johnson problem, *Vesti Akad. Navuk BSSR*, Ser. Fiz.-Mat. Navuk. 6, 99-101 (in Russian).

[18] R. Ramaswamy, and N. Chakravarti, 1995, Complexity of determining exact tolerances for min-sum and min-max combinatorial optimization problems, Report WPS-247/95, Indian Institute of Management, Calcutta.

[19] J. Picard, and M. Queyranne, 1978,The time-dependent traveling salesman problem and its application to the tardiness problem in one machine scheduling, *Oper. Res.*, 26, 86-110.

[20] M. Pinedo, 1995, *Scheduling. Theory, Algorithms, and Systems*, Prentice-Hall, Englewood Cliffs, New Jersey.

[21] D.R. Shier, and G. Witzgall, 1980, Arc tolerances in shortest path and network flow problems, *Networks*, 10, 277-291.

[22] Yu.N. Sotskov, 1989, The stability of high-speed optimal schedules, *U.S.S.R. Comput. Maths. Math. Phys.*, 29, 57-63.

[23] Yu.N. Sotskov, 1991, Stability of an optimal schedule, *European J. Oper. Res.*, 55, 91-102.

[24] Yu.N. Sotskov, 1993, The stability of the approximate Boolean minimization of a linear form, *Comput. Maths. Math. Phys.*, 33, 699-707.

[25] Yu.N. Sotskov, N.Y. Sotskova, and F. Werner, 1996, Stability of an optimal schedule in a job shop, Preprint 15/96, Otto-von-Guericke-Universität Magdeburg, FMA.

[26] Yu.N. Sotskov, and F. Werner, 1995, On the calculation of the stability radius of an optimal or an approximate schedule, Preprint 23/95, Otto-von-Guericke-Universität Magdeburg, FMA.

[27] Yu.N. Sotskov, V.K. Leontev, and E.N. Gordeev, 1995, Some concepts of stability analysis in combinatorial optimization, *Discr. Appl. Math.*, 58, 169-190.

[28] B. Sussmann, 1972, Scheduling problems with interval disjunctions, *Oper. Res.*, 16, 165-178.

[29] V.S. Tanaev, Yu.N. Sotskov, and V.A. Strusevich, 1994, *Scheduling Theory, Multi-Stage Systems*, Kluwer Academic Publishers.

[30] R.E. Tarjan, 1982, Sensitivity analysis of minimum spanning trees and shortest path trees, *Inform. Processing Letters*, 14, 30-33.

[31] G.R. Wilson, and H.K. Jain, 1988, An approach to postoptimality and sensitivity analysis of zero-one goal programs, *Naval Res. Log.*, 35, 73-84.

5

SCHEDULING JOBS WITH SEQUENCE DEPENDENT SETUPS ON COUPLED LINES

E. C. Sewell and J. C. Tan*

Department of Mathematics and Statistics
Southern Illinois University at Edwardsville
Edwardsville, Illinois 62026-1653
USA

**Department of Systems*
University of Pennsylvania
Philadelphia, Pennsylvania 19104
USA

ABSTRACT

This paper introduces the problem of scheduling a set of jobs with sequence dependent setup times on a machine with a pair of coupled processing lines, i.e., both lines must be operated simultaneously. A heuristic procedure is developed to find a good schedule which then is used as a starting point for a tabu search. A mixed integer programming formulation of the problem is also given. Computational results on real-world problems are presented.

1 INTRODUCTION

This paper investigates the problem of scheduling a set of jobs with sequence dependent setup times on a machine with a pair of coupled processing lines. The novel feature of this machine is that both lines must be operated simultaneously, that is, neither line is allowed to process a job while the other line is idle. One ramification of coupled lines is that if a setup is required for a job on one of the lines, then the processing of the job on the other line must be halted until the setup is completed. Furthermore, according to the job specifications, the machine can process different jobs at different speeds. However, as a direct

G. Yu (ed.), *Industrial Applications of Combinatorial Optimization*, 109–124.

consequence of the lines being coupled, both lines must operate at the same speed. Hence both lines must operate at the speed required for the slower job.

The particular version of this problem that is addressed in this paper arises from the packaging industry, where machines with coupled lines are used to print cartons for beverages such as milk or juice. We mention that while we only address the case of machines with two coupled lines, packaging companies also operate machines with four coupled lines.

Each job consists of printing a specified number of identical cartons, where the cartons require up to four distinct colors. The number of jobs to be scheduled ranges from 15 to 75 or more, while the number of cartons to be printed for a job can range from as few as 1,000 to as many as 500,000. A job cannot be split into two parts and run on both lines simultaneously because each job requires special tooling that is too expensive to duplicate.

Each color for a carton requires a different printing plate and each printing plate requires five minutes to set up. The setup time of a job also depends on the job that immediately precedes it in two different ways. First, since many of the jobs represent orders from the same customer and since different cartons from the same customer may be similar, two jobs may have one or more printing plates in common. The common printing plate(s) does not need to be set up again for the second job. Second, if the job currently being processed on a line requires fewer than four colors, then some of the printing plates for the next job can be set up in the unused positions while the machine is operating. For example, if a job that requires three colors is immediately followed by a job that requires two colors, then one printing plate can be set up while the machine is running and the machine will only have to be stopped for five minutes to set up the other printing plate. As a final note on setups, we mention that if two jobs are scheduled to start on the two lines at the same time, then their setups can be performed simultaneously, thus the downtime of the machine will be equal to the maximum of the two setup times. In the jargon of the industry, a simultaneous setup for the two lines is called a "double-make-ready."

The processing time of a job is determined by the number of cartons that are printed, the number of colors that are required, and the number of colors required by any jobs that are processed on the other line simultaneously. As mentioned above, both lines must operate at the same speed, so a job will be processed at a slower speed if it is processed simultaneously with a job that requires more colors.

The objective is to find a schedule that processes the jobs in the least amount of time, i.e., minimize the makespan. The quality of a schedule is determined by three factors. First, the total time spent on processing the jobs can be decreased if jobs that are processed simultaneously require the same number of colors. Second, the total amount of time spent on setting up the machine can be decreased by choosing an appropriate sequence of jobs for both lines. Third, the total downtime of the machine can be decreased by using double-make-readys. Of course, these three factors may conflict with each other, so trade-offs will be necessary.

Clearly, it will not be possible to have many double-make-readys unless many pairs of jobs have exactly the same number of cartons. In order to match more pairs together, thus permitting more double-make-readys, the contracts allow the packaging company to deliver within $\pm 10\%$ of the requested number of cartons. (We will call this contractual clause the 10% rule.) However, it is also desirable to alter the orders of as few jobs as possible, so alteration should be done only when it results in an improved schedule.

It is easy to see that this problem is theoretically quite difficult. First, if all of the jobs are processed at the same speed and have equal processing times, then the problem reduces to finding the sequences that minimize the total setup time and is equivalent to the traveling salesman problem, which is well-known to be a difficult NP-complete problem. Second, if the setup times are all zero and all the jobs are processed at the same speed, then the problem reduces to assigning the jobs to the two lines such that the total processing times on both lines are as close to equal as possible, which is equivalent to the partitioning problem and is also NP-complete. Thus, we do not attempt to solve the problem optimally, but instead rely on heuristics to find good solutions.

A large number of papers have addressed the problem of scheduling jobs on parallel machines under various assumptions regarding the objective function, setup times, release times, precedence constraints, due dates, preemption policy, and the similarity of the machines. However, we are not aware of any papers that address the problem of scheduling parallel machines that are coupled together. See [2] for a recent review of parallel machine scheduling.

In Section 2, a formal statement of the objective function is given. Section 3 presents two heuristics for this problem. The first heuristic produces a good feasible schedule which then is used as a starting point for a tabu search. Section 4 presents a mixed integer programming formulation of the problem and develops a lower bound that allows us to evaluate the quality of the schedules produced by the heuristics. Section 5 presents computational results on problems arising from real-world data. For these problems, the value of the best schedule was within 8.2% of the lower bound, on average, indicating that the heuristics performed quite well, despite the theoretical difficulty of the problem.

2 OBJECTIVE FUNCTION

In this section we discuss the choice of an objective function for the problem.

Typically, the schedule that is created will take at most two days to execute and all of the jobs will be completed on time, regardless of their position in the schedule. Thus, the completion times of the jobs are immaterial and objectives such as minimizing the number of late jobs or tardiness costs are inappropriate. The real goal here is to find a schedule that uses the machine as efficiently as possible.

There are many possible objective functions to measure the efficiency of a schedule. One natural choice is to minimize the sum of the processing times plus the sum of the setup times. However, this objective function does not take into account the benefit of double-make-readys, since the setup times of both jobs will be included in the cost, instead of the maximum of the two setup times. Of course, this difficulty can be resolved by modifying the objective function to use the total downtime of the machine due to setups rather than the sum of the setup times. This can then be restated as minimizing $t_1 + t_2$, where t_i is the completion time of the last job on line i, $i = 1, 2$.

Now, if t_1 is required to be equal to t_2, due to the lines being coupled, then minimizing $t_1 + t_2$ is equivalent to minimizing $\max(t_1, t_2)$, i.e., minimizing the makespan. In fact, however, it is not practical to require $t_1 = t_2$, since it may not even be possible to partition the jobs in such a way as to achieve $t_1 = t_2$ and, even if it is possible, the resulting schedule may be very inefficient. Thus, in practice $t_1 \neq t_2$ is permitted, with the understanding that several of the jobs near the end of the schedule (or possibly a portion of the last job) may not actually be processed with this set of jobs, but will be rescheduled to be processed with other orders that arrive later. Of course, we want to reschedule as few job as possible, so it is desirable to have $|t_1 - t_2|$ as small as possible. Therefore, minimizing the makespan is a better objective than minimizing $t_1 + t_2$, since it will penalize large values of $|t_1 - t_2|$. Note that minimizing the makespan will automatically ensure that at most one job will be scheduled by itself at the end of the schedule, which means that at most one job will have to be rescheduled, since a schedule that doesn't meet this condition can be improved by moving one of the jobs that is scheduled by itself to the other line.

There still remains one more issue to address regarding the objective function, namely, that minimizing the makespan inherently favors schedules where the 10% rule has been used to decrease the number of cartons actually produced. To prevent this, we modify the objective function by including an adjustment factor, α_i, for use of the 10% rule on job i. Let

$$\alpha_i = (c_i - c_i')/r_i,$$

where c_i is the number of requested cartons for job i, c_i' is the number of cartons actually produced, and r_i is the maximum rate at which job i can be processed. (The maximum rate at which a job can be processed is determined by the number of colors that it requires and is given in Table 1.) Thus $|\alpha_i|$ is the time required to process the difference between the number of requested cartons and the actual number of cartons produced. Now α_i needs to be added to the objective function for every job i scheduled on the longest line. For computational convenience, we take a simpler approach by adding $\sum_{i=1}^{n} \alpha_i$ to twice the makespan. Thus our objective function is

$$T = 2\max(t_1, t_2) + \sum_{j=1}^{n} \alpha_j.$$

Note that it is necessary to use twice the makespan in this formula, otherwise the objective function would inherently favor schedules where the 10% rule has been used to increase the number of cartons actually produced. With this objective function, the cost of a schedule does not decrease by simply using the 10% rule to decrease the number of cartons that are produced, which supports our subobjective of modifying as few jobs as possible. Furthermore, two different schedules can be accurately compared to each other, even though the 10% rule might have been applied differently in the two schedules. This is important in developing the lower bound presented in Section 4.

Table 2 provides the data for an example with five jobs that we will use to illustrate the definitions. Consider a schedule where jobs 1, 3, and 5 are placed on the first line, in that order, and jobs 2 and 4 are placed on the second line. The actual number of cartons produced for job 2 can be reduced to 9,000 so that jobs 1 and 2 finish simultaneously, thus permitting a double-make-ready. Thus $\alpha_2 = (9,750 - 9,000)/375 = 2$. The setup time for job 3 is 5 minutes

Table 1 Maximum Processing Rate.

Number of Colors	*Number of Cartons per Minute*
1 or 2	600
3	500
4	375

Table 2 Data for an example with five jobs.

Job	*Cartons*	*Colors*
1	9,000	3
2	9,750	4
3	3,000	2
4	9,000	2
5	7,500	4

since one of its printing plates can be set up while job 1 is running. The setup time for job 4 is 10 minutes. Hence the setup time for the double-make-ready is 10 minutes. When job 3 finishes, job 4 must be stopped to permit job 5 to be set up, which requires 10 minutes since two of its printing plates can be set up while job 3 is running. Then $t_1 = 69$ minutes, $t_2 = 61$ minutes, and

$$T = 2\max(69, 61) + 2 = 140.$$

The schedule is depicted in Figure 1.

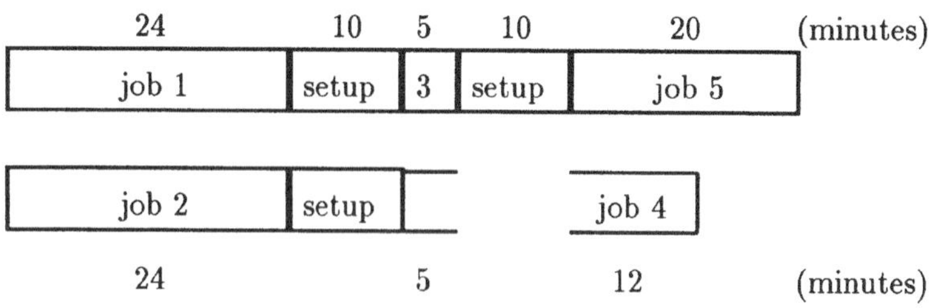

Figure 1 Schedule for the example.

3 HEURISTICS

We attempted to obtain good schedules by developing two different heuristics. The first heuristic, MPT, uses methods based on heuristics for matching, partitioning, and the traveling salesman problem. The schedule obtained from MPT is used as the starting point for a tabu search, in an attempt to further improve the schedule.

MPT consists of three phases that roughly correspond to matching some of the jobs together, using a partitioning heuristic to assign the unmatched jobs to the two lines, and using a traveling salesman heuristic to form a sequence of matched jobs on the two lines.

The first phase separates the jobs into groups of jobs that can be matched together, by which we mean that all of the jobs in a group require the same number of colors and for any pair of jobs in a group, the number of requested cartons can be adjusted to be equal by using the 10% rule. Thus, if a pair of jobs from a group are scheduled to start at the same time, then this matching will ensure that (i) the printing speed of one job will not have to be reduced to accommodate the other job, (ii) the setups for the two jobs can be performed simultaneously and (iii) the pair of jobs will end at the same time.

The second phase schedules the unmatched jobs, with these jobs being grouped into sets according to the number of required colors. Initially, both lines are empty; subsequently, the jobs from one of the sets are added iteratively to the lines, where the unscheduled job with the most number of cartons is added to the line that currently has the least number of cartons. Before proceeding to the next set of unmatched jobs, an attempt is first made to balance the number of cartons on the two lines by using the 10% rule and, if necessary, by adding jobs from the matched groups onto the shorter line.

The third phase schedules the matched jobs, one pair at a time, and places the resulting schedule prior to the schedule obtained in phase 2. The goal of this phase is to separate the jobs into pairs of jobs that require the same number of cartons (using the 10% rule, if necessary) and the same number of colors, in such a way that the total setup time is minimized. We use a simple heuristic that essentially creates a sequence that alternates between pairs of jobs that require a large number of colors and pairs of jobs that require a small number of colors. This alternation allows many of the printing plates to be set up while the machine is still running. As previously mentioned, the schedule obtained from this heuristic was then used as a starting point for a tabu search in an attempt to improve it.

Tabu search is a popular heuristic procedure for finding good solutions to optimization problems. We only give a brief description of the method here — see [3, 4, 5], and [6] for a detailed description. Tabu search searches the set of solutions of the problem; at each iteration it "moves" from the current solution to another solution, where a "move" must be defined for the particular problem that is being solved. In order to decide what move should be made in a given iteration, a set of candidate moves are evaluated and the one yielding the best solution is selected. Unlike simple descent methods that continue to move until no further improvement is possible, tabu search avoids being trapped in a local minimum by making the best candidate move, even if that move increases the value of the objective function. To avoid cycling back to a solution that has just been considered, tabu search creates a list of moves that remain forbidden for a specified number of iterations.

For our problem, a move was defined as swapping the positions of two jobs in the current pair of sequences for the two lines. This results in new sequences for the two lines, which then must be converted into a complete schedule (i.e., a starting time for each job must be defined) in order that the objective function can be evaluated and compared to that of other candidate moves. If the number of cartons for each job was fixed, then the pair of sequences for the two lines would completely specify the schedule. However, the number of cartons for each job can be adjusted by using the 10% rule, with the goal of increasing the number of double-make-readys. Since the problem of determining the optimal number of cartons for each job, given a pair of sequences for the lines, is nontrivial, and since this must be performed for each candidate move in each iteration of the tabu search, we used a simple heuristic to apply the 10% rule. This heuristic adds the jobs to the lines iteratively, where the next line to receive a job is the one with the fewest cartons. Each time a job is added to a line, it is determined whether or not the number of cartons on the two lines can be made equal by applying the 10% rule to the jobs that have been added since the last double-make-ready. At each iteration of the tabu search, every possible pair of job swaps was evaluated and the best nontabu move was selected.

We mention that we originally tried to use moves defined by swapping two blocks of jobs, where a block of jobs is a consecutive set of jobs on one of the lines such that the setup for the first job in the block is part of a double-make-ready, and similarly, the setup following the last job in the block is part of a double-make-ready. The motivation for this type of move was that it would allow the schedule to be modified without destroying any double-make-readys.

Unfortunately, we found that block moves didn't provide good results because it did not perturb the schedule enough to allow the tabu search to escape from local minimums.

In our computational studies, the tabu search was run for 100 iterations. We tried tabu lists of length 7, 10, and 15, and observed that using 15 provided the best results. We mention that this is a very simple implementation of tabu search since it does not include an aspiration function, long term diversification strategies, nor search intensification strategies. Nonetheless, the tabu search was able to improve the schedules from MPT by 3.6%, on average, thereby reducing the gap between MPT and the lower bound by 29.9%, on average (see Section 5). [1] provide an excellent review of more complex tabu searches applied to production scheduling problems.

4 INTEGER PROGRAM AND LOWER BOUNDS

In this section we present an integer programming formulation of the problem and a lower bound.

The problem can be formulated as a mixed integer program by defining a set $P = \{1, \ldots, n\}$ of positions to which the jobs can be assigned. Two jobs (or portions of two jobs) are assigned to each position, one of the jobs for the first line and one of the jobs for the second line. Thus a position represents two jobs (or portions of jobs) that are processed simultaneously. Each position is preceded by a setup. If both of the jobs for that position must be set up, then a double-make-ready occurs. Let

$$y_{ikl} = \begin{cases} 1 & \text{if a portion of job } i \text{ is assigned to position } k \text{ on line } l, \\ 0 & \text{else,} \end{cases}$$

c_i =number of cartons requested for job i (parameter),

c'_i = actual number of cartons produced for job i,

c''_{ikl} = number of cartons produced for job i in position k on line l,

s_{ij} = setup time for job j if it immediately follows job i (parameter),

s_{0j} = setup time for job j if j is the first job on a line (parameter),

$s_j^{\max} = \max_{i \in J} s_{ij}$,

S'_{kl} = setup time for the job in position k on line l,

S_k = setup time for position k,

$$f_{ikl} = \begin{cases} 1 & \text{if job } i \text{ is completed on or before position } k \text{ on line } l, \\ 0 & \text{else,} \end{cases}$$

r_i = maximum processing rate for job i (cartons/minute) (parameter),

t_{kl} = time needed to process the cartons assigned to position k on line l,

T_k =time needed to process the cartons assigned to position k,

α_i = adjustment factor for job i.

Then the problem can be formulated as:

$$\min 2\sum_{k=1}^{n} T_k + 2\sum_{k=1}^{n} S_k + \sum_{i=1}^{n} \alpha_i \tag{5.1}$$

$$.9c_i \le c'_i \le 1.1c_i \qquad i \in J \tag{5.2}$$

$$c'_i = \sum_{l=1}^{2} \sum_{k=1}^{n} c''_{ikl} \qquad i \in J \tag{5.3}$$

$$c''_{ikl} \le 1.1c_i y_{ikl} \qquad l = 1,2; i \in J; k \in P \tag{5.4}$$

$$\sum_{i=1}^{n} c''_{ik1} = \sum_{i=1}^{n} c''_{ik2} \qquad k = 1,\ldots,n-1 \tag{5.5}$$

$$\sum_{i=1}^{n} y_{i1l} = 1 \qquad l = 1,2 \tag{5.6}$$

$$\sum_{i=1}^{n} y_{i,k+1,l} \le \sum_{i=1}^{n} y_{ikl} \qquad l = 1,2; k = 1,\ldots n-1 \tag{5.7}$$

$$\sum_{l=1}^{2} \sum_{k=1}^{n} y_{ikl} \ge 1 \qquad i \in J \tag{5.8}$$

$$S'_{1l} \ge \sum_{i=1}^{n} s_{0,i} y_{i1l} \qquad l = 1,2 \tag{5.9}$$

$$S'_{kl} \ge s_j^{\max} y_{jkl} - s_j^{\max} + \sum_{i=1, i \ne j}^{n} s_{ij} y_{i,k-1,l} \qquad l = 1,2; j \in J; k = 2,\ldots,n \tag{5.10}$$

$$S_k \ge S'_{kl} \qquad l = 1,2; k \in P \tag{5.11}$$

$$t_{kl} = \sum_{i=1}^{n} c''_{ikl}/r_i \qquad l = 1,2; k \in P \tag{5.12}$$

$$T_k \ge t_{kl} \qquad l = 1,2; k \in P \tag{5.13}$$

$$f_{ikl} \ge y_{ikl} - y_{i,k+1,l} \qquad l = 1,2; i \in J; k = 1,\ldots,n-1 \tag{5.14}$$

$$f_{i,k+1,l} \ge f_{ikl} \qquad l = 1,2; i \in J; k = 1,\ldots,n-1 \tag{5.15}$$

$$f_{in1} + f_{in2} = 1 \qquad i \in J \qquad (5.16)$$

$$y_{ikl} \leq 1 - f_{i,k-1,l} \qquad l = 1,2; i \in J; k = 2, \ldots, n \qquad (5.17)$$

$$\alpha_i = (c_i - c'_i)/r_i \qquad i \in J \qquad (5.18)$$

$$y_{ikl}, f_{ikl} \text{ binary} \qquad l = 1,2; i \in J; k \in P \qquad (5.19)$$

$$c'_i, c''_{ikl}, S'_{kl}, S_k, t_{kl}, T_k, \alpha_i \geq 0 \qquad (5.20)$$

The constraints (5.2) ensure that the actual number of cartons produced for a job is within ±10% of the requested number of cartons. Constraints (5.3) set c'_i equal to the total number of cartons assigned to the various positions for job i. Constraints (5.4) permit the assignment of cartons for job i to position k on line l only if y_{ikl} has been set equal to one. Constraints (5.5) force the same number of cartons to be assigned to the two lines in position k, except for the last position. Constraints (5.6) require that jobs are assigned to the first position on both lines, (5.7) allow position $k+1$ to be used only if position k is used, and (5.8) ensure that every job is assigned to at least one position.

Constraints (5.9), (5.9), and (5.11) determine the setup time for the jobs that are assigned to position k. Constraints (5.9) determine the setup time for the special case of the first position, where there is no job preceding the job to be set up. Constraints (5.9) determine the setup time for line l at position k. Finally, constraints (5.11) force the setup time for position k to be the maximum of the setup times for the two lines at position k. Similarly, constraints (5.12) and (5.13) determine the time to process the cartons that are assigned to position k. Constraints (5.12) determine the time to process the cartons that are assigned to line l at position k, and constraints (5.13) force the time for position k to be the maximum of the processing times for the two lines.

The variables f_{ikl} have been defined for the sole purpose of ensuring that a job is not split into two or more parts, i.e., that all of the cartons for a job must be assigned to consecutive positions on the same line. Constraints (5.14) determine at which position a job is finished. Constraints (5.15) require that if a job has finished by position k, then it has also finished by position $k+1$. Constraints (5.16) force each job to finish on only one of the two lines, thus each job can be processed on only one of the two lines. Constraints (5.17) do not allow cartons for a job to be assigned to position k if the job has already finished by position $k-1$.

Constraints (5.18) determine the adjustment factors for the jobs.

A few remarks regarding the integer program are in order. First, the variables c_i' and c_{ikl}'' should be integer variables, since it is impossible to produce a fractional number of cartons. We have decided to drop this requirement since the number of cartons that must be produced is so large that it is quite simple to round to a feasible solution with essentially the same objective function value. Second, as mentioned previously, the optimal schedule may not assign an equal number of cartons to the two lines. This possibility is permitted by the integer program since constraints (5.5) are not included for the last position. This means, however, that such a schedule must assign cartons to all n positions, which in turn requires $2n$ setups. This appears to conflict with the fact that there must be exactly n setups, one for each job. The integer program resolves this difficulty by permitting fictitious setups of length zero whenever a job is assigned to two consecutive positions.

In Section 5, the linear programming relaxation (LPR) of the integer program will be used to compute a lower bound on the cost of an optimal schedule. The linear programming relaxation is obtained by replacing the binary constraints on y_{ikl} and f_{ikl} by $0 \leq y_{ikl} \leq 1$ and $0 \leq f_{ikl} \leq 1$.

A different lower bound can be obtained from the data by summing the minimum processing time needed to produce the requested number of cartons for each job and adding to that sum a lower bound for the setup times obtained by solving an assignment problem. It is to be expected that this lower bound is not very tight since it implicitly assumes: 1. The jobs are processed at their maximum processing rate, which is equivalent to assuming that the jobs are perfectly matched together on the two lines of the machines; 2. Every setup is part of a double-make-ready; and 3. An equal number of cartons is assigned to both lines.

5 COMPUTATIONAL RESULTS

In this section we report our computational experience on six real-world problems. The problems were obtained from a packaging company and are proprietary. Table 3 summarizes the main characteristics of the test problem set. The jobs are divided into categories based on the number of colors that are required. For each category, the column labeled Jobs contains the number of jobs in that category and the column labeled Cartons contains the total number of cartons in that category.

Table 3 Characteristics of the test problems.

	1 Color		*2 Colors*		*3 Colors*		*4 Colors*	
Prob.	*Jobs*	*Cartons*	*Jobs*	*Cartons*	*Jobs*	*Cartons*	*Jobs*	*Cartons*
1	0	0	2	90,000	6	783,500	9	256,000
2	0	0	10	304,000	10	418,000	1	90,000
3	1	14,000	12	427,500	12	306,000	1	16,000
4	1	20,000	5	240,000	13	578,000	12	600,000
5	6	199,600	5	684,000	21	647,400	20	726,450
6	8	141,000	29	1,054,800	24	507,500	10	217,800

Lower bounds for the test problems were obtained by solving the linear programming relaxation (LPR) of the integer programming formulation of the problem and by the alternative method described in Section 4. The integer programs were generated by using GAMS 386 Version 2.25. GAMS was unable to generate integer programs for the two largest problems due to the large number of variables and constraints. The LPRs were solved by using the dual algorithm from CPLEX 4.0. Problem 4 exceeded the memory capacity of the computer that we used.

Table 4 presents the lower bounds for the test problems. The columns labeled Rows and Cols contain the number of rows and columns in the integer programming formulation. The columns labeled Pivots and CPU contain the number of dual pivots and the cpu time in seconds (Intel P90 processor) required to solve the linear programming relaxation. The column labeled z contains the optimal value of the linear programming relaxation rounded up to the next largest integer. The column labeled Min Process contains the minimum time to process all of the requested cartons. The column labeled Setup LB contains the lower bound on the setup time obtained by solving an assignment problem. The column labeled LB is the alternative lower bound and is equal to the sum of the previous two columns.

The lower bound provided by LPR is approximately the same as the alternative lower bound for problems 2 and 3, but it is substantially weaker for problem 1. The reason for this is that LPR provides a very weak lower bound on the setup times since constraints (5.9) allow the setup time for position k on line l to be set equal to zero whenever the values of y_{ikl} are less than one half. Thus LPR does not provide as good a lower bound for problem 1 since the setup time is large (compared to problems 2 and 3). For this reason, we did not pursue generating and solving LPR for problems 4, 5, and 6 any further.

Table 4 Lower bounds for the test problems.

	LP Relaxation					*Alternative LB*		
Problem	*Rows*	*Cols*	*Pivots*	*CPU*	*z*	*Min Process*	*Setup LB*	*LB*
1	2,977	1,871	2,270	53	2,420	2,361	140	2,501
2	4,517	2,815	3,472	207	1,603	1,581	20	1,601
3	6,892	4,265	6,190	1,711	1,401	1,385	20	1,405
4	9,767	6,015				3,182	140	3,322
5						4,690	215	4,905
6						3,574	50	3,624

As can be seen from Table 4, the LPRs are fairly large and can require a large amount of cpu time to solve. This effectively eliminates any possibility of finding the optimal solution by using branch and bound. Indeed, we attempted to solve the smallest problem by using the default settings of CPLEX, except that the dual algorithm was applied to the root node. CPLEX was given a time limit of ten hours, during which time it solved 316 nodes but neither improved the lower bound provided by LPR nor found a better solution than the one found by tabu search.

The tabu search was implemented in Turbo Pascal 5.5; no special optimization options were used to compile the program. Table 5 presents the results. The column labeled LB is the lower bound, T is the value of the objective function in minutes, Gap is the relative difference between T and LB, and CPU is the cpu time in seconds on a personal computer with an Intel P90 processor. For the tabu search, the column labeled Best Iter is the iteration in which the best schedule was found.

Table 5 Performance of MPT and tabu search.

			MPT		*Tabu Search*			
Problem	*Number of Jobs*	*LB*	*T*	*Gap*	*T*	*Gap*	*CPU*	*Best Iter*
1	17	2,501	2,773	10.9%	2,691	7.6%	3	3
2	21	1,601	1,916	19.7%	1,755	9.6%	7	17
3	26	1,405	1,595	13.5%	1,523	8.4%	11	7
4	31	3,322	3,671	10.5%	3,605	8.5%	19	41
5	52	4,905	5,434	10.8%	5,342	8.9%	86	5
6	71	3,624	3,933	8.5%	3,859	6.5%	212	57

MPT produced schedules with an average gap of 12.3% and that were always within 19.5% of the lower bound. The tabu search found improved schedules with an average gap of 8.2% and that were always within 9.5% of the lower bound. Thus tabu search reduced the average gap from 12.3% to 8.2%. Given the complexity of the problem (recall that it contains both the traveling salesman problem and the partitioning problem as special cases), an average gap of 8.2% appears to be pretty good. Furthermore, the gap is measured relative to a lower bound, not the true optimal solution. Since the lower bound that we have used is rather simplistic, we believe that a significant proportion of that gap is due to the lower bound, which further leads us to believe that the schedules produced by the tabu search are very good. Note that the largest problem required only three and a half minutes of cpu time, which is quite reasonable considering that the schedules require at least 23 hours to execute.

The current shop-floor practice is to find schedules manually using simple "rules-of-thumb," such as trying to maximize the number of double-make-ready's. We did not have access to the actual schedules that were created manually for the test problems, so we cannot estimate how much improvement the company could expect to obtain by using the tabu search.

Finally, notice that the best schedule found by tabu search was usually found very early during the search. This indicates that 100 iterations was sufficient to allow the tabu search to thoroughly search the solution space. Indeed, we ran tabu search with 350 iterations on several of the smaller problems without any further improvement in the objective function.

Acknowledgments

The authors thank Henry Shum for introducing this problem to us. The authors gratefully acknowledge Chung-Lun Li for participating in many useful discussions and Drew Spangler for helping with the computer implementation of the tabu search.

REFERENCES

[1] J. W. Barnes and M. Laguna, 1993, A tabu search experience in production scheduling, *Annals of Operations Research*, 41, 141-156.

[2] T. C. E. Cheng and C. C. S. Sin, 1990, A state-of-the-art review of parallel-machine scheduling, *European Journal of Operations Research*, 47, 271-292.

[3] F. Glover, 1989, Tabu search, part i, *ORSA Journal on Computing*, 1, 190-206.

[4] F. Glover, 1990a, Tabu search - part ii, *ORSA Journal on Computing*, 2, 4-32.

[5] F. Glover, 1990b, Tabu search: a tutorial, *Interfaces*, 20, 74-94.

[6] F. Glover, 1995, Tabu search fundamentals and uses, Technical report, University of Colorado.

6

OPTIMIZING THE RADIAL COMPONENT INSERTION OPERATIONS ON PCBS

Soo Y. Chang, Hark-Chin Hwang, and Katta G. Murty*

Department of Industrial Engineering
Pohang University of Science and Technology
Hyoja Dong San 31, Pohang, 790-784, Kyungbook
SOUTH KOREA

**University of Michigan*
Ann Arbor, Michigan 48109-2117
USA

ABSTRACT

We consider the problem of generating an optimal plan for radial components insertion on PCBs. The problem is modeled as an integer programming model, but it can easily be shown to be NP-hard. Hence, we propose a heuristic method and test its performance through computational experiments on the actual PCB data obtained from S-electronics company which assemble PCBs for their high-resolution computer display devices. This heuristic method with an easy-to-use graphic user interface is implemented and being used currently.

1 INTRODUCTION

PCBs are usually green plastic rectangular boards up to 10" by 15" with many holes in which electronic components are to be inserted and fixed. PCBs are used extensively in electronic devices such as high-resolution computer display devices. There are literally thousands of types of PCBs with new types being created daily.

The S-Electronics Company in South Korea ("SEC" for short) produces about ten thousand units of assembled PCBs per month. In every month, about forty

G. Yu (ed.), Industrial Applications of Combinatorial Optimization, 125–141.

different types of PCB's are manufactured, and about four to five new models are introduced into the manufacturing system for the first time.

When a new model is introduced into the manufacturing system for the first time, SEC must come up with "good" NC (Numerical Control) code for processing that PCB model on the various automated manufacturing machines from the design specification given in the form of the electronic CAD data by its customer. The efficiency of this NC code as well as its prompt derivation are vital to SEC's competitive edge against many small scale computer display manufacturers which could respond to the customers relatively faster than SEC, since they are free from much of the organizational overhead that the bigger companies like SEC must deal with.

For each new model, the NC codes must be generated for the machines for the axial part insertion and several different types of machines for the radial part insertion. Radial insertion machines are expensive, and SEC has six different types of them. In this paper, we will focus our discussion on one type of radial machines called the "RHU" machine, since these machines are the most popular machines for the radial part insertion, and they are the "bottle-neck" which determines the overall productivity of the PCB lines at SEC.

In the following sections, we give detailed description of the "RHU" machine and the problem definition along with its mathematical model. Then we propose a heuristic method and summarize our computational experience.

2 PROBLEM DEFINITION

A schematic diagram of the "RHU" machine is given in Figure 1. The machine consists of an X–Y table on which the board is fixed. On top of the table there is an insertion gun. During the insertion operation, the X–Y table moves the PCB to position each hole right under the gun. The insertion gun itself never moves. There is a linear strip through the insertion gun containing a row of stations. Each station on the linear strip is allowed to be loaded with only one type of components arranged serially on a long tape. There could be 20 - 70 different types of radial components on each PCB. A PCB may need to have upto about 5 components of each type. Each component goes into a hole. The holes are in predetermined positions on the PCB as illustrated in Figure 2. The number of stations on the strip is always greater than or equal to the number of different types of radial components to be inserted.

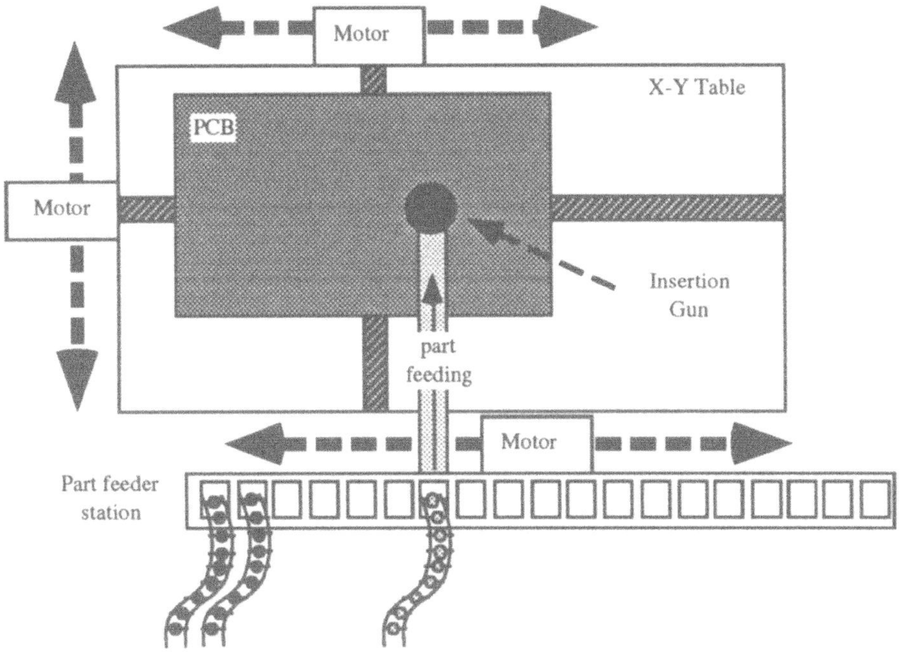

Figure 1 The schematic diagram of the typical radial part insertion machine.

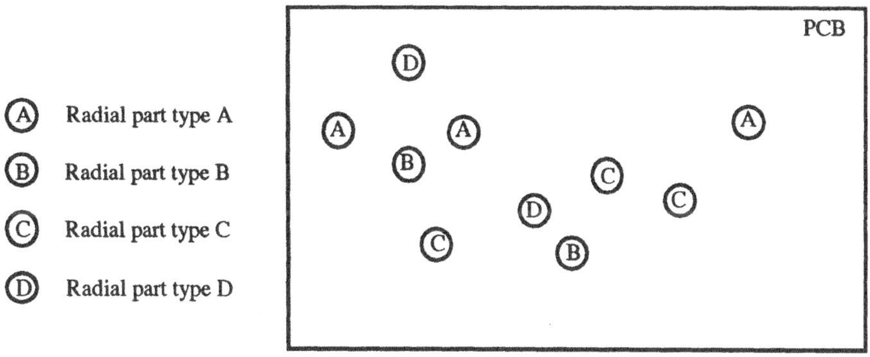

Figure 2 The schematic diagram of a PCB with 4 different types of radial parts.

The generation of NC codes for manufacturing a PCB involves two decisions: (*a*) the insertion sequence, i.e., the order in which the holes in the PCB are filled, and (*b*) the assignment of component types to the stations on the linear strip. Collectively, these two decisions determine the ***process plan*** for processing the PCB on the RHU machine.

Each insertion of a component in a hole is carried out through the following steps.

(1) The X–Y table moves from its present position so as to bring the hole right under the insertion gun. While this is going on, the linear strip in the insertion gun moves so that the station containing the type of component to be inserted into the hole comes into position in the insertion gun. When both these moves are complete, the component is inserted by the gun right into the hole. The machine then moves on to the next hole in the sequence.

(2) The X–Y table moves both in the X and Y directions simultaneously until the correct position is reached at a constant speed of d inches per time unit in each direction. Hence after inserting a component at the hole whose Cartesian coordinates are (x_1, y_1) on the board, the time that it takes for the X–Y table to position the hole at the point (x_2, y_2) under the insertion gun is: $(1/d)(\max\{|x_1 - x_2|, |y_1 - y_2|\})$ time units. As an example, if holes 1 and 2 have their centers at points (5, 7) and (12, 3) on the board, and the speed d is 3 inches/time unit, then bringing hole 2 under the insertion gun after filling hole 1 will take the X–Y table $(1/3)(\max\{|5-12|, |7-3|\})$ $= 7/3$ time units.

The linear strip in the insertion gun can move left or right to bring the necessary station into the insert position, but its speed depends on the length of the move. To move a length of r stations, it takes t_r time units, where t_r is monotonically increasing with r and all t_r's are given. Assume that the stations are numbered 1,2,... from left to right on the strip. Thus if station k_1 is currently in the insert position, it takes the strip $t_{|k_1-k_2|}$ time units to move so as to bring station k_2 into the insert position.

Suppose the hole with its center at (x_1, y_1) has just been filled with the component stocked at station k_1; and the next hole to be filled has its center at (x_2, y_2), and the component that goes into this hole is stocked in station k_2 on the linear strip. Then the time lapse before the next part insertion is executed is $\max\{(1/d)(\max\{|x_1 - x_2|, |y_1 - y_2|\}), t_{|k_1-k_2|}\}$ time units, since the X–Y table and the linear strip of stations move simultaneously and independently.

3 MATHEMATICAL MODEL

We present the mathematical model for the problem of finding the optimum process plan for radial component insertion.

Assume that the PCB needs p different types of radial components numbered $k = 1, \ldots, p$; and that there are m stations numbered $i = 1, \ldots, m$, where $m \geq p$. Suppose we have n holes numbered $j = 1, \ldots, n$ on this PCB. For the jth hole, we are given the coordinates of its center (x_j, y_j), and the number of the type of component, $f(j)$, that goes into it.

Then the problem of generating the optimum process plan for this PCB is: finding the *insertion sequence* of the n holes; as well as the *station assignment*, i.e., the assignment of the p component types to the m stations on the linear strip (each station can store at most one component type, and each component type must appear on at least one station), to minimize the total time for the entire radial component insertion operation. The decision variables are:

- $\pi = (\pi_1, \ldots, \pi_n)$: the insertion sequence, permutation of $\{1, \ldots, n\}$.
- $A = (a_{ki})$, the $p \times m$ incidence matrix of radial part type assignment to the stations, i.e.,

$$a_{ki} = \begin{cases} 1 & \text{if } k \text{ th component type is stored in the} \\ & i\text{th station on the linear strip} \\ 0 & \text{otherwise} \end{cases}$$

The mathematical model for the problem is easier in the special case when $p = m$, since in this case the matrix $A = (a_{ki})$ is a square matrix and a permutation matrix (as there are m stations and exactly m types of components in this special case, each type of component gets stored in exactly one station) and hence we present this case first. The mathematical model for the special case is:

$$\text{Minimize} \quad \sum_{j=1}^{n-1} \max\{(1/d)|x_{\pi_j} - x_{\pi_{j+1}}|, (1/d)|y_{\pi_j} - y_{\pi_{j+1}}|, t_{|s(j)-s(j+1)|}\}$$

$$\text{subject to} \quad \sum_{i=1}^{m} a_{ki} = 1 \qquad \text{for } k = 1, \ldots, p$$

$$\sum_{k=1}^{p} a_{ki} = 1 \qquad \text{for } i = 1, \ldots, m$$

$$s(j) = \sum_{i=1}^{m} i a_{f(\pi_j)i} \qquad \text{for } j = 1, \ldots, n$$

$$\pi = (\pi_1, \ldots, \pi_n) \qquad \text{is a permutation of } \{1, \ldots, n\}$$

$$a_{ki} \geq 0 \qquad \text{and integer for all } k = 1, \ldots, p; \quad i = 1, \ldots, m$$

Here $s(j)$ denotes the station number on the linear strip where the component type $f(\pi_j)$ that goes into the jth hole filled in the sequence π, is stored.

There are two things to be noted about this model. The first is the fact that there are two different decisions involved in it, the station assignment (defined by the variables a_{ki}), and the insertion sequence (defined by the variables π). If one of them is fixed, the model becomes a bit simpler. We refer to this overall model with both these decisions to be determined as MODEL A.

The second thing to be noted about this model is the fact that the problem becomes the shortest Hamiltonian path problem in 2-dimensional space if we ignore the cost associated with the movement in the linear strip, namely, if we set all $t_r = 0$. In this case the model reduces to:

$$\text{Minimize} \qquad \sum_{j=1}^{n-1} \max\{(1/d)|\, x_{\pi_j} - x_{\pi_{j+1}}|, (1/d)|\, y_{\pi_j} - y_{\pi_{j+1}}|\}$$

$$\text{subject to} \qquad \pi = (\pi_1, \pi_2, \ldots, \pi_j, \ldots, \pi_n) \quad \text{is a permutation of } \{1, 2, \ldots, n\}$$

This special case of MODEL A will be referred to as MODEL B. It is a shortest Hamiltonian path problem with L-infinity norm which is known to be NP-hard. Therefore, MODEL A is clearly an NP-hard problem also.

It is useful to note, however, that the optimum objective value for MODEL B is a lower bound to the optimum objective value of MODEL A, since;

$$\max\{(1/d)|\, x_{\pi_j} - x_{\pi_{j+1}}|, (1/d)|\, y_{\pi_j} - y_{\pi_{j+1}}|\} \leq$$
$$\max\{(1/d)|\, x_{\pi_j} - x_{\pi_{j+1}}|, (1/d)|\, y_{\pi_j} - y_{\pi_{j+1}}|, t_{|s(j)-s(j+1)|}\}.$$

Model A becomes more complicated when $m > p$. In this case, a component type can be stored in more than one station, and some stations may not be used at all. In fact, efficient use of the stations when $m > p$ is a very important part of the heuristic method that we present later. In this case, if the linear strip has to move from its present position to a station containing the kth component type that is stored in more than one station, it will move to the one closest to its present position. To present Model A in this general case, when T is an index set over which an index t can vary, and b_t is a real valued function defined over T, we define

$$\arg \ \min\{\min_{t \in T}\{b_t\}\}$$

to be the set of indices $t \in T$ that correspond to the smallest value of b_t over T. The model is:

$$\begin{array}{ll}
\text{Minimize} & \sum_{j=1}^{n-1} \max\{(1/d)|x_{\pi_j} - x_{\pi_{j+1}}|, (1/d)|y_{\pi_j} - y_{\pi_{j+1}}|, t_{|s(j)-s(j+1)|}\} \\
\text{subject to} & \sum_{i=1}^{m} a_{ki} \geq 1 \qquad \text{for } k = 1, \ldots, p \\
& \sum_{k=1}^{p} a_{ki} \leq 1 \qquad \text{for } i = 1, \ldots, m \\
& s(1) \in \{i : a_{f(\pi_1),i} > 0\} \\
& s(j+1) \in \arg\min\{\min\{t_{|s_j - i|} : \text{over } i \text{ such that } a_{f(\pi_{j+1}),i} > 0\}\}, \\
& \text{for } j > 1 \\
& \pi = (\pi_1, \ldots, \pi_n) \qquad \text{is a permutation of } \{1, \ldots, n\} \\
& a_{ki} = \quad 0 \text{ or } 1 \text{ for all } k = 1, \ldots, m; \quad i = 1, \ldots, m
\end{array}$$

This model can be transformed into standard 0-1 linear mixed integer programming by defining additional 0-1 variables, as shown below. Define 0-1 variables z_{ji} to be 1 if the component to be filled in the jth hole is drawn from the ith station, or 0 otherwise. Then, the model becomes;

$$
\begin{aligned}
\text{Minimize} \quad & \sum_{j=1}^{n-1} \max\{(1/d)|x_{\pi_j} - x_{\pi_{j+1}}|, (1/d)|y_{\pi_j} - y_{\pi_{j+1}}|, t_{|s(j)-s(j+1)|}\} \\
\text{subject to} \quad & \sum_{i=1}^{m} a_{ki} \geq 1 \qquad \text{for } k = 1, \ldots, p \\
& \sum_{k=1}^{p} a_{ki} \leq 1 \qquad \text{for } i = 1, \ldots, m \\
& s(j) = \sum_{i=1}^{m} i z_{\pi_j i} \qquad \text{for } j = 1, \ldots, n \\
& \sum_{i=1}^{m} z_{ji} = 1 \qquad \text{for all } j = 1, \ldots, n \\
& z_{ji} \leq a_{f(j)i} \qquad \text{for all } j = 1, \ldots, n \text{ and } i = 1, \ldots, m \\
& \pi = (\pi_1, \ldots, \pi_n) \qquad \text{is a permutation of } \{1, \ldots, n\} \\
& a_{ki} \quad \text{is 0-1 integer for } k = 1, \ldots, p; \quad i = 1, \ldots, m \\
& z_{ji} \quad \text{is 0-1 integer for } j = 1, \ldots, n; \quad i = 1, \ldots, m
\end{aligned}
$$

4 HEURISTIC METHOD

The heuristic method that we propose in this paper consists of the alternating series of insertion sequencing and the station assignment procedures.

The insertion sequencing procedure tries to find the best insertion sequence while keeping the station assignment fixed. If the station assignment is fixed, the problem becomes very much like the shortest Hamiltonian path problem with cities in 3-dimensional space where the distances among cities are measured by the objective function of MODEL A.

On the other hand, if the insertion sequence is fixed, getting the station assignment becomes similar to the quadratic assignment problem(QAP) and the station assignment procedure seeks the best station assignment for the given insertion sequence.

The station assignment is not, however, a one-to-one assignment of component types to stations since we may have more stations than the number of radial component types. In such cases, the effective use of stations plays an important role. Figure 3 demonstrates how the effective use of stations could improve the efficiency of the solution.

In Figure 3, we have two different station assignment plans for a PCB with 4 different types of radial components. Both of the plans in Figure 3 use the same insertion sequence. Only difference is that the plan in part (a) uses 4 stations and the plan in (b) uses 5 stations. As one can see from the example, the use of extra available station yields less total number of jumps among stations. The less total number of station jumps, however, does not necessarily mean the plan is more efficient, since the overall efficiency of the plan is determined by the objective function as defined in MODEL A. Nevertheless, a plan with less total number of station jumps will yield better or same efficiency as the plan with more total number of station jumps, if both of the plans use the same insertion sequence.

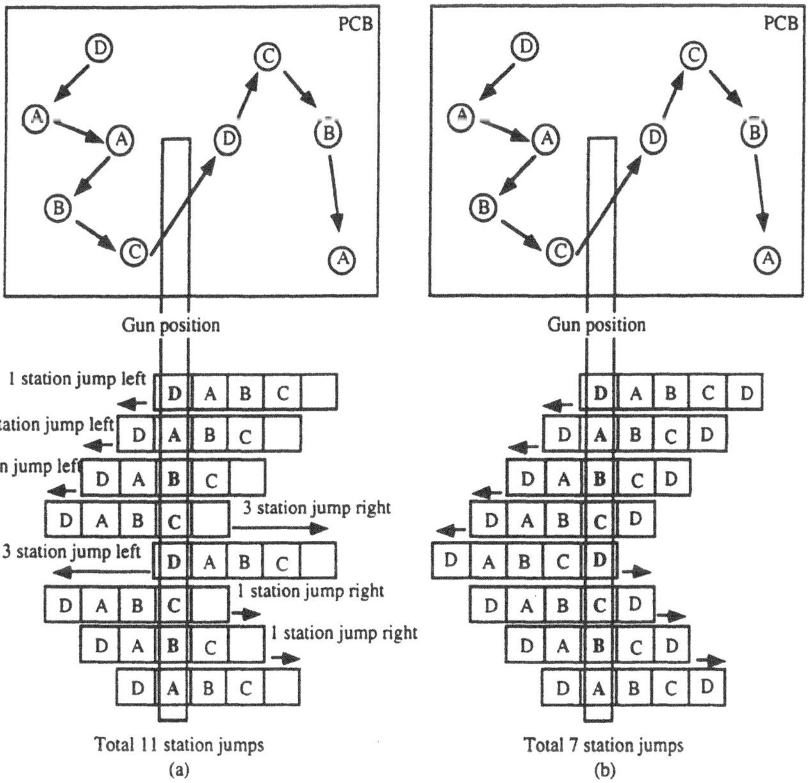

Figure 3 Station assignment examples.

The station assignment is done by a greedy type procedure. It starts out with a tentative station assignment assuming unlimited number of available stations, then reduces the number of stations used if necessary. In the process of reducing the number of used stations, however, we tend to introduce more station jumps than before. The following example illustrates this reduction process.

Figure 4 shows a PCB with 16 components with 5 different types and 9 available stations on the machine. Also, the predetermined insertion sequence is shown on the board.

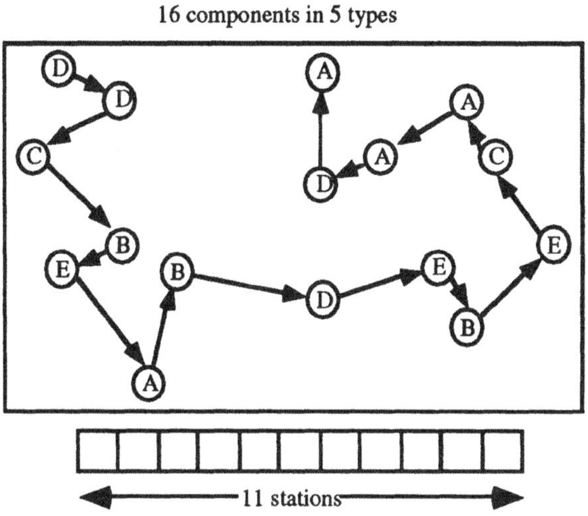

Figure 4 A PCB example.

In this example, the insertion sequence starts from type D component and moves on to C, B, E, A, B, D, E, B, E, C, A, D and then to A. In this sequence, there are 13 component type changes. If we must keep this insertion sequence, 13 station jumps is clearly inevitable. Hence, if we had unlimited number of stations, we could have used 14 stations and come up with the optimum station assignment. However, we only have 11 available stations. Therefore, we have to merge three stations with others. Namely, we have to choose a pair of stations which are assigned the same type of the radial component and merge them, and repeat this three times. In our heuristic method, we choose the particular pair of stations assigned the same type of radial component which incurs the least additional station jumps when they are merged.

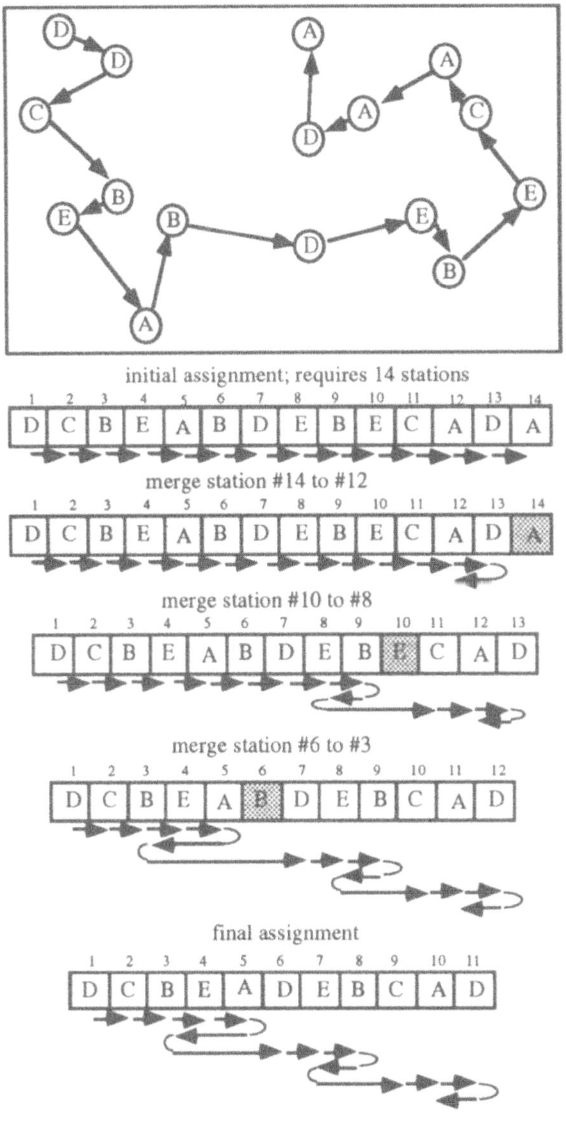

Figure 5 Station assignment procedure.

Suppose we have decided to merge the 14-th station containing the part type A with the station number 12. Then the movement on the linear strip will end at the station number 12 instead of 14. In this particular case of merge, there is no increase in the total number of station jumps. After this merge, we still have 13 stations, so further reduction must be done. Suppose we have chosen to merge the station number 10 containing the component type E with the station number 8. In this case, however, we introduce one more station jump than before. Finally, if we have chosen to merge the station number 6 containing the component type B with the station number 3, we introduce three additional station jumps and we have the final station assignment using all 11 available stations.

A station assignment gives the sequence of component types stored in the various stations on the linear strip. We now describe the procedure for computing the number of station jumps associated with a station assignment.

PROCEDURE 1: To compute the number of station jumps given a station assignment and insertion sequence:

Let the permutation $\pi = (\pi_1, \ldots, \pi_n)$ of $\{1, \ldots, n\}$ define the insertion sequence. Let the station assignment be $C_1, C_2, \ldots, C_t$ occupying t consecutive stations on the linear strip ($C_!, \ldots, C_t$ are the component types stored in stations $1, \ldots, t$). Define $s(j)$ to be the position of station from which the the component for the jth hole is picked up, computed by:

$s(1)$ = the left most station among those storing component type $f(\pi_1)$.

$s(j+1)$ = the station containing component type $f(\pi_{j+1})$ that is closest to station s_j (break ties arbitrarily), for $j \geq 1$.

Then the number of station jumps corresponding to the present station assignment and insertion sequence is $\sum_{j=1}^{n-1} t_{|s(j)-s(j+1)|}$.

We now present the greedy type method outlined above for finding the station assignment, given the insertion sequence, in pseudocode form.

BEGIN

1. Let the permutation $\pi = (\pi_1, \ldots, \pi_n)$ of $\{1, \ldots, n\}$ define the insertion sequence. Write the sequence of component types in this sequence $f(\pi_1), \ldots, f(\pi_n)$ and merge consecutive entries which are the same into

a single entry. The resulting sequence is the initial tentative assignment. Go to 2.

2. Let the present tentative station assignment be $C_1, \ldots, C_t$ occupying t stations. If $t \leq m$, terminate with this as the station assignment.

   If $t > m$, go to 3.

3. Let the present tentative station assignment be $C_1, \ldots, C_t$ occupying t stations.

   For each pair of stations in this assignment that have the same component type, say stations r and $s > r$; using Procedure 1 find the number of station jumps corresponding to the two tentative station assignments

   $$C_1, \ldots, C_{r-1}, C_{r+1}, \ldots, C_t \quad \text{and} \quad C_1, \ldots, C_{s-1}, C_{s+1}, \ldots, C_t$$

   i.e., the two tentative station assignments occupying $t - 1$ stations obtained by deleting one of the two entries in the pair.

   Among all these tentative station assignments occupying $t - 1$ stations evaluated above, make the one having the smallest number of station jumps as the new present tentative station assignment, breaking ties arbitrarily; and go to 2 with it.

END

How to Find the Best Insertion Sequence Given the Station Assignment

Suppose the station assignment is fixed. Now we consider the problem of finding the best insertion sequence. We consider two cases.

Case 1: Each component type is stored in only one station in the station assignment

In this case, each component type is available in a unique station, and the problem of finding the best insertion sequence can be modeled as the first Model A discussed above corresponding to $p = m$. It is a shortest Hamiltonian path problem of covering all the holes 1 to n by a simple path when the cost matrix is $c = (c_{ij})$ where

$$\begin{aligned} c_{ij} &= \text{time lapse to fill hole } j \text{ immediately after filling hole } i \\ &= \max\{(1/d)|x_i - x_j|, (1/d)|y_i - y_j|, t_{|s(i)-s(j)|}\} \end{aligned}$$

Since $s(i)$, the station number where component type $f(i)$ that goes into hole i is stored, is unique and well defined for all i in this case, this cost matrix $c = (c_{ij})$ can be computed unambiguously in this case.

We will discuss how to solve the shortest Hamiltonian path problem with the given cost matrix, after presenting Case 2.

Case 2: Some component types are stored in more than one station

In this case some components can be obtained from more than one station. As an example, if the station assignment is $D\ A\ B\ C\ B\ A$, component type A can be obtained either from stations 2 or 6, B from either stations 3 or 5; but C, D from the unique stations 4, 1 only, respectively. In this case the problem of finding the best insertion sequence can be modeled using the general Model A corresponding to the case $p > m$. It is a shortest Hamiltonian path problem, but the cost function is not linear, it is sequence dependent. So, in this case we find the insertion sequence using a heuristic approach which preselects and assigns a station index number for getting the component type needed in each hole. For $j = 1$ to n, here we define

$s(j)$ = Station number in the station assignment, containing component type $f(j)$ that goes into hole j, which was used in Procedure 1 to compute the number of station jumps corresponding to the present station assignment

and fix this as the station index number for station j; i.e., we assume that the component to insert into hole j will be fetched from station j for $j = 1$ to n. With $s(j)$s defined as here, we can approximate the time lapse to fill hole j immediately after filling hole i by c_{ij} as defined in Case 1. With this approximation, the problem of finding the insertion sequence again becomes a shortest Hamiltonian path problem with cost matrix $c = (c_{ij})$.

The shortest Hamiltonian path problem can be easily transformed to a traveling salesman problem (TSP) by introducing an additional city, say city 0, to the problem. Namely, we add city 0 having a constant distance say α to and from all the existing cities. After solving the TSP for the augmented problem, we write the optimum tour beginning and ending with city 0. Then, we could recover the shortest Hamiltonian path in the original problem simply by deleting the added city from this tour. To solve the TSP, we use the algorithm that was developed by R. Jonker and T. Volgenant in [TSP 1,2,3].

When we first start our heuristic method, however, we don't have any station assignment. So, we solve the shortest Hamiltonian path problem with only the given x and y coordinates of each hole on the PCB. This 2-dimensional problem is used not only to give the initial insertion sequence, the optimum objective value in it is used as a lower bound for the optimum objective value in our problem to give an approximate evaluation of the quality of the final solution.

5 COMPUTATIONAL EXPERIMENT

For the insertion sequencing procedure, we have used the TSP code in Pascal called "tsp1" for personal computers developed by R. Jonker and T. Volgenant in [TSP 2,3,4]. The station assignment procedure is implemented in C language on a SPARK 20 UNIX workstation.

We show the results obtained when the method is applied to determine the process plan for 14 types of PCBs. These were the major types of PCBs being manufactured at SEC when we did this work for them.

The PCB data are from two product groups. One of them (the bottom 7 examples in Table 1) is processed on a single radial insertion machine with 62 stations.

The PCBs from the second group (the top 7 examples from Table 1) are big, and none of the RHU machines at SEC have enough number of stations on their linear strip to handle them. So, they are processed on two seperate radial insertion machines serially, each processing about half the holes on the board. Even though we find the process plan for these PCBs treating the total available stations on the two machines as one linear strip on an imaginary machine with 136 combined stations, the process plan is later partitioned into two seperate plans one for each of the two machines to be used in actual production.

The following table summarizes the computational experiment. The first column contains the code name of the PCB board. The second through fourth columns contain the number of holes on the PCB board, the number of types of the radial components, and the number of the stations available.

The next column is the lower bound for the optimum objective value, that was obtained from the first insertion sequencing procedure, i.e. the result that we got from the first 2-dimensional problem (MODEL B). The next column is the MODEL A objective value of the final process plan obtained by the heuristic method. The final column is the ratio of the objective value of the heuristic solution to the lower bound.

Each station assignment procedure took less than 30 seconds in every case. We stopped the execution of "tsp1" when it reaches 300 iterations which took about 2 to 4 minutes on 486 personal computer, and in all of our cases, "tsp1" found either the optimum solution or a solution within less than 1% of the optimum.

We terminated our heuristic method whenever the solution from the insertion sequencing procedure does not improve. In all of our examples, it took less than 4 alternating executions of station assignment and the insertion sequencing to terminate.

The objective values were measured by the time unit which is about one tenth of a second. The X–Y table is assumed to move 1 inches per time unit. Each station jump is assumed to take 1.5 of the time unit.

Before the use of the system that we developed, the SEC had to spend about a week to come up with the NC code manually for each new PCB design whenever it is introduced to the production. The method that we developed is implemented on a UNIX workstation together with a graphic user interface which allows the user to easily edit the insertion plan on the screen. With this new system, it takes about a half day to complete the derivation of the entire NC codes for a new PCB model.

Table 1 Summary of the computational experiment.

PCB code Name	No. of holes	No. of types	No. of stations	Lower bound	Heuristic Solution	Solution divided by Lower Bound
cmh	186	77	136	882	892	1.01
ctg	188	73	136	868	927	1.07
cmq	206	78	136	939	1034	1.10
cmg	206	82	136	925	1005	1.09
fx	223	77	136	956	1212	1.27
cmb	232	86	136	970	1220	1.26
cma	246	89	136	1057	1429	1.35
g62	131	62	62	752	1099	1.46
h62	131	62	62	751	1117	1.49
b62	144	62	62	783	1144	1.46
t62	150	62	62	762	1204	1.58
q62	152	62	62	817	1213	1.48
a62	157	62	62	856	1359	1.59
x62	165	62	62	826	1360	1.65

REFERENCES

[1] R. Jonker, T. Volgenant, 1984, Nonoptimal edges for the symmetric traveling salesman problem, *Operations Research*, 32, 837-846.

[2] A. Volgenant, R. Jonker, 1982, A branch and bound algorithm for the symmetric traveling salesman problem based on the 1-tree relaxation, *European Journal of Operational Research*, 9, 83–89.

[3] A. Volgenant, R. Jonker, 1983, The symmetric traveling salesman problem and edge exchanges in minimal 1-trees, *European Journal of Operational Research*, 12, 394–403

7

COMBINATORIAL OPTIMIZATION IN A CATTLE YARD: FEED DISTRIBUTION, VEHICLE SCHEDULING, LOT SIZING, AND DYNAMIC PEN ASSIGNMENT

Moshe Dror and Janny M.Y. Leung

Management Information Systems Department
College of Business and Public Administration
University of Arizona
Tucson, Arizona, 85721
USA

ABSTRACT

In this chapter describes various interesting combinatorial optimization problems which constantly present themselves in an operation of a large cattle yard. These problems include feed distribution, distribution vehicle scheduling, feed lot sizing (mixing) operation, and reassignment of cattle to pens. We examine the mathematical models and the corresponding solution methodologies for these problems, modeling them as a combination of arc routing, machine scheduling, lot sizing, and very large general 0-1 integer programming problems. Partial results and the considerable insight gained by our analysis is presented.

1 INTRODUCTION

This work examines in detail the daily and weekly operational issues of managing a very large cattle yard. The daily operations consist primarily of producing and distributing the different kinds of feed in a coordinated fashion, under rather restrictive feeding-time requirements, among the different categories (age/weight) of the cattle in the yard. The weekly managerial problem is that of a cost-effective assignment of the cattle lots to the over 600 pens in the yard as they arrive, their redistribution as the cattle mature and the release of pens as mature cattle lots are sent to their final destination.

G. Yu (ed.), Industrial Applications of Combinatorial Optimization, 142–171.

In this chapter we describe the various interesting combinatorial optimization problems which constantly present themselves in such a large cattle yard operation. These combinatorial problems, whilst not formulated explicitly as such, are solved daily "on line" by the management team (cowboys), to provide for a smooth, efficient, and profitable operation of the cattle yard. We examine the mathematical models and the corresponding solution methodologies of these operational problems, indicating considerable insight gained by our analysis, and present some partial results. The model for feed distribution presented here was first outlined in Dror and Mullaseril (1996). This chapter provides a more exhaustive list of the other various combinatorial problems encountered in a large cattle yard.

We start with the description of the cattle yard in this section. Next in Section 2, we focus on the feed distribution activities, presenting some mathematical models and solutions. After that, in Section 3, the resulting vehicle scheduling problem is described in a general framework, along with a description of the particular setting encountered on the cattle yard. The feed production problem, which can be modeled as a single-machine lot-sizing problem with side constraints, is discussed in Section 4. The last issue examined, in Section 5, is that of the dynamic assignment of the cattle lots to the different pens. We conclude the chapter with a short summary of the models and the results.

1.1 The Cattle Yard

The company that motivated this study is in the business of growing cattle for slaughter and has about 100,000 head of cattle in its yard at any one time. These animals are brought onto the premises by rail or by truck, usually when they are calves, and are placed in pens where they are fed and fattened. When mature, the cattle are taken, again by rail or trucks, for slaughter. Thus, the "added value" provided by the cattle yard is in the additional weight, and the improved meat quantity and quality, of the fattened cattle. Hence the overall goal of the cattle yard is to minimize feed production and delivery costs while maximizing the profit from the sale of the fattened cattle.

For the purpose of feed production, the company has a feed mill on the premises which is a continuous-process mill where different types of feeds are prepared. Usually, there are five different types of feeds required at any given time. The input materials to the mill consist of solids (e.g. hay, rolled grain, straw, alfalfa), minerals, micro ingredients, and liquids. There are five bins of fixed capacity in the plant to hold the solids to be fed into the feed production process and a bin can hold only a single solid type at any one time. The solids are weighed by a belt scale before their introduction to the mixing plant. Depending on the type of raw material and its relative proportion in the feed type being produced, there is a lower and upper bound to the flow rates due to the measurement accuracy of the belt scale. The other ingredients are added to the mix as required to form the finished product (i.e., a given feed type). The feed output of a production run in the mill is taken from the mill by a conveyor belt to temporary holding bins, called hoppers, of fixed volume capacity. Currently, six such hoppers are available. Since the relative proportion of the ingredients are different for the different feed types, and only certain flow rates are allowed (to assure accurate weighing), the production rate of the mill is different for the different feeds and the holding capacity (by weight) of the temporary storage hoppers is different for each feed. Furthermore, each hopper can only hold one type of feed, so no more than 6 different type of feeds can be held in temporary storage. These storage hoppers are about 25 feet above the ground, so that a truck may be driven under them to be loaded with feed, which is then delivered to the cattle in the appropriate pens. The loading time for a truck is approximately 5 minutes.

The cattle are housed in pens, which are generally rectangular in shape and approximately 50 feet by 150 feet each in size, that are arranged in rows. Figure 1 is a map of the cattle yard. The pens are accessed by both paved and unpaved roads. The feeding trough for a pen runs alongside the entire front edge of the pen that borders the service road. Currently, the pens are partitioned into five groups — one for each driver (truck), and each driver is responsible for the timely delivery of all types of feeds to the pens assigned to him.

The trucks have been specifically designed for carrying and discharging feed to the pens. Each truck may hold only one type of feed at a time. There are presently five trucks, four of which can hold about 25,000 lbs (12.5 tons) of feed and the fifth about 32,000 lbs (16 tons). Since the densities of the feed vary, the amount of feed carried, in tons per truckload, is different for each type of feed. The trucks are specially equipped to discharge the feed directly into the feeding trough of the pens from a chute fixed to the left (driver's) side of the truck while the truck moves at a pre-determined speed. The driver can control the flow rate by manipulating the truck speed and the size of the opening of the discharge chute. This feeding process necessitates the feed-discharging operation to be conducted in one direction only. When a truck has discharged all its load, it returns to the temporary storage hoppers at the feed mill to refill for the next run.

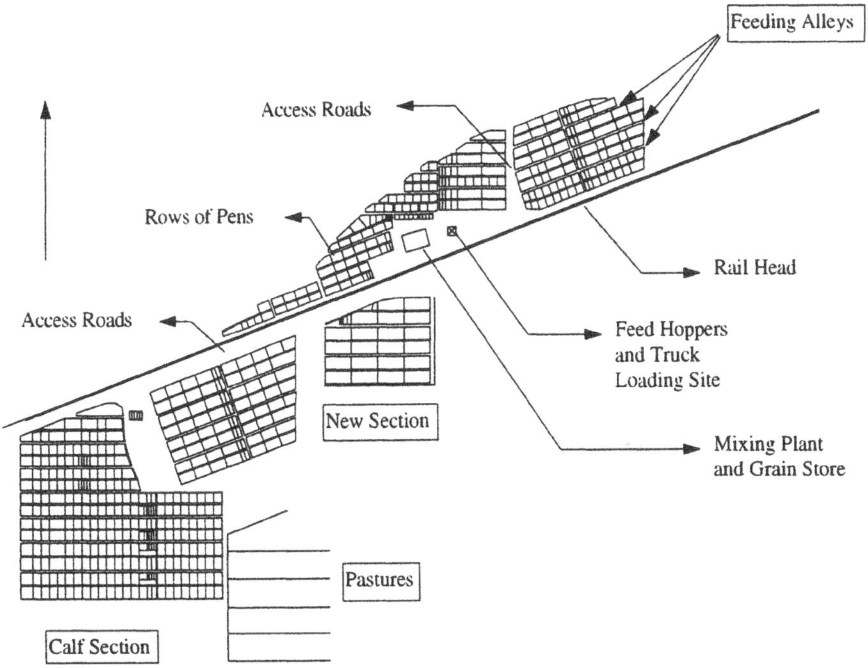

Figure 1 Map of the cattle yard.

The livestock feeding policy in the past was to feed the cattle once a day (with exceptions noted later), and except for 'small' variations, a given cattle lot in a pen must be fed at the same time each day. The daily demand and delivery quantity for each pen is not constant, and is estimated using a 7-day moving average of the daily consumption. Feeding exceptions arise when the feeding troughs of some pens are 'slick', that is, completely empty. Early each morning, the pens are inspected and a limited amount of feed (later deducted from the daily total) needs to be distributed to the pens with slick troughs in order not to leave those cattle hungry, since that might cause overeating later with some health-related detrimental consequences. The present feeding policy is to feed the cattle twice a day with a preset interval of 6 to 10 hours between the two feed deliveries, so now two feeding trips must be scheduled for each pen.

In addition to the management of feed production and distribution, one of the main activities in the yard involves the movement of cattle groups. When a new cattle lot arrives, it must be assigned to pens. As cattle mature, they must be moved to pens with facilities appropriate for larger animals. Periodically, when cattle are to be vaccinated or medically (hormonally) treated, they must be moved to one of the two confinement "service" areas. When cattle are shipped for slaughter, they must be loaded onto trucks. The cattle yard manager has stated that between the cattle weight loss and the labor involved, each cattle movement within the yard costs about 6 to 10 dollars per head, which is quite expensive. Furthermore, since cattle of different age and weight require different feeds, the pen assignments affect the location of the cattle requiring the same feed type and thus significantly impact the travel time and distance, and hence the cost, of feed delivery. Thus, minimizing the expected total number of cattle head movements within the yard over an extended period (for example, a year) would be a very desirable goal!

In the remainder of this chapter, we provide a more mathematical description of these various problems which must be tackled routinely in the course of operating the yard. As can be seen in the following sections, these problems are large combinatorial optimization problems. Terminology from graph theory, integer programming, scheduling theory, etc., will be introduced as the various problems are described, and their relation to more readily recognizable canonical optimization problems will also be discussed.

2 THE FEED DISTRIBUTION PROBLEM

The entire feed distribution operation in the yard is driven by the cattle's needs. In this section, we first focus on the problem of delivering the right feed to the cattle at the right time, assuming that the appropriate feed can be produced and will be ready for delivery as needed.

The cattle feed delivery activity is modeled as a sequence of Capacitated Rural Postman Problems (CRPP) with some additional constraints. In order to characterize our problem within a larger family of related problems, we start by defining the Rural Postman Problem (RPP).

2.1 The Rural Postman Problem and its Adaptation to Cattle Feed Distribution

The Rural Postman Problem (RPP) is the problem of finding a minimum length/cost traversal of a given subset of arcs $R \subseteq A$ in a connected graph $G = (N, A)$ with an associated distance/cost function defined on the arcs A. The set R is usually referred to as the *required arcs.*

The Rural Postman Problem (RPP) was proven strongly NP-hard by Lenstra and Rinnooy Kan (1976) even for completely directed and completely undirected graphs (see also Garey and Johnson, 1979, and Papadimitriou, 1976). Many problems such as mail delivery, garbage collection, street cleaning, street inspection, road maintenance, snow-plowing, residential utilities meter reading, torch path determination in flame cutting, and others can be modeled as a variation of RPP. See, for example, Bodin and Kursh (1978), Manber and Israni (1984), Stern and Dror (1979), and in Assad and Golden (1995), Eiselt, Gandreau and Laporte (1995) for a recent survey.

In our application, the required arc set R is further partitioned by the different feed types (i.e., $R_1, \ldots, R_{\bar{f}}, R_i \cap R_j = \emptyset, i \neq j$) since a vehicle can carry only one feed type on a single route, and there are also capacity constraints on the trucks. We describe our model of the feed delivery problem as a Capacitated Rural Postman Problem (CRPP) in more detail below.

The cattle yard is represented by a directed graph $G = (N, A)$. The nodes in N are of two kinds:

1. nodes representing intersections of roads (any service or connecting roads),
2. nodes which correspond to the boundary between the feeding troughs of two adjacent pens, and are of in-degree one and out-degree one.

The arcs of G represent segments of the service roads, either corresponding to the front (feeding border) of a pen, or road segments. Clearly, each arc in the subset R corresponds to the feeding trough for a single pen. The arcs that require traversal are partitioned into feed categories from 1 to $\bar{f}$ indexed by f. Thus, for each $f, 1 \leq f \leq \bar{f}$, let G_f denote the graph induced by the arcs in R requiring feed type f. Note that $G_f = (N_f, R_f)$ need not be a connected subgraph. Disregarding at this point the issue of the timing of traversals for each arc in R, we formulate the CRPP for the graph G with $R = R_f$ for a given feed type f. This formulation is introduced in Dror and Mullaseril (1996).

2.2 Mathematical Formulation of the CRPP

Since we solve a different distribution problem for each feed type, we omit the index f for notational simplicity. Let

$$
\begin{aligned}
q_{ij} &= \text{the demand along arc } (i,j) \in R, \\
W_v &= \text{the vehicle capacity of trip } v, \\
c_{ij} &= \text{the distance (length) of an arc } (i,j) \in A. \text{ (Note } c_{ij} \geq 0, \forall (i,j) \in A), \\
x_{ijv} &= \text{the number of times the arc } (i,j) \in A \text{ is traversed in trip } v, \\
V &= \text{the upper bound on the number of vehicles}, \\
y_{ijv} &= \begin{cases} 1 & \text{if the feed is discharged along the arc } (i,j) \in R \text{ in trip } v, \\ 0 & \text{otherwise.} \end{cases}
\end{aligned}
$$

Node 1 is designated as the depot and represents the truck loading area at the holding bins.

In actual feed delivery, the vehicles traverse the arcs in R at two different speeds — discharging speed and deadheading speed. However, since the discharging speed (for a given feed and a given pen) adds a fixed time to any solution, we can disregard it initially in the following CRPP formulation which finds a minimum distance distribution plan (ignoring time-window constraints):

$$(CRPP): \qquad \min \sum_{(i,j)\in A} \sum_{v=1}^{V} c_{ij} x_{ijv}$$

subject to

$$\sum_{k\in N} x_{kiv} - \sum_{k\in N} x_{ikv} = \quad 0, \qquad \forall i \in N, v = 1,2,\ldots,V, \tag{7.1}$$

$$\sum_{v=1}^{V} y_{ijv} = \quad 1, \qquad \forall (i,j) \in R, \tag{7.2}$$

$$\sum_{(i,j)\in R} q_{ij} y_{ijv} \leq \quad W_v, \qquad v = 1,\cdots V, \tag{7.3}$$

$$x_{ijv} \geq \quad y_{ijv}, \qquad \forall (i,j) \in R, \tag{7.4}$$

$$M \sum_{i\notin N[S], j\in N[S]} x_{ijv} \geq \sum_{(j,k)\in S} x_{jkv}, \qquad \forall S \subseteq R,\ 1 \notin N[S], v = 1,\ldots,V, \tag{7.5}$$

$$y_{ijv} \in \quad \{0,1\}, \qquad \forall (i,j) \in R, v = 1,\ldots,V, \tag{7.6}$$

$$x_{ijv} \in \quad Z^+, \qquad \forall (i,j) \in A, v = 1,\ldots,V, \tag{7.7}$$

where M is a large constant greater than or equal to sum of traversals of arcs in any given $S \subseteq R$, and $N[S]$ is the set of nodes incident to the arc set S. In this formulation, the index v denotes a trip, and V is the maximum number of trips allowed. Time compatible trips can be assigned to the same vehicle, and we will discuss this assignment problem in Section 3.

The objective function represents the total distance traveled by all the vehicles. Note that arcs are allowed to be traversed an integer number of times (that is, more than once). The first set of constraints are the common 'flow conservation' constraints for network-flow formulations. The second set of constraints require that at least one traversal is made for each of the arcs in R to assure that all animals are fed. The third set of constraints are the capacity constraints for the vehicles. The next set of constraints require that vehicle v traverses the arc $(i,j) \in R$ if it delivers feed to this arc. The fifth set of constraints are subtour-elimination constraints ensuring each trip is connected to the depot. Note that this formulation of the CRPP is different than the one given in Golden and Wong (1981) with respect to subtour elimination constraints and the integrality requirement on the x_{ijv} variables.

Essentially, the above CRPP problem has to be solved separately for each feed type, as represented by its subset of required arcs R_f. In the livestock feed problem that motivated this study, one type of feed is more often used than the others. Since only one out of the five vehicles present in the yard is of a larger capacity, we assume that this truck will always be assigned to distribute the more dominant feed type; this, in turn, simplifies the other CRPPs somewhat by limiting their formulation to a single truck capacity. Thus, for one feed we allow two types of vehicles, and for the other feed types the vehicle fleet is homogenous (i.e., with a single capacity).

In the above mathematical formulation of the CRPP, the feeding time constraints were completely ignored and instead the focus is on generating solutions with minimum total distance. In the next subsection we re-introduce the issue of the feed times and its relation to the CRPP solution.

Time Window Constraints for Arc Routing

In this section, we examine time-window constraints for the CRPP and point out the difficulty of formulating the CRPP with time windows, which turns out to be a much more difficult modeling task than its node-routing (VRP) counterpart.

Associate with each arc $e \in R$ a time window $[a_e, b_e]$ within which delivery must be completed, and a positive duration t_e for the traversal of arc e at the discharging speed. With each arc in R, we associate a delivery starting time and completion time, which indicate the times when the vehicle is at the tail of the arc at the beginning of the delivery and at the head of the arc at the completion of the delivery.

In the vehicle routing literature for node-routing problems with time windows, the original "physical" network can be augmented by adding an arc (i, j), with its arc-length equal to the 'shortest-path' distance, whenever there is a path from i to j in the original network. With this augmented network, we can then make the assumption that a node is visited no more than once, and the time-window restrictions can be modeled by the constraints:

$$\begin{aligned} x_{ijv}(T_{iv} + t_{ij} - T_{jv}) &\leq 0 \quad \forall (i,j) \in A,\ \forall v \\ a_i \leq T_{iv} &\leq b_i \quad \forall i \in N,\ \forall v \end{aligned}$$

where T_{iv} is a decision variable representing the time when vehicle v arrives to deliver to node i, $[a_i, b_i]$ is the allowed time window for node i and t_{ij} is the traversal time for arc (i, j). (See Desrosiers, Dumas, Solomon and Soumis, 1995.)

In an arc-routing problems such as CRPP, each arc $e \in R$ is serviced exactly once but can be traversed an additional number of times in a deadheading mode if so required by the minimum-distance objective. Hence, we cannot associate a unique starting and completion time for an arc $e \in A$. Moreover, it is not possible to augment the network in a manner analogous to node-routing problems to get an equivalent formulation where arcs (or nodes) are visited only once.

We do note that the CRPP can be transformed from arc-routing into an equivalent node-routing (VRP) problem (Pearn, Assad and Golden, 1987). Such a transformation replaces each arc in R with three vertices each having one-third of the original demand, with the same time-window $[a_e, b_e]$ for each of the three nodes. Solving this much larger node-routing with time-windows problem would give a solution to the corresponding arc-routing with time-windows problem. However, there is no simple reverse transformation back to the original arc-routing formulation — the unique starting time for the node-routing problem generate multiple 'first-visit' times for the same node in the arc-routing problem. Thus, somewhat surprisingly, this transformation for arc-routing problems with time windows is one-directional. Since the size of the corresponding VRPs is multiplied by a factor of three in comparison with the original CRPP, "the interest of these transformations is mostly formal and their algorithmic value has yet to be demonstrated" (Eiselt, et al., 1995).

Discussion of the Subtour Elimination Constraints

In Golden and Wong (1981) there are two versions of the subtour elimination constraints for CRPP. The first version consists of the set of constraints:

$$\sum_{i,j \in S} x_{ijv} \leq |S| - 1 + |N|^2 u_v^S$$

$$\sum_{i \in S} \sum_{j \notin S} x_{ijv} \geq 1 - w_v^S$$

$$u_v^S + w_v^S \leq 1$$

$$u_v^S, w_v^S, x_{ijv}, y_{ijv} \in \{0, 1\}$$

$$\forall S \subseteq N \setminus \{1\}; S \neq \emptyset; v = 1, \cdots, V$$

The interpretation of the variables u_v^S, w_v^S is as follows: $w_v^S = 0$, if and only if the vehicle v enters the subset S. If $w_v^S = 0$ for a subset S, then the corresponding value of the u_v^S variable can be 1 thus allowing deadheading (traversing an arc without delivering any feed) inside the subset S. This set of subtour elimination constraints is not applicable for the CRPP where the variables x_{ijv} are not restricted (as in Golden and Wong, 1981) to take binary values only. In CRPP, a vehicle v might be required to traverse the arc (i, j) numerous times as illustrated in the example shown in Figure 2. In this example, all arcs are required, and the arc $(1, 2)$ must be traversed over and over again.

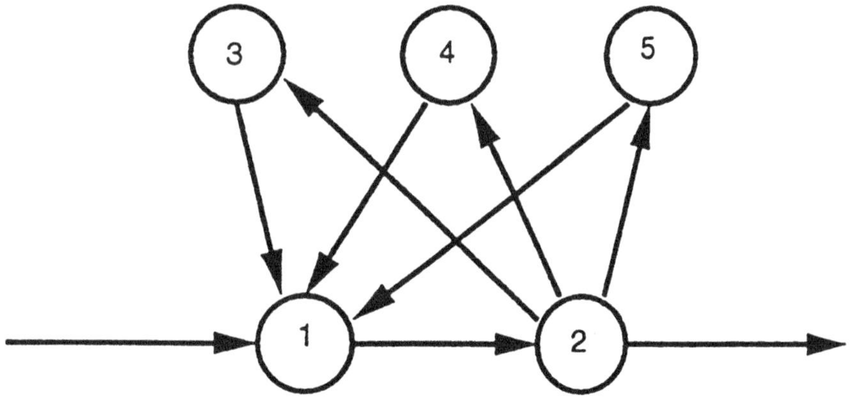

Figure 2 Example showing that arcs may be traversed repeatedly.

The second version of subtour elimination in Golden and Wong (1981) is of the following form:

$$\begin{aligned} \sum_k f_{ikv} - \sum_k f_{kiv} = & \sum_j y_{ijv} && i \in N \setminus \{1\};\ \forall v \\ f_{ijv} \leq & \ |N|^2 x_{ijv} && \forall (i,j) \in A;\ \forall v \\ f_{ijv} \geq & \ 0 && \forall v \end{aligned}$$

Essentially the f_{ijv} are flow variables and every node is assigned an external demand equal to the number of required arcs incident from it. Since the depot (node 1) is the only source, flow must emanate from it and the x_{ijv} variable cannot be zero if there is flow on arc (i,j).

If there were no capacity constraints, a much simpler version of subtour elimination constraints for the (uncapacitated) rural postman problem (RPP) which allows x_{ijv} to be integer-valued (and not just binary) is the following:

$$R[S] \sum_{i \in S} \sum_{j \notin S} x_{ijv} \geq \sum_{(i,j) \in R[S]} y_{ijv},\ \forall S \subseteq N \setminus \{1\}; S \neq \emptyset; S \cap N[R] \neq \emptyset; v = 1, \cdots, V$$

where $N[R]$ denotes the subset of nodes incident to arcs in R, and $R[S]$ denotes the set of required arcs incident to or from a node in S. Both the second version of Golden and Wong (1981) and our version of subtour elimination constraints above fail to eliminate subtours on a subset of nodes which are not incident to the required arcs; however, if the arc costs are non-negative, then the objective function will ensure that such subtours do not occur in the optimal solution.

Split Delivery for Arc Routing

If the demand of an arc can be served by more than one vehicle, then we have a split-delivery arc-routing problem. To allow for split delivery, we can replace the 0-1 constraint (7.6) in the (CRPP) formulation by

$$y_{ijv} \geq 0, \qquad \forall (i,j) \in R, \forall v.$$

Some properties of the optimal solution for the vehicle routing problem (VRP) with split deliveries (see Dror and Trudeau, 1989, Dror and Trudeau, 1990, and Dror, Laporte and Trudeau, 1994) can be extended to the split deliveries for arc-routing in the following manner.

Definition : Let $S = \{e_1, \cdots, e_k\} \subseteq R$ and $k > 1$. If there exist k vehicle routes such that e_1 and e_k are on the same route, and e_t and e_{t+1} are on the same route for every $t = 1, \cdots, k-1$, then S is a k-split cycle.

Proposition : If the distance matrix used satisfies the triangle inequality, there always exists an optimal solution not containing k-split cycles.

The proof is similar to the one in Dror and Trudeau (1990) for VRP with split deliveries. We note that if a shortest-path distance matrix is used, the triangle inequality is always satisfied.

Extending the discussion of the above CRPP to include the split delivery option is not made here just because of academic interest. This is in fact the delivery mode on the cattle ranch in question. It is partially dictated by the fact that the truck loading process is imprecise and the vehicle might run out of the feed in the middle of a delivery to a particular pen, so non-split deliveries, even if intended, cannot be realized. As already noted in Mullaseril, Dror and Leung (1996), allowing for split deliveries represents a generalization of the corresponding routing problem which is significantly harder to solve if exact solutions are required.

2.3 Solution Procedures for Feed Distribution

Having modeled the feed distribution problem as a sequence of capacitated arc-routing problem with time-windows, we experimented with a solution strategy which was adapted from the heuristic procedures for split-delivery node-routing problems described in Dror and Trudeau (1990).

Our heuristic approach consists of two phases:

1. Route Generation - a set of trips are generated which satisfies all the demand and their corresponding time-window requirements and the vehicle capacity constraints, but no demand is split among different trips.
2. Split-Delivery Introduction and Improvement — heuristics are applied which generate split-deliveries that reduce the total distance traveled without violating time windows.

A full description of the route generation and route improvement procedures is given in Mullaseril, Dror and Leung (1996). Here we just summarize the heuristic principles and their corresponding routing results. To generate the initial capacitated arc-routing solutions in phase 1, two heuristic procedures were used:

1. A Path Scanning procedure — which construct a route adding the path segment to one demand arc at a time until the vehicle capacity is reached (introduced by Golden, DeArmon and Baker, 1983) and,
2. An Augment-Merge heuristic — introduced by Golden and Wong (1981) which is based on a savings principle similar to Clark and Wright (1964) for node routing.

Both procedures were modified for routing with arc time-windows (see Mullaseril et al., 1996).

In phase 2, the heuristics applied are k-split generation and k-route addition. Both are adapted from the heuristics developed by Dror and Trudeau (1990) for split-delivery node-routing. In Dror and Trudeau (1990) a *k-Split* procedure for node-routing was introduced which tests if delivery to a node can be split across k other candidate routes in such a way that the highest savings is obtained. Another concept introduced in Dror and Trudeau (1990) for split-delivery in node-routing was that of *route addition*. This procedure searches over nodes whose demand is split among several routes and considers consolidating them into a new route which will realize net savings in distance traveled. We do not describe the algorithmic details of the adapted solution procedures in this chapter and the interested reader is referred to Mullaseril et al. (1996).

In our computational experiments on data obtained from the cattle ranch, these heuristics proposed were able to lower the total distance covered by the trucks substantially, when compared to the existing practice. For all five different feed types, the heuristic procedures proposed outperformed the current methods by at least 25%, and as high as 50% on a denser graph.

3 ASSIGNMENT OF ROUTES TO VEHICLES: THE SCHEDULING PROBLEM

Given that we have solved the Feed Distribution Problem described in the previous section, we are now faced with another combinatorial optimization problem — that of assigning the trips to the vehicles (delivery trucks) in the yard and scheduling their timely execution.

From a solution to the CRPP for each feed type, we obtain a set of vehicle trips starting and ending at the depot (i.e., at the six temporary storage hoppers) which cover at least once each arc in R. As noted earlier, the total number of such trips will normally exceed the number of vehicles in the yard since each vehicle is planned to make a number of feed deliveries during a work day. Given this set of trips, the vehicle scheduling (dispatching) problem consists of the assignment of starting times and specific vehicles for each trip, so that trips do not overlap in time for each vehicle, and demand is delivered to each arc $e \in R$ within its feeding interval $[a_e, b_e]$ on the corresponding trip. We assume that the feed with the highest usage has the exclusive use of the larger truck, and all the other feeds are delivered by the four remaining trucks of smaller capacity. In our computational testing, the trips generated by the procedures described in Section 2 for the different feed types were generated according to these truck capacities.

For each trip j, we compute its duration t_j by adding the traversal time at the deadheading speed for each arc in the trip to the incremental time of feed-discharge for each of the arcs in R serviced by this trip. (This incremental time is feed and pen/arc dependent.)

To compute the latest completion time of a trip, we assume that it is started as late as possible so that it finishes serving its first required arc at the end of its allowed time window. Next, we examine the start and completion times for feed delivery to the pens along to the traversal sequence of the trip and check that the feeding times are feasible in relation to the trip's travel time.

If a time-window for a subsequent required arc is violated, the entire trip is shifted to begin earlier so that that time-window constraint is just met. The completion time of this realization of this trip is the latest completion time of the trip. We can compute the earliest start time for the trip by assuming that the trip is started at a time so that it arrives at the first pen at the beginning of its time window and adjust for the traversal of the remainder of the trip in a similar manner.

Given that the vehicles are all of the same capacity, the problem of assigning start-times and vehicles to these trips is equivalent to finding a feasible solution to the scheduling problem with m parallel machines (corresponding to the vehicles) and n jobs (corresponding to trips) with release times and deadlines where no preemption is allowed. This problem (known as *Sequencing Within Intervals*) is strongly NP-hard even for one machine (see Garey and Johnson, 1979, Dempster, Lenstra, and Rinnooy Kan, 1982). However, if preemption is allowed, then the problem is solvable in polynomial time as a network flow problem (see Horn, 1974).

We first present a formulation of the problem in the next subsection and discuss some of its properties in the remainder of this section.

3.1 Mathematical Formulation for Trip Assignments

We formulate the trip assignment problem as that of determining a feasible solution to the interval scheduling problem of n jobs on m parallel identical machines ($n > m \geq 1$). In the case of the cattle yard we studied, since there are four identical trucks and one bigger truck, the feasibility trip scheduling problem is partitioned into two separate problems. One of the problems is a single machine (larger truck) problem and the other corresponds to constructing a feasible schedule on four identical parallel machines. There is no setup times since there is no need to clean the trucks when switching between the deliveries of different feed types. We also note that the split delivery policy used in route construction does not affect the subsequent scheduling problem.

Let $\mathcal{J}$ denote the set of trips for all feed types handled by the smaller-sized trucks. For each trip j, let t_j denote its duration, r_j its earliest start time and d_j its latest completion time. (We assume that the work day has been partitioned into time intervals of equal lengths and the data (t_j, r_j and d_j) have been discretized according to these time intervals. We adopt the interpretation that if job j has $r_j = 4$ and $d_j = 7$, then the trip can be started in period 4 or later and must be completed *prior* to period 7.) Let

$$s_{vjk} = \begin{cases} 1 & \text{if trip } j \text{ starts at period } k \text{ on machine } v \text{ (vehicle } v\text{), and} \\ 0 & \text{otherwise.} \end{cases}$$

To solve the vehicle dispatching problem, we seek a binary solution vector $s = (s_{vjk})$ to the following system:

$$s_{v\bar{j}\bar{k}} + \sum_{k=\bar{k}+1}^{\bar{k}+t_j-1} \sum_{j} s_{vjk} \le 1, \quad v = 1,\ldots,m; \bar{j} = 1,\ldots,n; \quad \bar{k} = r_j,\ldots,d_j - t_j, \tag{7.8}$$

$$\sum_{v=1}^{m} \sum_{k=r_j}^{d_j-t_j} s_{vjk} = 1, \quad j = 1,\ldots,n, \tag{7.9}$$

$$s_{vjk} = 0 \ \ or \ 1, \quad v = 1,\ldots,m; j = 1,\ldots,n; \quad k = r_j,\ldots,d_j - t_j, \tag{7.10}$$

For the case of the larger capacity vehicle, its feasible schedule is found by finding a feasible solution to (7.8) – (7.10) where $\mathcal{J}$ now denotes the set of trips for this larger capacity vehicle and m, the number of vehicles, is equal to 1.

NP-completeness

In our application, preemption is not allowed. Vehicles cannot stop delivery in the middle of a trip (with the truck somewhere in the cattle yard) and *immediately* start another trip from the depot or continue delivery of a different trip. Without preemption, the scheduling problem as formulated by (7.8) – (7.10) is NP-hard, even on one machine with one 'nested' job, as can be shown by the following reduction from the PARTITION problem.

In the PARTITION problem, we are given a set $U = \{a_1, \cdots, a_n\}$, with 'size' $s(a_i)$ for element a_i such that $s(a_i) \in Z^+$ and $\sum_{i=1}^{n} s(a_i) = 2B$. The question is to determine if the set U can be partitioned into U_1 and U_2 such that $\sum_{i \in U_1} s(a_i) = \sum_{i \in U_2} s(a_i) = B$.

Given an instance of a PARTITION problem, construct a set of $n+1$ jobs $J = \{a_1, \cdots, a_n, a_{n+1}\}$ to be scheduled on a single machine, where the processing time for job a_i, is equal to $s(a_i)$ for $i = 1, \cdots, n$, and the processing time for job a_{n+1} is equal to one unit. The release times and the deadlines are: $r_i = 1, d_i = 2B+2$, for $i = 1, \cdots, n$ and for the job $a_{n+1}, r_{n+1} = B+1, d_{n+1} =$

$B + 2$. Clearly, our problem has a nonpreemptive solution iff the answer for the PARTITION problem is YES (see also Garey and Johnson, 1979).

We also note that, without preemption, if the set of release times and deadlines for the trips have a certain strict non-nestedness property, (i.e., no trip has its earliest starting time and its latest completion time contained in the time interval defined by another trip's earliest start and latest completion time) and the duration of all the trips are the same, then this feasibility scheduling problem can be solved by a simple list processing algorithm, as described in Section 3.2.

If preemption is allowed, then the problem of finding a feasible schedule can be modeled as a network flow problem. Let the set of release times and deadlines of all the jobs be ordered as:

$$e_1 \leq e_2 \leq \cdots \leq e_K$$

Let $e_0 = 0$. Let I_j be the interval $[e_{j-1}, e_j)$ for $j = 1, \cdots, K$, and let $\Gamma = \{I_1, \cdots, I_k\}$. Let $N = \{1, \cdots, n\}$ denote the set of jobs. Consider a network with $\{s\} \cup N \cup \Gamma \cup \{t\}$ as its node set, and with arcs of the following three types:

1. (s, i) for all $i \in N$ with capacity $= t_i =$ processing time of job i,
2. (j, t) for all $j \in \Gamma$ with capacity $m(e_j - e_{j-1})$ representing total available machine time in interval I_j.
3. (i, j) for all $i \in N, j \in \Gamma$ if $r_i \leq e_{j-1}$ and $d_i \geq e_j$ for job i, with capacity $(e_j - e_{j-1})$ since a job can be processed on only one machine at a time.

Note that this network has a very special structure where each node $i \in N$ is incident to "consecutive" nodes in Γ. Horn (1974) showed that a preemptive schedule exists if the above network has a feasible flow.

3.2 Heuristic Generating a Feasible Trip Covering Solution - List Algorithm

Given that there exists a feasible solution to the system (7.8) – (7.10) above, we describe a simple list processing algorithm for assigning trips to the vehicles and their starting times. As we mentioned earlier, we can view the vehicles as parallel identical processors and list the trips according to the nondecreasing earliest start time. Then we apply a list processing algorithm which assigns the trips in order to the first available vehicle, breaking ties according to the lower vehicle index.

For the larger vehicle, its dispatch schedule is easily found. It simply executes its trips in order, according to nondecreasing earliest start time. For the smaller-sized vehicles, we consider the following two algorithms:

Algorithm 1: (backward algorithm)
Select the trip l with deadline $d_l = \max_{i \in \mathcal{J}'}\{d_i\}$ where $\mathcal{J}'$ is the set of trips not assigned to any vehicles. ($\mathcal{J}'$ includes all the trips initially.) Break ties primarily in favor of the latest release time, and secondarily by longest processing time. Ties between trips with identical release times, deadlines and processing times are broken arbitrarily. Assign this trip (job) to the vehicle (machine) with least load (or lowest index in the case of ties) at time $d_{max} - t_l$ or as late as possible. Delete trip l from the set $\mathcal{J}'$. Continue selection and assignment until all trips are assigned vehicles and start-times.

Algorithm 2: (forward algorithm)
Select the trip l with the release time $r_l = \min_{i \in \mathcal{J}'}\{r_i\}$. Break ties primarily in favor of the earliest deadline, and secondarily by shortest processing time. Ties between trips with identical release times, deadlines and processing times are broken arbitrarily. Assign this job to the machine with least load (or lowest index in the case of ties) at time r_l or as early as possible. Delete the trip l from the set of trips $\mathcal{J}'$. Continue selection and assignment until all trips are assigned vehicles and start-times.

Polynomially Solvable Case

Claim: In the case that all the trips (jobs) are of equal duration and the release time and deadlines are non-nested in the sense that no trip (job) has a [release-time, deadline] interval that strictly contains another's, then Algorithm 1 and Algorithm 2 constructs a feasible assignment if one exists.

Proof : We prove the claim for Algorithm 1 since the proof for Algorithm 2 is very similar.

The proof is by contradiction. Assume all jobs are of equal duration $t > 1$ (the case of $t = 1$ is trivial) and the release times and deadlines satisfy the conditions of the claim. Suppose that a job $k-1$ in the sequence according to Algorithm 1 has been feasibly assigned to a machine at time s_{k-1} where $s_{k-1}+t \leq d_{k-1}$, and we cannot find a feasible assignment for the kth job in the sequence. Without loss of generality, assume that job $(k-1)$ is assigned to machine 1.

Case 1: $d_{k-1} > d_k, r_{k-1} \geq r_k$
The completion time for the $(k-1)$-th job is equal to $c_{k-1} \equiv s_{k-1} + t \leq d_{k-1}$. Denote by $\bar{S}$ the maximum start time for the jobs assigned last on each machine, other than machine 1, before the $(k-1)$-th job. Then, $\bar{S} \leq c_{k-1}$ (according to the selection rule for Algorithm 1), and $\bar{S}$ is attained on a machine other than machine 1. Note that at each new job assignment, the job is assigned to a machine so that it starts as late as possible. From the assumption that job k cannot be feasibly assigned, we must have $\bar{S} - t < r_k$. Since all the release times of the jobs already assigned are greater than or equal to r_k and since all the jobs have identical processing times, no job can be switched to the k-th position in this sequence to obtain feasibility. Thus, no feasible assignment exists in contradiction to the statement in the claim.

Case 2: $d_{k-1} \geq d_k, r_{k-1} > r_k$.
This case is symmetric in argument to Case 1. □

When the trip times are not identical, Algorithms 1 and 2 are heuristics and there is no guarantee that they will find a feasible set of assignments even if one exists. We run both Algorithms 1 and 2, and select the schedule that minimizes the infeasibility (as measured by $\sum_j (r_j - s_j)^+$ for Algorithm 1 and $\sum_j (c_j - d_j)^+$ for Algorithm 2). We should point out that even if both algorithms generate feasible solutions, they need not be the same.

3.3 Preliminary Computational Results

In checking for feasibility with respect to time windows, the route generation heuristics assumed a deadheading speed of 15 miles per hour and a discharging speed of 1 mile per hour. For the data set we tested, a feasible assignment of the trips to the 5 vehicles is possible using Algorithm 1. As mentioned above, neither Algorithms 1 nor 2 are guaranteed to give feasible solutions. The speed of the vehicles turns out to be a critical factor for determining whether feasible assignments are found, as shown by the results of a computational 'sensitivity' study reported in Table 1. In particular, the discharging speed is very important; when it is below 0.4 mph, no feasible schedule is found for our test data unless the deadheading speed is increased substantially, whereas when it is above 0.6 mph, feasible schedules are found.

Table 1 Relationship between Vehicle Speed and Time-window Constraint Violations

Speed (mph)		Forward Algorithm (2)		Backward Algorithm (1)	
Deadheading	Discharging	No. of Vehicles with Time-window Violations	Maximum Violation (minutes)	No. of Vehicles with Time-window Violations	Maximum Violation (minutes)
10	1.0	0	0	0	0
10	0.65	0	0	0	0
20	0.6	0	0	0	0
10	0.60	0	0	1	1
12	0.50	0	0	1	9
10	0.50	0	0	1	18
9	0.50	3	7	4	29
8	0.50	4	14	4	36
7	0.50	5	29	5	42
10	0.40	5	56	5	58
12	0.40	5	33	4	40
24	0.40	3	3	3	9
26	0.40	1	1	1	4
30	0.40	0	0	1	1
32	0.40	0	0	0	0

4 THE FEEDMILL OPERATION - SYNCHRONIZING PRODUCTION WITH DISTRIBUTION

A solution to the feed distribution problem presented in Sections 2 and 3 produces a feeding schedule for the different pens by specifying the time, the vehicle, and the route which deliver the feed to the cattle. To execute such a schedule, the feedmill has to produce the appropriate amounts of the different feeds (5 feed types in our test data) in coordination with this distribution schedule. Since the distribution plan is driven by the cattle's needs, the feed production schedule has to maintain the inventory levels of the different feed categories, while also focusing on the important aspect of feed quality in its feed mixing process. The quality of feed mixing objective can be expressed in terms of the length of a production run of a feed type. The longer the production run, the greater the control over the feed-mix in meeting specifications. Thus, minimizing the number of total production startups for all feed types, while satisfying the demand of feed-requirements as prescribed by the distribution plan, serves as an appropriate surrogate objective function for optimizing quality. A similar problem exists in a paper mill where a paper is wasted until a machine can be fully adjusted when changing from one grade of paper to another. This problem of short-term production schedule for a facility that processes one product at a time and incurs a changeover penalty when changing from one product to another has received considerable attention in the literature. Next, we examine a mathematical model, similar to one in Magnanti and Vachani (1990) for determining the start times and run length of the production run for the different feed types which enable the subsequent execution of the feed distribution solution. Similarly stated problems and an overview can be found in Maes and Van Wassenhove (1988) and Salomon, Kroon, Kuik and van Wassenhove (1991), where the problem is called a Discrete Lotsizing and Scheduling Problem (DLSP). One important difference between our problem and that described in Magnanti and Vachani (1990), is the fact that the storage facilities for the different feed types (in finished product) are of finite capacity and each storage facility can store only one feed type at a time.

We make the following assumptions:

1. The changeover time is zero, that is, the changeover preparations can be done off-line before the changeover has to take place. Machine setup can be maintained even if machine is idle in a particular period.
2. The changeover costs are not sequence dependent, but only depend on the product for which the facility is to be setup.
3. The production level at any period is either set to the required, feed specific, production rate, or is zero (i.e., discrete production policy).

4.1 Mathematical Formulation for Feed Production as a Capacitated Changeover Scheduling Problem (CCSP)

We present a formulation of the problem of production scheduling and lot sizing at the feed mill, given the feed delivery requirements as determined by the feed distribution plan. Let

y_{ij} = 1 if feed type i is produced in period j and is zero otherwise,

z_{ij} = 1 if the production of feed i is started in period j, and is zero otherwise,

b_{ij} = number of temporary storage hoppers allocated for feed i in period j.

Other parameters are:

$$
\begin{aligned}
K_i &= \text{the startup cost for feed } i, \\
r_i &= \text{the production rate for feed } i, \\
q_{ij} &= \text{the demand for feed } i \text{ in period } j, \\
t_i &= \text{the minimum length of a production run for feed } i, \\
\bar{Q}_i^0 &= \text{the maximum capacity of a storage hopper for feed type } i, \\
\bar{B} &= \text{the total number of storage hoppers available.}
\end{aligned}
$$

We assume that the storage hoppers are initially empty.

$$(CCSP): \qquad \min \sum_{i=1}^{\bar{f}} \sum_{j=1}^{T} K_i z_{ij}$$

subject to

$$
\begin{aligned}
\sum_{j=1}^{t} r_i y_{ij} &\geq \sum_{j=1}^{t} q_{ij} & 1 \leq t \leq T, & \qquad (7.11) \\
\sum_{i=1}^{\bar{f}} y_{ij} &\leq 1 & \forall j, & \qquad (7.12) \\
z_{ij} &\geq y_{ij} - y_{i,j-1} & \forall i, j, & \qquad (7.13) \\
t_i z_{ij} &\leq \sum_{t=j}^{j+t_i} y_{it} & \forall i, j, & \qquad (7.14) \\
\sum_{j=1}^{t} (r_i y_{ij} - q_{ij}) &\leq \bar{Q}_i^0 b_{it} & 1 \leq t \leq T, & \qquad (7.15) \\
\sum_{i=1}^{\bar{f}} b_{it} &\leq \bar{B} & \forall t & \qquad (7.16) \\
z_{ij} &\in \{0,1\} & \forall i, j, & \qquad (7.17) \\
y_{ij} &\in \{0,1\} & \forall i, j, & \qquad (7.18) \\
b_{ij} &\geq 0 \text{ integer} & \forall i, j & \qquad (7.19)
\end{aligned}
$$

The objective function represents the total cost of machine setups. Constraints (7.11) require that the cumulative production for any one of the feeds is sufficient to satisfy the demand for that feed. Constraints (7.12) state that no more than one feed type can be produced at any given time. Constraints (7.13) is used to count the number of startups for each feed. Constraints (7.14) state that each production run has a minimum duration. Constraints (7.15) represent the storage hopper capacity constraints for each feed type and constraints (7.16) represents the limit on the total number of storage hoppers. Note that if one feed type has been completely distributed, its storage hopper becomes available for storage for some other feed.

The fact that (CCSP) is NP-hard follows from a result of Bruno and Downey (1978). They showed, via a reduction from the (equality) knapsack problem, that the problem is NP-hard even in the special case where feed is required in only two specified periods and there are no capacity restrictions.

The uncapacitated changeover scheduling problem (UCSP) is obtained by omitting constraints (7.14), (7.15), (7.16) and (7.19) from (CCSP). Magnanti and Vachani (1990) studied a variant of the (UCSP) where the setup costs are time-dependent and there are inventory holding costs, and reported that Orlin (private communication, 1986) proved that variant of (UCSP) to be NP-hard in the strong sense. Still, since Orlin's result is not in the public domain we present here a simple strong NP-hardness proof for the special case of (UCSP)where only one item has a time-dependent setup cost.

Claim : The (modified) Uncapacitated Changeover Scheduling Problem is NP-hard in the strong sense.

Proof : By reduction from the NP-complete problem **3-PARTITION**:
Let $B \in \mathbf{Z}^+$, let $A = \{a_1, a_2, \ldots, a_{3m}\}$ be a set of $3m$ elements, with size $s(a_i) \in Z^+$ for each $a_i \in A$, such that $B/4 < s(a_i) < B/2$ and $\sum_{a_i \in A} s(a) = mB$. Can A be partitioned into m disjoint sets $A_1, \ldots, A_m$ such that for $1 \leq i \leq m, \sum_{a_j \in A_i} s(a_j) = B$?

Assume $3m+1$ different feed types and $mB+m-1$ production periods. Assume that the number of production periods required for one unit of feed i is given by $s(a_i)$ for $i = 1, 2, \ldots, 3m$. For feed $3m+1$, one production period is required to produce one unit. For $i = 1, 2, \ldots, 3m$, the feed requirement for period j, q_{ij}, is 1 if $j = mB + m - 1$ and zero otherwise. For feed $3m + 1$, $q_{3m+1,j} = 1$ for $j = B + 1, 2B + 2, \ldots, (m - 1)B + (m - 1)$. Further, let $K_i = 0$ for $i = 1, 2, \ldots, 3m$, and let the setup cost for feed $3m + 1$ in time period j be denoted by $K_{3m+1,j}$, where

$$K_{3m+1,j} = \begin{cases} 0 & \text{for } j = B+1, 2B+2, \ldots, (m-1)B + (m-1), \\ m+1 & \text{otherwise.} \end{cases}$$

Then, in a solution to this version of (USCP) with objective value $\leq m$, we must have feed $3m+1$ produced in periods $B+1, 2B+2, \ldots, (m-1)B+(m-1)$. Clearly, there is a feasible solution to this modified Uncapacitated Changeover Scheduling Problem with objective value $\leq m$, if and only if there is a solution to the corresponding 3-PARTITION problem.

□

4.2 Discussion of Solution Approaches to CCSP

Many authors have investigated problems similar to the one presented in this section. Some of the latest literature on this topic include Cattrysse, Salomon, Kuik and van Wassenhove (1993) and van Hoesel, Kuik, Salomon and van Wassenhove (1994). Most of the other related important references are mentioned in these two papers.

In Cattrysse et al. (1993), the authors discuss an uncapacitated version of the Discrete Lot-sizing and Scheduling Problem (DLSP) and formulate it as a Set Partitioning Problem (SPP). In their formulation, they include also inventory holding costs and production costs which are omitted from our formulation of (CCSP), but they did not address storage capacity limitations in their formulation. We are optimistic that the solution methodology of dual ascent and column generation which they have developed for their SPP formulation would be a promising approach for (CCSP). Cattrysse et al. (1993) report computational results on 420 test problems where 82% of the test problems were solved to optimality and the problems which were not solved to optimality had a duality gap of at most 1.5%. They compare their computational results to the ones obtained by a heuristic described in Fleischmann (1990). The speed of Fleischmann's heuristic was found to be superior but the resulting solution somewhat worst than the that of Cattrysse et al. (1993).

In van Hoesel et al. (1994), the DLSP problem is examined and two solution procedures are described in some detail. The first procedure is based on a "reformulation of the DLSP as a linear programming assignment problem, with additional restriction to reflect the specific (setup) cost structure. The second procedure is based on dynamic programming". The authors do not report computational results for the two procedures.

We have not conducted any significant computational experimentation to solve our problem. This topic is one of our future research directions, and we plan to adopt and modify the solution methodology of Cattrysse et al. (1993).

5 CATTLE MOVEMENT

In this section we describe the problem of modeling the assignment and movement of cattle lots among the pens in the yard. On average, each head of cattle spends about 200 to 250 days in the cattle yard before being sent to a slaughterhouse. During that period, each head of cattle has to be assigned an initial pen, followed by reassignments to different pens as the cattle matures, until it is shipped away. On average, each head of cattle experiences about 2 to 4 moves during its stay in the cattle yard. With 100,000 head of cattle in the yard and an estimated cost of 6 to 10 dollars per head per move, such costs quickly add up to a significant amount. For proper feeding, the maximum number of cattle allowed in a pen is proportional to the length of the feeding trough and this ratio is dependent on the age of the cattle. Thus, the capacity of a pen is different for cattle of different ages. As cattle matures, they may outgrow the capacity of the pens they are housed in and need to be moved to a different, larger pen. The perception of the management is that the decision problem regarding the optimal assignment of cattle lots to pens and their subsequent moves within the yard is a very hard problem to manage effectively.

In this section we present a mathematical formulation for this cattle assignment and movement problem. We do not propose here a solution methodology for this very large integer-programming problem. We also do not know of any other study regarding problems of this type, size, and structure. We only view this formulation as an initial attempt to model the problem and provide a starting point for future research.

We assume that the information regarding the number, time period, and the exact age of the young calves entering the yard is known in advance and the number, age, and time period of the cattle leaving the yard is also known. That is, we model this problem as a deterministic problem although, in reality, there are stochastic variations in the shipment dates. Depending on the number of time periods included in the model, we can repeatedly solve this problem using a rolling horizon. We also assume that every movement occurs at the beginning of the time period and has no significant duration. Given

$$
\begin{aligned}
c_{ij} &= \text{the cost of moving a single cow from pen } i \text{ to pen } j, \\
In_t(l) &= \text{the number of cattle of age } l \text{ entering the yard at time period } t, \\
Out_t(l) &= \text{the number of cattle of age } l \text{ leaving the yard at time period } t, \\
O_s &= \text{the set of pens from which cattle can be shipped out}, \\
L_o &= \text{the set of ages of the cattle which can be shipped out}, \\
I_o &= \text{the set of pens for the entering cattle}, \\
L_i &= \text{the set of ages for the entering cattle}, \\
q_i(l) &= \text{the capacity of pen } i \text{ for cattle of age } l, \text{ and} \\
N^t(l) &= \text{the total number of cattle in the yard of age } l,
\end{aligned}
$$

let

$$
\begin{aligned}
x_{ij}^t(l) &= \text{the number of cattle of age } l \text{ moved from pen } i \text{ to pen } j \text{ in period } t, \\
Y_i^t(l) &= \begin{cases} 1 & \text{if at time } t \text{ cattle of age } l \text{ is in pen } i, \\ 0 & \text{otherwise.} \end{cases}
\end{aligned}
$$

The cattle movement problem can be modeled as:

$$
(CMP): \qquad \min \sum_l \sum_t \sum_i \sum_{j:j \neq i} c_{ij} x_{ij}^t(l) \tag{7.20}
$$

subject to

$$
\sum_{j \in I_0} x_{0j}^t(l) = In_t(l), \quad t = 1, \cdots, T; l \in L_i, \tag{7.21}
$$

$$
\sum_{i \in O_s} x_{is}^t(l) = Out_t(l), \quad t = 1, \cdots, T, l \in L_o, \tag{7.22}
$$

$$
\sum_j x_{ji}^t(l) \leq q_i(l) Y_i^t(l), \qquad \forall i, t, l, \tag{7.23}
$$

$$
\sum_l Y_i^t(l) \leq 1, \qquad \forall i, t \tag{7.24}
$$

$$
\sum_i x_{ij}^t(l) - \sum_h x_{jh}^{t+1}(l+1) = 0, \qquad \forall l, t, j \tag{7.25}
$$

$$
\sum_j \sum_i x_{ji}^t(l) = N^t(l), \qquad \forall l, \tag{7.26}
$$

$$
x_{ij}^t(l) \in Z^+ \qquad \forall i, j, t, l \tag{7.27}
$$

The objective function (7.20) expresses that we are minimizing the total cost of cattle movement in the yard over the T time periods. One can view this optimization statement as a "rolling horizon" type of a statement. Constraints (7.21) express that all entering cattle has to be assigned pens in some subset of pens designated for the younger calves. Constraints (7.22) state that the mature cattle must be moved to some designated pens before being shipped out. Constraints (7.22) simply states that the number of cattle in pen i of age l can not exceed the capacity of the pen at any time for cattle of that age. Constraints (7.24), which state that only cattle of the same age can reside in the same pen. Constraints (7.25) are the "conservation" of flow constraints stating that cattle entering a pen has to be equal to the cattle leaving the pen or remaining in the pen. Constraints (7.26) state that all cattle of a certain age has to be accounted for, and constraints (7.27) are the integrality constraints.

In modeling livestock rearing, it may be appropriate to acount for the natural loss of cattle by a "loss" factor $g(l)$, representing the fraction of the cattle of age l that do not survive the current time period. With a loss model, constraint (7.25) is modified as

$$g(l) \sum_i x_{ij}^t(l) - \sum_h x_{jh}^{t+1}(l+1) = 0$$

and the variables $x_{ij}^t(l)$ are then usually modelled as continuous rather than integer-valued variables.

In the problem we studied, there are over 600 pens and about 40 different age groups for the cattle. With 52 weeks in the planning period, (CMP) is an integer program where the number of integer variables is on the order of 10^8. Clearly, solving a problem of such complexity directly to optimality is out of the question. In our future research, we plan to examine heuristic methods for this problem, with the goal of providing the cattle yard manager a tool that would be used to generate cattle movement plans a few times each month, for example, when new shipments of cattle are identified or new significant information becomes available. The real life problem is more stochastic in nature in terms of stock availability and demand, and cattle yard data is periodically updated.

6 CONCLUSIONS

In this chapter, we have described several problems encountered in the operations of a cattle yard. Many of these problems can be formulated as large combinatorial optimization problems. We have developed some algorithms for feed distribution planning which appear to be effective. On the other hand, the model for some other operational problems (e.g. cattle movement) are of a size that would challenge the state of the art in integer-programming methodologies. In the present daily operation, none of the problems described is solved using the mathematical representation and solution methodology stated here. In this paper, we have presented the feed distribution problem, the vehicle scheduling problem and the feedmill lot-sizing problem sequentially, suggesting that they would be solved in that order. In fact, the three problems are inter-related and a "one-pass" solution procedure would be used only for expediency and may lead to sub-optimal solutions. A research challenge would be to investigate how the three models might be linked and to develop an overall optimization procedure for when the objectives are combined. Much research is needed to investigate problems of this nature so that effective optimization-based decision-support tools can be offered to cattleyard managers.

REFERENCES

[1] A. Assad and B.L. Golden, 1995, Arc routing methods and applications, in M.O. Ball, et al. (eds.) *Handbooks in Operations Research and Management Science*, 8, Elsevier Science, Amsterdam, 375-483.

[2] L.D. Bodin and S.J. Kursh, 1978, A Computer-Assisted System for the Routing and Scheduling of Street Sweepers, *Operations Research*, 26, 525-537.

[3] J. Bruno, and P. Downey, 1978, Complexity of Task Sequencing with Deadlines, Set-up Times and Changeover Costs, *SIAM Journal of Computing*, 7, 393-404.

[4] D. Cattrysse, M. Salomon, R. Kuik, and L.N. Van Wassenhove, 1993, A Dual Ascent and Column Generation Heuristic for the Discrete Lotsizing and Scheduling Problem with Setup Times, *Management Science*, 39, 477-486.

[5] G. Clarke and J.W. Wright, 1964, Scheduling of vehicles from a central depot to a number of delivery points, *Operations Research*, 12, 568-581.

[6] M. Dempster, J.K. Lenstra, and A.H.G. Rinnooy Kan, 1982, *Deterministic and Stochastic Scheduling*, Reidel, Dordrecht.

[7] J. Desrosiers, Y. Dumas, M.M. Solomon, and F. Soumis, 1995, Time Constrained Vehicle Routing and Scheduling, in M.O. Ball, et al. (eds.) *Handbooks in Operations Research and Management Science, Vol. 8*, Elsevier Science Publishers B.V., 35-139.

[8] M. Dror, G. Laporte, and P. Trudeau, 1994, Exact Solutions for Split Delivery Routing, *Discrete Applied Mathematics*, 50, 239-254.

[9] M. Dror and P.A. Mullaseril, 1996, Live stock feed distribution: Capacitated Rural Postman Problem with Time Windows, Working Paper, MIS, University of Arizona, Tucson, Arizona.

[10] M. Dror and P. Trudeau, 1989, Savings by split delivery routing, *Transportation Science* 23, 141-145.

[11] M. Dror and P. Trudeau, 1990, Split delivery routing, *Naval Research Logistics* 37, 383-402.

[12] H.A. Eiselt, M. Gandreau, and G. Laporte, 1995, Arc routing problems - Part II: The rural postman problem, *Operations Research*, 43, 399-414.

[13] B. Fleischmann, 1990, The Discrete Lotsizing and Scheduling Problem, *European Journal of Operational Research*, 44, 337-348.

[14] M.R. Garey and D.J. Johnson, 1979, *Computers and Intractability: A Guide to the theory of NP-completeness*, Freeman, San Francisco, CA.

[15] B. Golden, J. DeArmon, and E. Baker, 1983, Computational experiments with algorithms for a class of routing problems, *Computers & Operations Research*, 10, 47-69.

[16] B.L. Golden and R.T. Wong, 1981, Capacitated Arc Routing Problems, *Networks*, 11, 305-315.

[17] S. van Hoesel, R. Kuik, M. Salomon, and L.N. Van Wassenhove, 1994, The Single-Item Discrete Lotsizing and Scheduling Problem: Optimization by Linear and Dynamic Programming, *Discrete Applied Mathematics*, 48, 289-303.

[18] W. Horn, 1974, Simple Scheduling Algorithms, *Naval Research Logistics Quarterly*, 21, 177-185.

[19] J.K. Lenstra and A.H.G. Rinnooy Kan, 1976, On General Routing Problem, *Networks*, 6, 273-280.

[20] J. Maes and L.N. Van Wassenhove, 1988, Multi-Item Single-Level Capacitated Dynamic Lot-Sizing Heuristics: A General Review, *Journal of the Operational Research Society*, 39, 991-1004.

[21] U. Manber and S. Israni, 1984, Pierce point minimization and optimal torch path determination in flame cutting, *Journal of Manufacturing Systems*, 3, 81-89.

[22] T.L. Magnanti and R. Vachani, 1990, A Strong Cutting Plane Algorithm for Production Scheduling with Changeover Costs, *Operations Research*, 38, 456-473.

[23] P.A. Mullaseril, M. Dror, and J. Leung, 1996, Split-Delivery Routing Heuristics in Livestock Feed Distribution, (forthcoming in *Journal of Operational Research Society*).

[24] W.L. Pearn, A.A. Assad, and B.L. Golden, 1987, Transforming arc routing into node routing problems, *Computers & Operations Research*, 14, 185-208.

[25] C.H. Papadimitriou, 1976, On the Complexity of Edge Traversing, *Journal of the Association for Computing Machinery*, 23, 544-554.

[26] M. Salomon, L.G. Kroon, R. Kuik, and L.N. Van Wassenhove, 1991, Some Extensions of the Discrete Lotsizing and Scheduling Problem, *Management Science*, 37, 801-812.

[27] H. Stern and M. Dror, 1979, Routing electric meter readers, *Computers & Operations Research*, 6, 209-223.

8

OPTIMIZING A RETAIL PROMOTIONAL CALENDAR BY MIXED INTEGER, LINEAR PROGRAMMING

Stephen A. Smith

University of Santa Clara
Santa Clara, California
USA

ABSTRACT

This paper formulates a linear programming model for determining an optimal promotional calendar for retail seasonal planning. The objective is to maximize the total season profit, subject to given resource constraints and seasonal variations, by selecting from a finite set of possible promotion policies that can be used each week.

1 INTRODUCTION

1.1 Overview

Retail buyers who do merchandise planning in department store chains face a complex optimization problem in preparing for each new season. In developing the seasonal promotional calendar for each class of merchandise, they must decide the type of advertising and promotional markdown, if any, that should be used for each week of the upcoming season. This decision must take into account promotional budgets, seasonal fluctuations in customer demand patterns, and some preset rules regarding the maximum frequency of promotions and their coordination with storewide events.

Even among the largest retailers, this decision making process relies largely on subjective judgment. Because of the complexity of the problem and the difficulty of incorporating all the factors that affect sales, buyers tend to simply reuse seasonal plans from previous years, with some minor modifications.

G. Yu (ed.), Industrial Applications of Combinatorial Optimization, 172–185.

Success or failure is generally measured by comparing a season's sales and profitability results to targets that were set before the season. There is no objective analytical way to judge the effectiveness of the promotional schedule itself, or to estimate the potential profit contribution of additional promotional resources. In addition, as customers' tastes change over time or new products are introduced, promotional schedules that were effective in the past may no longer be optimal for the new market conditions.

Sales forecasting methods for predicting customers' response to temporary price markdowns, advertising and other promotional variables have been developed and validated by marketing researchers. For background material on market response models, the reader is referred to Blattberg and Neslin (1990) and Hanssens, et al. (1990). Further details on the model forms used in this paper are found in Achabal, et al. (1990), Kalyanam (1996) and Smith, et al. (1996). Currently available methods can estimate the change in sales that will result from modifying the planned features of a given promotion, or estimate its incremental contribution to gross margin. However, no optimization models have been developed to schedule the placement of promotions on the seasonal calendar, taking into account budget constraints and seasonal variations in demand. The goal of this paper is to formulate a linear programming model for optimizing this discrete choice problem.

1.2 Retail Planning Environment

A retail buyer is typically responsible for developing a seasonal plan for a particular class of merchandise for a 13 to 26 week season, which will be implemented at all stores in the chain. A merchandise class, for example, might be men's casual slacks, which would then be further divided into subclasses, sometimes called merchandise "programs," for planning purposes. For the discussions in this paper, we will use the term "product" in place of "program," because it is a more familiar term to most readers.

The weekly unit sales for slacks are determined by factors such as seasonal effects, price and advertising. Retailers who use a marketing strategy of periodic promotions plan markdowns for each product during certain weeks of the season. Ads of various sizes, appearing as supplements in the Sunday paper, are also used to promote products periodically, usually in conjunction with a price markdown. Constraints may be in the form of resource budgets, the number of markdowns or ads of a particular type that can be used and the minimum spacing that is permitted between promotions during the season.

Retail buyers typically use trial and error to determine the seasonal plan, but many factors complicate this solution approach. For example, it may not be possible to find a solution that simultaneously meets a sales target and satisfies the resource constraints. This situation may arise when promotional budgets change from one year to the next, for example. When it appears that the resources are not adequate, the buyer may try to negotiate a revised sales target or obtain additional promotional resources.

A computer based optimization model can potentially make the following contributions to the retail seasonal planning process:

1. It can facilitate the identification and evaluation of better promotional plans. Even if the optimal solution itself is not used, studying the solutions developed by such a model may provide insights regarding general characteristics of good solutions.
2. The optimization algorithm can determine the maximum achievable gross margin, given a particular complement of promotional resources. This establishes an upper bound on the gross margin that can be achieved by any feasible plan and determines how much potential profit is being sacrificed by implementing an alternative plan.
3. The optimization model can be used to estimate the potential impact of increasing or decreasing the promotional resources for certain products.

2 MODEL FORMULATION

The proposed model will be stated in terms of weekly sales, since many retail department stores change prices and advertising on a weekly basis. Retail sales tend to be highly seasonal, with the Christmas season and back-to-school season being the strongest sales periods for many classes of items, while the months of January and February tend to be slow selling periods. Seasonal fluctuations are captured by the time indexed vector of parameters

$S(t) =$the seasonal multiple for week t,

where each $S(t)$ is expressed relative to a multiple of 1 for a "normal week," for convenience.

Price markdown and advertising effects are often captured by "response functions," that predict the multiplicative increase in sales that will result from a given markdown price or ad. Two commonly used response functions for temporary price markdowns are defined as follows

$$\text{Price Response Factor} = \left(\frac{p_0}{p}\right)^{\epsilon} \quad \text{Constant Elasticity Model} \tag{8.1}$$

$$\text{Price Response Factor} = e^{\epsilon(p_0 - p)/p_0} \quad \text{Exponential Response Model} \tag{8.2}$$

where

ϵ =the price response parameter

p_0 =the regular price of the item

p =the markdown price of the item.

For the constant elasticity model, it is the ratio p_0/p of the regular price to the promotional price that determines the price response, while for the exponential model it is the markdown percent $(p_0 - p)/p_0$ that determines the magnitude of the price response. Kalyanam (1996) compares the effectiveness of multiplicative models and linear response functions.

For advertising response, the constant elasticity response function can be written as:

$$\text{Advertising Response Factor} = \left(\frac{A}{A_0}\right)^{\alpha}, \quad \text{or } = 1 \text{ if there is no ad} \tag{8.3}$$

where A =the size of the ad (or the cost of the ad)

A_0 =a normalizing constant no greater than the smallest ad size (or cost)

α =the advertising response parameter.

Assuming simple multiplicative effects, the forecasted unit sales in week t, given a temporary markdown price p and an ad of size $A > 0$, can be then expressed as

$$U(t) = BS(t)\left(\frac{p_0}{p}\right)^{\epsilon}\left(\frac{A}{A_0}\right)^{\alpha} \tag{8.4}$$

where $U(t)$ =unit sales for week t and

B =the constant "base" weekly sales for the product.

This multiplicatively separable model of price markdown and advertising responses has the added advantage that the parameters ϵ and α can be estimated with linear regression, by taking the natural logarithm of the sales data and then performing the appropriate transformations to obtain the parameters. See Kalyanam (1996) or Smith, et al. (1994) for more discussion.

3 SPECIFYING THE OPTIMIZATION MODEL

In practice, retail buyers are usually restricted to a relatively small set of candidate price markdowns and ads. That is, if percentage markdowns are used, they are typically restricted to a small set of possibilities, e.g., 15%, 25%, 33% and 40% off. Alternatively, some retailers use "price points," i.e., certain prices that are believed to be attractive to customers, such as prices ending in $0.99. In either case, the selection of a markdown price for a particular week is a choice among discrete alternatives. Similarly, newspaper ads can be viewed as discrete choices from a set made up of alternatives such as: a quarter page ad, a half page ad, a full page ad, or possibly multiple page ads.

The buyer's decision problem can be expressed in terms of 0,1 variables in the following way. Let i be the index of the markdown type, $i = 0, \ldots, m$ and j be the index of the ad type, $j = 0, \ldots, n$. Define the following variables over these indices:

$Y_i(t) \quad = 1$ if markdown i is used in week t

$$
\begin{aligned}
&= 0 \ \text{otherwise} \\
Z_j(t) \quad &= 1 \ \text{if ad type } j \text{ is used in week } t \\
&= 0 \ \text{otherwise}
\end{aligned}
$$

p_i = the unit price when markdown i is used

A_j = the size of ad type j

No markdown corresponds to $i = 0$, while no ad corresponds to $j = 0$.

Multiplicative sales responses can then be captured for any discrete set of alternatives in the following way. For markdown i, define γ_i so that

$$\gamma_i = \left(\frac{p_0}{p_i}\right)^{\epsilon}, \quad \text{for } i > 0, \text{ and } \gamma_0 = 1. \tag{8.5}$$

Analogously, δ_j is defined as

$$\delta_j = \left(\frac{A_j}{A_0}\right)^{\alpha}, \quad \text{for } j > 0 \text{ and } \delta_0 = 1. \tag{8.6}$$

Thus, the unit sales in period t, reflecting both price markdowns and advertising effects, can be captured by the following summation and constraints.

$$U(t) = BS(t)\sum_{j=0}^{n}\sum_{i=0}^{m}\gamma_i\delta_j Y_i(t)Z_j(t) \tag{8.7}$$

$$\text{where } Y_i(t),\ Z_j(t) = 0,\ 1 \quad \text{and} \sum_{i=0}^{m} Y_i(t) = \sum_{j=0}^{n} Z_j(t) = 1. \tag{8.8}$$

Constraints (8.8) guarantee that exactly one markdown and one ad type are selected, including the possibility of no ad and no markdown. The revenue $R(t)$ can be expressed in terms of the $Y_i(t)$ and $Z_j(t)$ as well

$$R(t) = BS(t) \sum_{j=0}^{n} \sum_{i=0}^{m} p_i \gamma_i \delta_j Y_i(t) Z_j(t) \tag{8.9}$$

For simplicity, the cost model assumes that the merchandise is all purchased by the retailer before the start of the season and any remaining merchandise is salvaged at the end of the season. This would generally be the case for fashion and seasonal merchandise that must be cleared at the end of each season. [A more general cost model that includes inventory holding cost is described in Smith, et al., (1996).] The total cost for the season can be written as

$$C(I_0) = cI_0 - r\Big\{I_0 - \sum_{t=1}^{T} U(t)\Big\}, \tag{8.10}$$

where

I_0 =the initial inventory level (the number of units purchased)

T =number of weeks in the season

r =the residual value per unit at the end of the season

c =the unit purchase cost of the merchandise at the beginning of the season

Although the sales forecast is deterministic, I_0 is generally required to be greater than the total forecasted sales to provide adequate safety stock and to ensure that there is sufficient inventory available for an attractive store presentation of the merchandise. This can be stated as a constraint that I_0 must be $(1 + \alpha_s)$ times the forecasted sales, where α_s is a fractional value such as 10% to 20%.

The objective is to maximize the total gross margin for the season, which is given by $\sum_t R(t) - C(I_0)$.

3.1 Converting to a Linear Objective Function

The revenue function $R(t)$ and the unit sales $U(t)$ are both defined nonlinearly, since they are expressed as products of the decision variables. However, using results developed by Watters (1967) and Glover and Woolsey (1974), these forms can be expressed as linear functions by defining additional variables $X_{ij}(t)$ together with the appropriate constraints. The key result proved by Watters is that only the $Y_i(t)$ and $Z_j(t)$ need to be constrained to be 0,1. The variables $X_{ij}(t)$ are constrained only to be nonnegative. The equations are as follows

$$U(t) = BS(t) \sum_{j=0}^{n} \sum_{i=0}^{m} \gamma_i \delta_j X_{ij}(t), \tag{8.11}$$

$$\text{subject to (8.8) and } X_{ij}(t) \leq Y_i(t), \quad X_{ij}(t) \leq Z_j(t), \tag{8.12}$$

$$Y_i(t) + Z_j(t) \leq X_{ij}(t) + 1, \text{ for all } i, j, t.$$

3.2 Resource Constraints

Promotional resource constraints may be in the form of dollar budgets for the season, the total quantity of a resource available such as pages of advertising, or may be stated in terms of the number of each type of promotion that can be held during the season. Thus, the possible forms of the constraints can be specified as follows. Define

M_i = the number of markdowns of type i that are to be used

N_j = the number of ads of type j that are to be used

A_j = the size of ad type j in pages

M = the maximum total number of markdowns used during the season

T_A = the maximum total number of ad pages available for the season.

[Ad size could also be measured in dollar cost.]

Constraint equations of the following forms, or other variations, may be used:

$$\sum_{t=1}^{T} Y_i(t) = M_i \quad \text{for all } i \geq 1 \quad \text{number of markdowns of each type} \tag{8.13}$$

$$\sum_{t=1}^{T} Z_j(t) = N_j \quad \text{for all } j \geq 1 \text{ number of ads of each type} \tag{8.14}$$

$$\sum_{t=1}^{T} \sum_{i>0} Y_i(t) \leq M \qquad \text{total number of markdowns} \tag{8.15}$$

$$\sum_{t=1}^{T} \sum_{j=0}^{n} A_j Z_j(t) \leq T_A \qquad \text{total ad pages.} \tag{8.16}$$

3.3 The Linear Programming Model

The following mixed integer, linear programming model can now be stated for optimizing the promotional calendar. A more general model that includes inventory holding cost is described in Smith, et al., (1996).

$$max \sum_{t=1}^{T} R(t) + cI_0 - r\Big\{I_0 - \sum_{t=1}^{T} U(t)\Big\} \tag{8.17}$$

subject to

$$R(t) = BS(t) \sum_{i=0}^{m} \sum_{j=0}^{n} \gamma_i p_i \delta_j X_{ij}(t) \quad \text{for all } t \text{ revenue definition} \tag{8.18}$$

$$U(t) = BS(t) \sum_{i=0}^{m} \sum_{j=0}^{n} \gamma_i \delta_j X_{ij}(t) \quad \text{for all } t \text{ unit sales definition} \tag{8.19}$$

$$X_{ij}(t) \leq Z_j(t), \quad \text{for all } i, j, t \tag{8.20}$$

$$X_{ij}(t) \leq Y_i(t), \quad \text{for all } i, j, t \tag{8.21}$$

$$Y_i(t) + Z_j(t) \leq X_{ij}(t) + 1, \quad \text{for all } i, j, t \tag{8.22}$$

$$(1 + \alpha_s) \sum_{t=1}^{T} U(t) \leq I_0 \qquad \text{total inventory} \tag{8.23}$$

$$\sum_{i=0}^{m} Y_i(t) = 1 \;\; \text{for all } t \text{ one markdown selection per week} \tag{8.24}$$

$$\sum_{j=0}^{n} Z_j(t) = 1 \;\; \text{for all } t \quad \text{one ad selection per week} \tag{8.25}$$

$$\sum_{i>0} [Y_i(t) + Y_i(t+1)] \leq 1 \;\; \text{for all} \;\; t \leq T - 1 \;\; \text{markdown spacing} \tag{8.26}$$

$$Y_i(t), Z_j(t) = 0, 1 \text{ and } X_{ij}(t) \geq 0 \;\; \text{for all } i, j, t, \tag{8.27}$$

Constraints selected from (8.13)–(8.16), depending on the type of solution desired.

Based on the definition of unit sales $U(t)$ in (8.7), constraints (8.18) and (8.19) define the quantities $U(t)$ and $R(t)$ used in (8.17). Constraint (8.23) enforces a safety stock constraint for the total inventory for the season, while (8.24) and (8.25) ensure that exactly one combination of price and ad size is selected for each week. Constraint (8.26) enforces the rule that markdowns cannot be held back-to-back during the regular season.

3.4 Analysis of Test Runs

Test runs of the optimization model were made using historical data for eight products obtained from a major retailing chain. More complete details on the data and its analysis are presented in Smith, et al. (1996). The LINGO package solver developed by LINDO Systems was used to solve the linear programs. In applying the optimization methodology to these example situations in which the buyer's plan is already specified, it is interesting to consider alternative formulations that allow differing degrees of freedom for the model to improve the plan. Two cases were evaluated: (1) an optimal rescheduling of the buyer's selected promotions and (2) an optimized seasonal plan, subject to the existing resource constraints.

For each run, there were six possible ad size choices and five possible price markdown choices over a 26 week season, for a total of 780 $\{X_{ij}(t)\}$ variables, 130 $\{Y_i(t)\}$ variables and 156 $\{Z_j(t)\}$ variables. Thus 286 variables are constrained to be 0,1. In each case, the ordinary LP solution was first tested to see if the 0,1 constraints were automatically satisfied, which they were for ten of the sixteen cases. If not, the $\{Y_i(t)\}$ and $\{Z_j(t)\}$ variables were constrained to be 0,1 and the branch and bound option of LINGO was used to solve the problem.

Table 1 shows the number of iterations required to obtain an optimal solution in each case and whether or not the branch and bound method was needed to obtain a solution that satisfied the integer constraints. In Tables 1 and 2, "Rescheduled" refers to the case in which the buyers' promotions were merely rescheduled, while "Optimized" refers to the case in which an optimal solution was generated, subject to the same constraints on total promotional resources. This table suggests that computation time may become an issue for larger problems because of the wide variation in the number of iterations required.

Table 1 Number of Iterations Required for Solution

Product	Rescheduled	Optimized	Branch/ Bound
1	527	211	no
2	69,056	69,056	yes
3	380	271	no
4	1638	927	yes
5	42,758	33,690	yes
6	622	278	no
7	586	295	no
8	539	357	no

Detailed comparisons of the optimization model's results to those obtained with the buyers' plans, as well as analysis of a more general model with inventory holding cost, is presented in Smith, et al. (1996). The model's impacts on unit sales, total revenue and total gross margin for the season are summarized in Table 2, expressed as percentage changes relative to the results obtained with the buyers' plans. For these test cases, optimizing the seasonal plan left the total number of units sold roughly unchanged, increased the total revenue slightly, and increased the gross margin more significantly, as compared to the buyers' results. The gross margin changes are substantial in dollar terms because the combined total gross margin for these eight products was roughly $6 Million. Thus, optimizing only the schedule is projected to increase the gross margin by roughly $500,000 for the eight products, while the overall optimum increased it by roughly $700,000. Thus, by merely rescheduling the markdowns and ads in an optimal way, the model predicts that is possible to increase the gross margin significantly. These results are, of course, specific to these example products and assume that the model's sales forecasts are correct.

Table 2 Percentage Changes from the Buyer's Plan

	Rescheduled	Optimized
Total Units Sold	2.6%	−0.4%
Revenue	4.6%	4.0%
Gross Margin	8.6%	11.8%

4 CONCLUSION

This paper formulates a model for optimizing the schedule of promotions for retail products over a season, choosing from a discrete set of alternatives, subject to a variety of resource constraints. An integer programming formulation of this problem was applied to eight test runs to test the feasibility and effectiveness of the optimization approach. With limited computing resources, the optimal solution was obtained quickly in most cases, although certain cases required many iterations. The optimization model has four other operational benefits for the buyers. It can

1. Provide bounds for the gross margin that can be achieved by any seasonal plan

2. Test the feasibility of achieving sales targets for given levels of promotional resources

3. Evaluate the impact on gross margin of changing promotional resources

4. Provide insights regarding the general features of optimal plans.

The application of the linear programming model to these test examples demonstrates the feasibility of using this formulation approach. In many cases, buyers would like to be able to develop seasonal plans for multiple products in the same merchandise class simultaneously, subject to shared resource constraints, and perhaps using more than the two promotional variables of price and ad size. Conceptually, the mixed integer, linear programming formulation developed here is capable of solving this problem, subject to computational limitations. It is hoped that this article will stimulate further research on this problem that leads to improved optimization methods that can solve larger promotional planning problems.

Acknowledgments

The author is grateful to Professors Dale Achabal, Robert Collins, Kirthi Kalyanam and Shelby McIntyre for their valuable suggestions in the course of this research, and to the corporate members of the Retail Workbench at Santa Clara University for their financial support and guidance in defining this research topic. The author is fully responsible for the opinions expressed and any remaining errors.

REFERENCES

[1] D. Achabal, S. McIntyre, and S. Smith, 1990, Maximizing Profits from Department Store Promotions, *Journal of Retailing*, 66, 383-407.

[2] R. C. Blattberg and S. A. Neslin, 1990, *Sales Promotion: Concepts, Methods and Strategies*, Prentice-Hall, Englewood Cliffs, NJ.

[3] F. Glover and E. Woolsey, 1974, Converting the 0-1 Polynomial Programming Problem to a 0-1 Linear Program, *Operations Research*, 22, 180-200.

[4] D. M. Hanssens, L. Parsons, and R. Schultz, 1990, *Market Response Models: Econometric and Time Series Analysis*, Kluwer Academic Publishers, Boston, MA.

[5] K. Kalyanam, 1996, Pricing Decisions Under Demand Uncertainty: A Bayesian Mixture Model Approach, *Marketing Science*, 15, 207-221.

[6] S. A. Smith, D. Achabal, and S. McIntyre, 1994, A Two Stage Sales Forecasting Procedure Using Discounted Least Squares, *Journal of Marketing Research*, 31, 44-56.

[7] S. A. Smith, R. Collins, and S. H. McIntyre, 1996, A Discrete Optimization Model for Seasonal Merchandise Planning, Retail Workbench Working Paper W-RW96-01, Santa Clara University.

[8] L. J. Watters, 1967 Reduction of Integer Polynomial Programming Problems to Zero-One Linear Programming Problems, *Operations Research*, 15, 1171-1194.

9

AN IMPLEMENTATION OF A SYSTEM USING HEURISTICS TO SUPPORT DECISIONS ABOUT SHOP FLOOR SETUP CHANGES

Rahul Dé

Department of Business, Public Administration, and Economics
Bowie State University
Bowie, Maryland 20715
USA

1 INTRODUCTION

Shop floor scheduling remains as one of the hardest and most researched problems in management science. Heuristic approaches to solving such problems are widely acknowledged as a feasible way to proceed. This paper describes a shop floor problem at a real manufacturing facility, where a knowledge-based system was installed and used for supporting decisions related to scheduling. The specific problem dealt with scheduling the setup changes on machines that are at the final stage of a multi-stage processing sequence.

The system described in this paper is part of a larger research project in which a number of systems were designed and built to address various parts of the scheduling problem [3, 2, 6, 4, 1]. The research goals of the larger project had to do with representing the factory scheduling problem in such a manner that it could be reasonably addressed with a combination of knowledge-based and traditional operations research techniques. The problem was thus divided into a number of parts that could be tackled independently and then combined to present a comprehensive solution. The research reported in this paper emerged as one of those parts.

The issues addressed in this paper deal with defining the scope of the specific scheduling problem, identifying the heuristics needed to solve the problem, finding an adequate representation for the implementation of the heuristics, and the validation of the system. In the sections that follow, these issues are discussed as part of the description of the problem, the design of the system, its implementation and validation.

G. Yu (ed.), Industrial Applications of Combinatorial Optimization, 186–211.

2 OVERVIEW OF THE SETUP CHANGE SCHEDULING PROBLEM

The site where the planning problems were studied is the Westinghouse Specialty Metals Plant, Commercial Nuclear Fuel Division (in Blairsville, Pennsylvania). This plant manufactures zirconium alloy fuel tubes that are used in the nuclear fuel industry. The input raw material are zirconium alloy tubes that are relatively thick (where the outer diameter is close to three inches) and these are converted through various stages of processing to tubes whose outer diameter is a fraction of an inch. Different "products" at this facility are marked by the different final diameters to which they are reduced, although the input raw material is the same for all.

The principal operation conducted at this plant is that of pilgering, or drawing, the zirconium metal tubes down to finer diameters of given dimensions. This is a cold working process and the tubes are drawn in a number of passes or stages, usually four. After each pilgering pass the tubes are cut to smaller sizes, as they extend in length during pilgering, cleaned of the lubricating greases, and annealed in furnaces for several hours. The cleaning is done by immersing batches of tubes in an acid bath and then washing with water. This cleaning process is referred to as pickling. The last operation of annealing is done to relieve the tensions introduced in the metal during the pilgering process.

Consider the layout of the manufacturing facility shown in Figure 1. The figure shows how input raw material is converted to the final product through four passes, or stages, of processing. Each pass consists of the three processes — pilgering, pickling, and annealing.

The raw material is input in the form of *lots* that usually have between five and ten tubes. As these lots are processed through the various passes the diameter of the tubes is reduced, which lengthens them, and these are then cut, thereby increasing the number of tubes in the lots. By the final pass there are several hundred tubes in each lot. The increasing number of tubes increases the amount of time required to process the lots in each pass. For instance, the ratio of the time required to pilger the lots in the first pass to that of the time required to pilger in the final pass is 1:22. This means that a larger number of machines are required to pilger the lots in the final passes.

The arrows show the passage of the lots through the shop floor. Each arrow denotes a separate product. Products 1 and 2 are currently running through the entire facility. As opposed to these, products 3 and 4, have lots only in

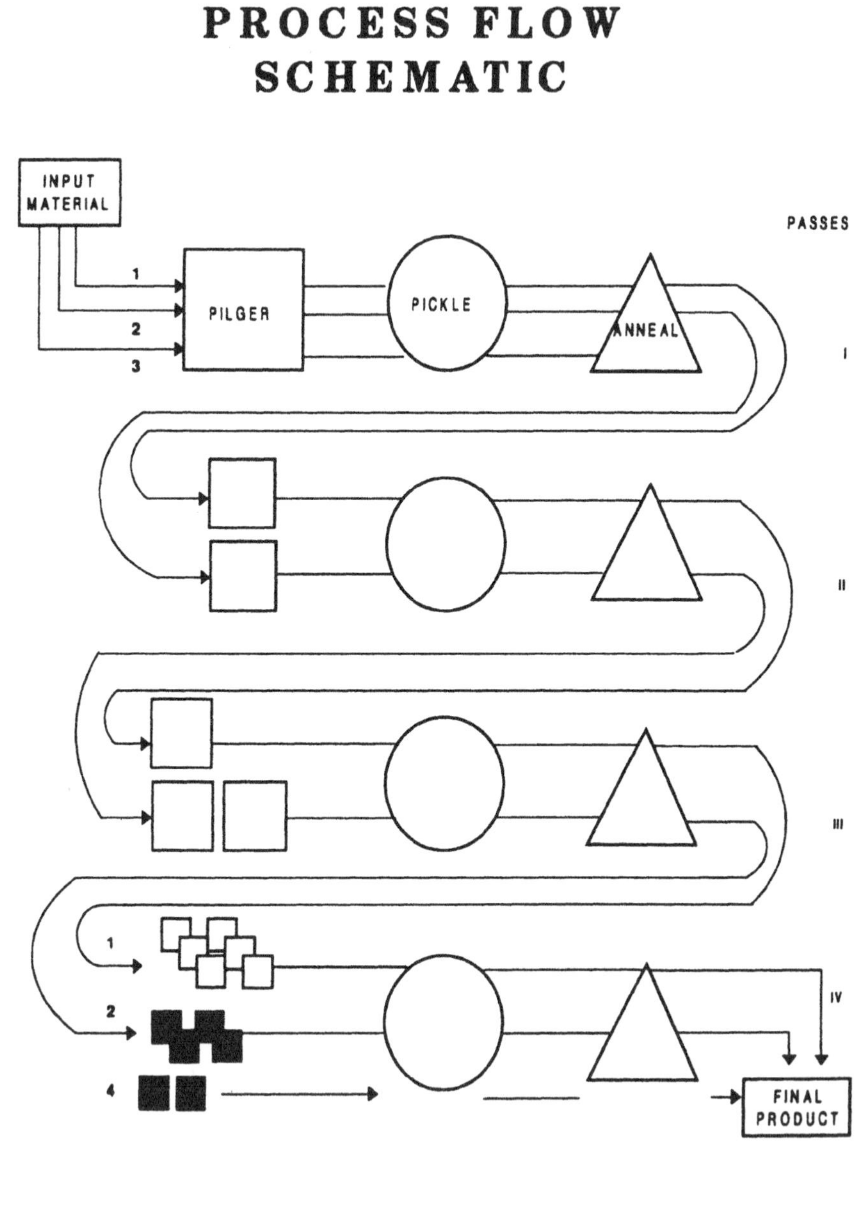

Figure 1 Layout of the Manufacturing Facility

some stages of the process. Product 3 is entering the facility and has lots that are still in the first pass. Product 4 is leaving the facility at the stage marked "Final Product," and has lots only in the final pass. These lots are about to complete; once the pilger machines have completed pilgering them, they will be assigned to the other products.

The setup requirements for the pilger machines (or pilger mills) at the earlier stages are such that most products can be pilgered on the same mills. At the later stages different mill setups are required for different products. This is especially important for the final pass where each product has to have different setups for the group of machines that are assigned to it. This is shown in the figure by the different shades of the boxes denoting the mills. Setup changes are expensive, in the amount of time required and the actual cost of the dies and quality testing required.

From the perspective of planning and scheduling the various tasks on the shop floor, the Westinghouse plant is a very complex environment. This complexity arises from the nature of the products, the manner in which the products are processed, the variability of process times, etc. Some of these factors are:

- The Westinghouse plant produces close to a hundred products, where each product is distinguished by its inner and outer diameters. The raw stock for all the products is the same. For most products the outer diameters to which they are reduced during the first three passes is the same. In the last, fourth, pass the final product diameters are produced. For other products the differences in the diameters begins from the second pass itself.
- The final products are made to very close tolerances. The internal and external diameters are quality tested carefully, both for the final tubes and for the intermediate tubes produced. (The Westinghouse Specialty Metals Plant is a recipient of the Malcolm Baldridge Award for manufacturing quality.)[1]. Thus, the machine setups and process parameters have to be monitored and controlled very closely.
- The plant can produce over a hundred different products. At any time there are usually five or six products being processed concurrently on the shop floor.
- The regular products at the plant are the fuel tubes. These constitute about 90% of the total production. The plant also produces "thimble"

[1] These tubes are used in nuclear reactors to hold the uranium fuel pellets. The amount of fuel in the core of the reactor is controlled by raising or lowering these tubes

and "instrument" tubes (used for different purposes in nuclear reactors) which are processed over the same machines, but with different settings and process times. These products are tracked separately for purposes of scheduling and planning.

- Each product is processed in batches, where each batch consists of several lots. The individual products might have anywhere between a few to over a hundred lots in each batch. Thus, some batches or runs last for several weeks at a time.

- A lot consists of a few pieces initially, but as it goes through the passes the number of pieces multiply and in the final pass it has about five hundred. Owing to this the pilgering and pickling time for each lot increases.

- Most products are processed over four passes, but some are moved through two, three, or five passes also.

- Processing times have high variability. The average pilgering times on the third and fourth stage machines have high variance. This is critical because the average time to process a lot at this stage is very high, as compared to the earlier passes.

- There are multiple bottlenecks on the shop floor. Depending on the build-up of queues in front of machines and their operating capacities and times, the bottlenecks may shift. For example, both the furnace and the final pass pilgers operate at very high utilization levels. Any breakdowns or variability in their runs will cause the bottlenecks to shift.

- Changing the setups of the pilger machines to process different products is expensive. Owing to the high precision requirements, every setup change requires qualification. At the third and fourth passes, this implies processing one-to-two lots through the recently changed dies, testing them for dimensional tolerances, and correcting for any discrepancies. Each setup change requires about 30% of the time required to process one job.

- Material is produced to meet demand. The demand forecast is made available by the management for several weeks into the future. Although, this sets a fairly clear guideline for planning the production activities, there are frequent and numerous fluctuations in the products ordered as well as the quantities. The penalties for missing the demand due dates are high and the scheduler ensures that such costs are rarely incurred.

- There are many products that require special handling. For example, one product requires that all its lots be annealed together in the same furnace. In this case, the scheduler has to ensure that all the lots arrive at the furnace in close succession so as to start the annealing without waiting much. Other products, such as experimental products, require that they be specially processed across all the machines.

These factors highlight the characteristics that make the Westinghouse facility a very complex domain. Although all the factors are not pertinent for any one decision they are, nevertheless, related.

2.1 Knowledge Elicitation

The author was part of a team of researchers that worked with the scheduler for several months to understand the planning and scheduling tasks that he performed. He introduced the team, gradually, to the complex organization and processes on the shop floor.

The team's approach was to first understand what he did to control the shop floor and also how he did this. Sessions were taped in which he would talk through the decisions he was making. He showed the information that he had gathered from the shop floor and then explained how he used this to make the planning and scheduling decisions. Later, transcriptions of the tapes were used to further understand the details of the decision making process. Once the team had a basic grasp of the heuristics he was using to make the decisions, they used the same information to try and imitate his decision making. He looked over their decisions and provided comments and criticisms, furthering their understanding of his decision processes.

The methodology of scheduling used in this document is based on the heuristics and strategies learned from the scheduler. His expertise in these tasks (he had worked at this facility for over 25 years) forms the basis for using this knowledge in designing, implementing and validating the scheduler.

2.2 Decision Tasks

The number of pilgering machines available in each pass is limited, and the scheduler had to balance the load between them carefully. He tried to make as few setup changes as possible in order to produce the requisite final products. The setups of the mills can be changed to pilger different products, but this is an expensive operation. An important goal of the scheduler was to avoid idleness on the expensive machines. He also tried to keep the queues in front of the different machines at a minimum so as not to increase the work-in-process inventory.

The scheduler made a number of decisions regarding the utilization of the various resources at the plant. The demand schedule set the targets for him and since on the average a lot of material takes about 4-6 weeks to go through all the operations in the plant, he had to plan for the resources with that time frame in mind.

An important decision is to determine the setups changes of the mills, from their current configuration, to process the various products on the shop floor. The setup changes may move one mill from one product line to another within the pass in which it is being used, or, in a few cases, the change may be from one pass to another, within the same product line or a different one. These decisions are governed by various factors such as the status of the shop floor, the stage of processing of the lot the mill is working on, the load on the mill etc.

Another important decision is that of the quantity of raw stock to start processing on a daily basis. The raw material is introduced in the form of lots that are marked for the final product they are going to be turned into. The scheduler knew the demand for the product, which is given in the form of number of pieces, and used this to estimate the number of lots of each product to start every day, keeping in mind the lead time required and the status of the shop floor.

3 HEURISTICS FOR SCHEDULING SETUP CHANGES

This section explains the manner in which the scheduler created the plans for controlling the resources on the shop floor. The details about how various tactics and heuristics were employed in different situations were obtained from the scheduler during various scheduling sessions. The format that was followed in these sessions was to work with the pickup sheet (the *pickup sheet* is a document that shows the snapshot of the shop floor at any point of time; for each product, it shows where the lots are and whether they are being processed on any machine (Table 1)) of that particular day (along with the source sheet[2] and the forecast sheet[3] in some cases) and discuss the schedules and plans that had to be created to address the situation on the shop floor. In any particular session the scheduler would develop plans for all the machines in the different passes, the furnaces, and also the priorities for the pickle houses. (The decisions made to schedule all these resources are not independent of each other.)

The following are some of the statements made by the scheduler, regarding machine setups (particularly for the final pass machines), that highlight the particular heuristic that he is using. These statements are extracted from the transcripts of three scheduling sessions. The manner in which these rules are specifically implemented in the system is described later.

1. " ...to find out where we are and where we want to be."

   This is the general philosophy of planning followed by the scheduler. The process of planning begins by assessing the current situation and then creating the plans to move towards certain goals or targets.

2. Whenever there are excess mills available for a product (in the final pass), these can be pulled down and assigned to another product line.

   If there is a light load on any product line because there are not enough lots to keep the mills running, and there is a possibility of some mills going idle, then mills should be removed from this line and assigned to another product. The load is gauged on the basis of the work available to the mills for the next few shifts.

3. For a product that already has final pass mills set up, it is advisable to move the backstream lots through at a priority to feed these mills.

   This is another heuristic for reducing idle times on final mills, as also, preventing unnecessary setup changes. This applies to lots in the second and third passes. Giving priority to the lots implies preferring their processing over all machines, particularly bottleneck machines such as the furnace, in the second and third passes.

[2] The *source sheet* shows the due date of the different lots. The due dates are indicated in the form of week-ending dates, where batches of lots are due in any given week. The scheduler uses this sheet to track the progress of individual lots in relation to their week-ending due dates.

[3] The *forecast sheet* shows the demand forecast for various products over a six month time period. This document is prepared twice a calendar year and is updated according to the changes that might occur in the orders placed. The demand is shown in the form of aggregate numbers of pieces of tubes, of different sizes, that are required. The number of lots that have to be processed can be calculated from this.

4. When the mills have good queues, changeovers are not done.

   This is a general rule to ensure that mills are not moved away from product lines that have material in the queues that need to be processed. The "good queues" here refers to a sufficient supply of lots over the time horizon for which the situation is being assessed.

5. If you had two shifts of work left on one mill and three shifts on another, you would drive the lots towards the one with the lesser work remaining.

   In a situation where, for the same product line, there are two mills set up to process the lots, the material arriving from backstream would be queued for the mill with lesser work.

6. If there is a sudden die change required at the two-shift machine (previous point), then this would have to be considered as one with more than two shifts of work remaining.

   If a problem is noticed in processed tubes during quality inspection, then die changes are recommended. In such a situation, the lots are not moved away from the mill while the die is being changed. A change of die may require anywhere from one-half to one shift of time.

7. For the same product, different batches could have different drawing-item numbers. These batches would have different cut- lengths for the tubes. For any machine, it is preferable to finish off lots belonging to any given cut-length, or a particular drawing-item number, before moving to another cut-length.

   A specific product is defined by its inner and outer diameters. Some tubes, although belonging to the same product, sometimes require that their lengths be different. This is referred to as the cut-length. The above rule specifies that all tubes in a given lot should have the same cut-length, and if a different cut-length is required an entirely different lot marked with a different "drawing-item number" should be used. This is highly preferred. Although, in some cases, this rule is violated.

8. You plan for additional mills on any line when you have sufficient lottage built up to carry the machines.

   This is a general rule where the caution is to assign a mill to a product line only when it is certain that sufficient lots are available in backstream processes. "Sufficient" here is determined by the amount of time for which the lots will be processed over the mills. If the number of lots are not sufficient then the current status of the mills will be maintained, that is, no new mills' setup will be changed over to this line.

9. "If I can't get anything to the mill in six shifts ... I make the changeover."

   This rule specifies a time horizon for which the scheduler considers the possibility of backstream lots arriving at a mill that is a candidate for a setup change. The scheduler can actively "drive" lots towards this mill, however if this is not possible then the mill setup has to be changed to prevent it from remaining idle.

10. You anticipate when a product line is about to terminate and plan to move all the mills over (to some other product lines).

    This situation occurs when all the lots for a particular product line are about to complete and move off the final pass mills. The times at which the mills are about to complete processing these lots is estimated and the setup changes are scheduled.

11. The due date you work towards is not a ship-date but an "off- mill" date.

    The scheduler works with a due date for any lot which is actually the date by which the lot has to be made available to the final quality testers. This is different from the actual shipping-date for the lot.

12. Bring up mills to meet the objectives of the common pilger week.

    The "common pilger week" is a week-ending date that designates the period by which a number of lots, a batch, has to be pilgered and moved off the final pass mills. The date is determined by the capacities of the quality testing machines (previous point). The scheduler tries to assign as many mills as are required to finish off the lots. The lots that have to be processed in the common pilger week are marked in the source sheet.

3.1 Plans for Starting Lots

The decisions for starting product lots are based primarily on the demand forecast information. The scheduler used the information on the forecast and source sheets to determine the quantity and type of material to be introduced for processing on the shop floor. The following excerpts from scheduling sessions show some general heuristics that he used to determine the input material.

1. "In the week of ... five weeks before ... we want to start bringing it in the system. What we want to do is look five weeks ahead and say they want 4000 pieces of A, 5000 pieces of B, and 4600 pieces of C."

The horizon for these decisions is five weeks. The scheduler determined the number and type of pieces required, from the demand information, five weeks ahead and from this determined the number of lots to start.

2. "What you do here is estimate ...number of lots required to make the run."

   The number of final tubes that were required was divided by the number of pieces expected in each final lot. This gave the number of lots that had to be started for processing. This number was an approximation because pieces were lost due to quality checks.

3. "What this also tells you is that in the pilger week we pilger this, we're going to have three different sizes going through the mills."

   Each "pilger week" is denoted by the products that are being processed on the final pass mills. The computation for the number of lots to start shows that five weeks from now, in the pilger week, the number of lots in the final pass will be, approximately, the number being started.

4. "You want to bring in 6 lots a day ...."

   This is an average for the number of lots to start on a daily basis to ensure that the final pass mills will have a steady supply of material to process.

Using the information about the number of pieces demanded for each product, the scheduler would calculate the number of lots that would have to come off the final pilger mills in a given week, the pilger week. This information would be used to calculate the total number of lots to start in a given week.

In starting the analysis the scheduler would first establish the status of completion of the existing product lines (here information from both the source and pickup sheets would be used). It invariably happens that the production never progresses in the manner in which it is depicted in the forecast sheet, and the lines are ahead, implying that more lots of a particular product have been completed than are due in the pilger week, or are behind, which is the opposite situation. In determining the completion status the scheduler would establish what more needs to be done, and the pace at which it has to be completed. In this sense, the completion status provides the point from which the goals of starting lots, as determined from the forecast, can be achieved.

In addition to the demand forecast and the completion status the scheduler would use a number of other factors to determine the input lots. These are as follows:

- The throughput of the mills at the final stage imposes a limit on the total number of lots that can be processed in a given amount of time.
- If some products have been identified to be priority products, then lots of these products would be preferred over others.
- When a product line is about to start, then a few lots, called the lead lots, are started earlier to qualify the machines at the various stages.
- When a line is about to complete then the last lots may be started earlier to complete the line and free the machines for processing other products.

4 DESIGN OF THE DECISION SUPPORT SYSTEM

It was evident from the scheduler's explanations that the final pass mills were bottleneck resources that had to be scheduled carefully. From this the research team concluded that one system had to directly address the problem of scheduling setup changes for the final pass mills. It was also clear that the defining parameter for the decision had to be the queues building up in front of the mills assigned to specific product lines. The choices made for the design of the system are discussed below.

Determination of machine setup changes, within the system, is based principally on projections of queues. The queues are projected forward for the machine (and product) groups based on processing and consumption patterns. The objective is to identify if the expected loads in front of the machines are going to be within a certain prespecified range. In case the loads are not within this range, machine setups are changed to bring them to this value. Thus, the strategy followed is to maintain a consistent load in front of the machines to reduce idle time as much as possible. Every setup change scheduled is based on the expected loads and variations in these expectations.

The decisions are made as follows: to begin, the current status of the floor that shows the size of the queues in front of the machine groups, as well as the status of all the jobs in the earlier stages, is considered. This is called the *snapshot*. The snapshots are then projected for n periods into the future. The projections are made by simulating the movement of material on the shop floor. The simulation progresses one time period at a time, in which, the lots are stepped through the various stages of processing. After a specific number of time units, one day in this case, a new snapshot is created that shows the current status. In this manner, snapshots for n periods are created.

From the projections the net sizes of the queues for the final pass mills for each product are calculated. For any time period, the queue projections are made by adding to the queue the number of jobs that are expected to arrive in the next period and subtracting from it those that will be consumed by the machines. To estimate the jobs that can arrive at the queue, all the possible jobs in a given product line which are in previous stages are considered and probability estimates are used to compute whether they will arrive at the queue. The probabilities are based on observed behavior on the shop floor. The estimates of consumption are based on average performance of the machines. Sizes of the queues are thus projected for the different product lines. New jobs that are introduced on the shop floor in the current time period are also stepped through in a similar manner. The net queue sizes thus obtained are compared to the desired range of values for the queue.

A "good" value for the size of the queue is determined from the schedulers heuristics listed above. Heuristics #4 and #9 indicate what the scheduler considered desirable for the size of the queues. If the lots that are feeding the queue are sufficient to keep the mills running for the next six shifts, then this is considered to be within the desired range.

However, if the number of lots available to the queue is in excess of the range or short of that range, then this is a situation in which a setup change is possible in order to provide an additional machine to a line or remove a machine from one, at a given period. For example, if it is noticed that product line 2, in Figure 1 is consistently heavy (where heavy denotes that the queue sizes are above the desired range) for more than two (or any fixed number of) periods, say periods t, $t+1$, $t+2$, and product line 1 is consistently light (where light denotes that the queue sizes are below the desired range), a decision about moving a machine over from 1 to 2 can be made, at period t which is the beginning of the heavy load for 2.

Once such a decision is made the system changes the configuration of the shop floor by moving one mill from product line 1 and adding it to product line 2. This change is made in the snapshot of period t.

The projections are then re-computed from time $t+1$ with the change in the status of the product lines included. This forms the second set of projections. Again, the decisions about heavy or light queues are made and setup changes are scheduled, provided some are required. The process of projecting queues,

Table 1 An Example of a Snapshot. Cell values are number of lots.

TIME-INDEX: t					
Product	*Stage 1*	*Stage 2*	*Stage 3*	*Stage 4 Q*	*Stage 4 M*
1	0	2	7	9	4
2	1	8	7	1	5
3	6	2	0	8	3

Legend: Q = Number of lots in the queue; M = Number of lots on machines.

making decisions about machine setup changes, executing the changes, and re-projecting continues until the queue sizes are all within the ranges specified, and no more machine changes are required, for the n time periods for which the projections are done. The schedules, the final decisions for setup changes made by the system, are the setup changes that were made, at various time periods, to average out the queues.

Table 1 depicts an example of an input snapshot. The snapshot is a summarized version of the pickup sheet.

The algorithm that the system uses to decide upon the setup changes, based on the logic described above, is provided in the Appendix. This algorithm was implemented in the Lisp language.

The output from the system is a document showing a series of setup changes that are required in different time periods. Each change indicates the source product and the target product. An example of the output from the system is shown in Table 2.

5 IMPLEMENTATION AND VALIDATION

The performance of the system is measured by two methods, one relies on the historical record of setup changes and the second relies on an assessment by the scheduler. By the first method, the schedule of machine setup changes is

Table 2 Output from the system.

Move	From	To	At Index
2 Machines	Product 2	Product 1	1
1 Machine	Product 4	Product 5	3
⋮	⋮	⋮	⋮

compared with the historical record of actual machine setup changes that were executed on the shop floor.

5.1 Comparison with Historical Record

The metric used for comparing the schedules is that of counting the matches for machine setup changes executed within a particular time frame. For a given input snapshot the system creates schedules for exchanges or additions of machines to various products in specific time periods. For the same snapshot, i.e. the same date, the historical record or pickup sheets are consulted, and if the machine exchanges or additions are similar, then these are considered matches. The machine setup changes that were actually executed and recorded on the pickup sheets, were done so as a result of the scheduling activities of the scheduler.

The matches are recorded as "hits" or "errors" and are counted as follows: If the system schedules an addition or removal of a machine to or from a product line in periods 0, 1, or 2, and a machine was actually added to the particular product line in either of the dates corresponding to periods 0, 1, or 2, then this is counted as a *hit*. If there was no actual execution corresponding to this schedule then this is counted as an *error* of "Type A."

If an actual execution of a machine changeover took place in periods 0, 1, or 2, and a corresponding addition or removal was not scheduled by the system in any of the three periods, then this is counted as an error of "Type B."

A tally of hits and errors for 34 snapshots out of the total data set are used for the analysis. These snapshots are selected randomly to examine some simple statistics (there are several years worth of data available, 34 were selected as a sample).

1. The percentage of snapshot dates where at least one hit is recorded is 85%.
2. The net hits (hits minus errors), for Type A errors, is positive for 76% of the snapshots.
3. The net hits for Type B errors is non-negative for 47.1% of the snapshots at a lookahead level of 4 time periods. (The "lookahead" parameter is used by the module that marks mills for setup changes (see Appendix)).
4. For the 34 snapshots considered there were actually 266 machine setup changes executed. The total number of hits for these 34 snapshots is 110, which represents 41.3% of the total count.
5. There are 35 Type A errors and 153 Type B errors.

The analysis shows that the system is far more likely to make Type B errors than Type A errors. In other words, it is far more likely to make errors of *omission*, by not scheduling setup changes when they should have been done, than make errors of *commission*, where it wrongfully schedules a setup change. A net hit level of 76% for Type A errors is thus a very good result for the system's performance.

5.2 Assessment by the Scheduler

The system was implemented and functioned at the Westinghouse facility in Blairsville, for over a year. After this period, the nature of operations at the facility were significantly re-engineered, making the system irrelevant. However, for the period of its deployment, it was used on a daily basis by the scheduler, as a support tool for making decisions about setup changes. The system was implemented in Lisp on a DOS-486 machine running GoldWorks (a Lisp package).

The scheduler verified the schedules produced by the system against what his own ideas were about setup changes required for the shop floor. Based on his comments we have some results that validate the performance of the system in a real-time, reactive scheduling situation.

The metric that the scheduler used to comment on the system's output was to qualify the schedules for the various lookahead levels as "Best," "Correct," or some comments such as "too early" or "too late" (for a particular lookahead level). These judgements are qualitative and based on the scheduler's assessment of the overall situation on the shop floor and the system's response to this. The scheduler selected the "Best" category for those lookahead levels that appeared to provide the best schedules for the given situation. He used the "Correct" label to characterize those lookahead levels that seemed most suitable for the situation. The schedules with this label are less suitable than the ones with "Best." For a given day he either marked one lookahead level as "Best" or as "Correct," thus designating the hierarchy.

In some cases he used neither of these labels and indicated that the schedules produced by the system were inadequate. In this situation he would indicate the particular problem with the setup change scheduled as "too early" or "too late." These schedules are clearly wrong and unacceptable to him.

The validation of the system was thus based on what the scheduler was *scheduling* for the current situation as opposed to a historical record.

1. Setup change schedules are available for 50 working days (i.e. for fifty snapshots) and the scheduler marked "Best" for 28 of these cases. Thus, for 56% of the cases the system's schedules were reasonable and acceptable.

2. For 16 days the scheduler marked the schedules as "Correct," which is 32% of the cases.

3. Thus, for 88% of the cases, the system's suggestions were acceptable to the scheduler.

It must be pointed out that the scheduler was being somewhat "lenient" in assessing the system in the second method of validation. There are a number of factors that are external to the considerations of queue loads and demand that go into scheduling machine setup changes that the system did not consider. These include considerations such as maintenance downtime, delays in loading input material, delays in testing the material after setup changes, end-of-month inventory balancing, and unexpected fluctuations in demand. The scheduler did overlook some of the misses by the system, knowing that the system would have no way of knowing about the situation and accounting for it. Inspite of this, the scheduler agreed that the performance of the system was impressive.

6 CONCLUSION

The research reported in this paper discussed the process of transferring some knowledge obtained from an expert scheduler to a system that uses these to provide assistance to the scheduler in making decisions. A number of issues were raised in the introduction that served as the objectives for this research. In this concluding section, these issues are raised again, to further elaborate on the contributions of this paper.

The team of researchers on this project spent several years working on the various modules and implementing them. A significant portion of this time was spent with the scheduler in trying to understand the scheduling domain as he viewed it, and manipulated it. Much of the effort in the initial stages was spent in defining the scope of the project, and specifically, the scope of the various components. All the decisions on the shop floor are inter-related, and it is hard if not impossible to make any decision entirely independently. Defining the scope of the independent parts then meant understanding clearly the input information required for each decision, the time varying nature of the information, and the consequent impact of the decision's implementation on other parts of the shop floor. However, the design of the modules had to be conducted independently, assuming static and *ceteris paribus* conditions.

While working on the knowledge elicitation stage, an implicit goal was to first extract a set of "rules" that could then be implemented within an expert system shell. This would provide a first cut approximation to the design of the final system (in other words, a design prototype was envisaged). Rule-based systems have the advantage of having a declarative representation that can be easily upgraded with additions of rules. However, after a first attempt at constructing a prototype rule-base it became obvious that a very large set of rules would be required to accomplish the tasks. A procedural approach would be much more efficient and easier to code. Given that the setup change module relies on a discrete-event simulator to simulate the conditions on the shop floor, on an aggregate basis, the procedural approach for constructing this simulator was the most appropriate.

Eventually, after the system was implemented and validated, the conditions at the Westinghouse facility changed and the module for setup change decisions became irrelevant. In retrospect, it cannot be concluded that a rule-based, declarative approach would have been the better way to proceed (as it would have enabled the system builders to add extra rules to account for the new processes). The changes made within the shop floor were structural in nature.

Entire processing stages were added or removed. Within a rule-based system the design would have had to be sufficiently modular to account for such structural changes. Parametric changes within the process, such as changes in process times, can be easily included with a rule-based representation. Thus, a rule-based system would not have been a better representation. However, this underscores the important issue that systems designed for shop floor scheduling should have the capability to expand and be flexible [5].

The system was validated with a historical record as well as with the scheduler's comments. Both methods showed the system's performance to be excellent. Either method, by itself, would have been incomplete. To understand this point we have to consider the complex nature of the shop floor, and the decision making for this. The scheduler would make his decisions, prior to having the system, on the basis of information available at hand. Between the time he made the decision and its implementation by shop supervisors, if there were any changes to the critical information, such as machines going down or other emergencies, the scheduler would make on-the-spot decisions for setup changes. The historical record shows a combination of all these situations. On the other hand, the scheduler's decisions at the initial stages were based on the larger issues of the demand, the requirements of priority mills, etc. The two methods of validation thus cover the entire range of possible decision making circumstances.

REFERENCES

[1] J. H. May and L. G. Vargas, 1996, SIMPSON: An Intelligent Assistant for Short-term Manufacturing Scheduling, *European Journal of Operational Research*, 88, 269-286.

[2] S.A. Slotnick, 1991, MULTEX: The Short-term Expert Scheduling System, *Northeast D.S.I. 1991 Proceedings*, 375-377.

[3] J.H. May and L.G. Vargas, 1991, MULTEX: A Multiple Expert/DSS System for Shop floor Scheduling, *Northeast D.S.I. 1991 Proceedings*, 372-374.

[4] R. Dé, 1993 Empirical Estimation of Operational Planning Horizons: A Study in a Manufacturing Domain, University of Pittsburgh.

[5] J. Liebowitz and P. Lightfoot, 1993, Scheduling Approaches in Expert Systems: A Study of their Applicability, *Expert Systems With Applications: An International Journal*, 6-3, 231-235.

[6] J.H. May, T.E. Morton, D. Pentico, and L.G. Vargas, 1992, MULTEX:A Multiple Expert/DSS System for Shop Floor Scheduling – One year later, *Northeast DSI 1992 Proceedings*, 356-358.

APPENDIX A

Figure A.1 shows the flowchart for the module that computes an initial trail of snapshots starting with the initial snapshot at time period $i(= 0$, initially). The snapshot at i is an input to the system along with decisions about the new lots being started into the system, and n, the number of periods for which the projections have to be made. The output from this module is a trail of snapshots that are indexed from $i + 1$ to n. The decisions about new lots are made by another module that is described later (Figure A.4). The algorithm works in a loop, where each loop computes the snapshot status for each time index. The status updates are done by a procedure that uses a probability matrix to move a fraction of the lots from their current stage to the next state of processing. The status updates include the new lots that are added to the system. At this point the load, which is the size of the queue for the different product lines, is also computed and added into the data structure of the snapshot. The sequence of snapshots are appended.

Figure A.2 depicts the module that marks the mills that may have their setups changed. The module takes as input the trail of snapshots from index 0 to n computed in the previous step. The output from the system is a trail of snapshots which has products marked for addition or removal of mills. For each product the system checks whether the load remains heavy or light for k consecutive periods (this is called the *lookahead* parameter). If such is the case then the product is marked as an "add-mill" or a "remove-mill" product. If a product has no final pass mills allocated to it and it has some lots in its queue, then it is marked as a "no-mill" product. The parameter k is decided by conditions on the shop floor. Heuristics #4, #8, #9 and #10, as described above, are used to design this module.

Figure A.3 depicts the algorithm for the module that computes the decisions for the setup changes. Broadly, the algorithm works by finding the first index where setup changes are required, from the trail of snapshots. It changes the snapshot at this index to account for the setup changes and then calls the module described above to create a new set of projections for the changed situation. This is referred to as backtracking. A new set of projections are

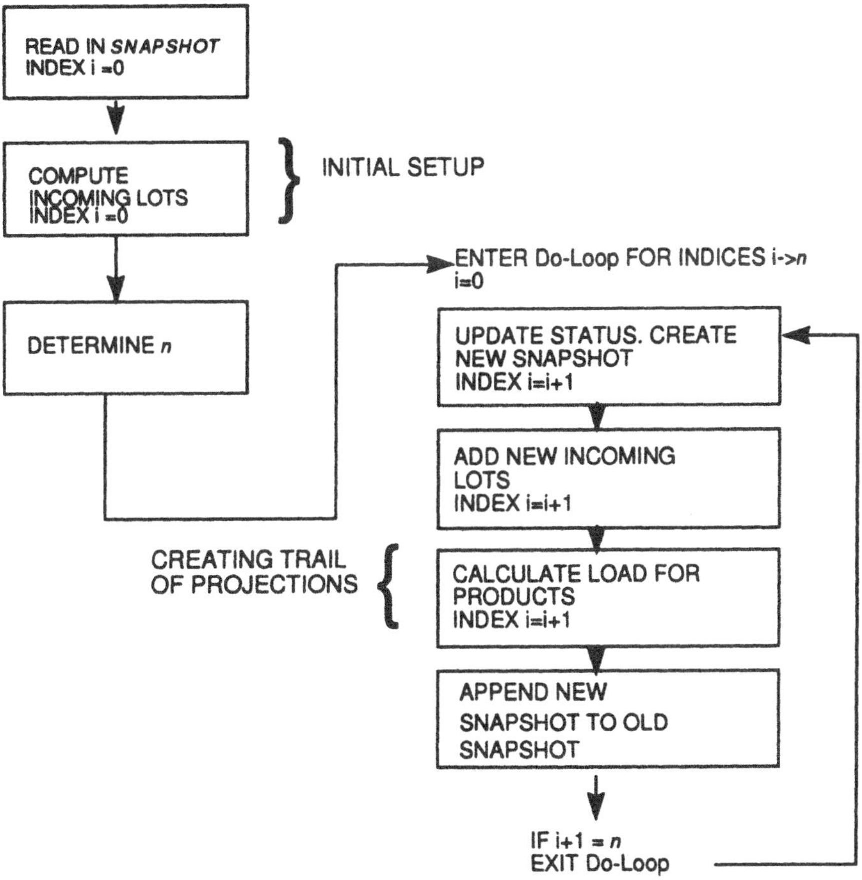

Figure A.1 Creating Snapshots for Periods i through n.

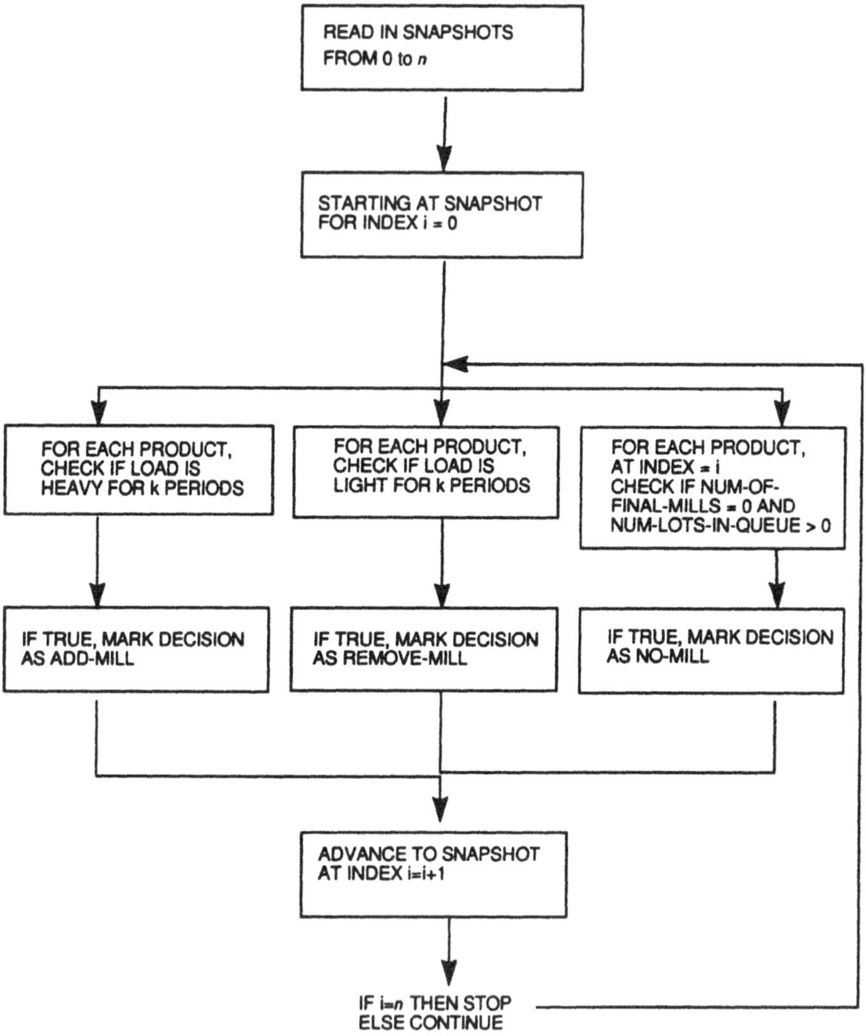

Figure A.2 Marking Setup Change Decisions

obtained which are again marked for setup changes. The algorithm then calls itself recursively to continue the process. It terminates when there are no more indices at which setup changes are required.

As shown in the figure, at a particular index, where mill setup changes are required, we can encounter a number of situations: one, where the products from which mills have to be removed is more than the ones to which a mill has to be added; two, the reverse of the previous situation where the add-mills products exceeds the remove-mills products; and three, the situation where the mills to be added and removed are the same. In the first situation, the algorithm prioritizes the remove-mills products according to specified criteria (point #10 in Section 3), and the highest priority products are used to remove mills from. In the second situation, mills are removed from all the products in the remove-mills list (and added to the add-mills products). The factory cannot have more than 15 mills in the final pass, so this is imposed as a constraint. Machines are added to the add-mills products from prior passes, as long as the total does not exceed 15. In the third case the mills are simply exchanged. Heuristic #2 is used to design this module.

Figure A.4 depicts the logic for starting new material on the shop floor. The rationale for this is as described before in Section 3.1. Heuristics #7, #11 and #12 are used to design this module.

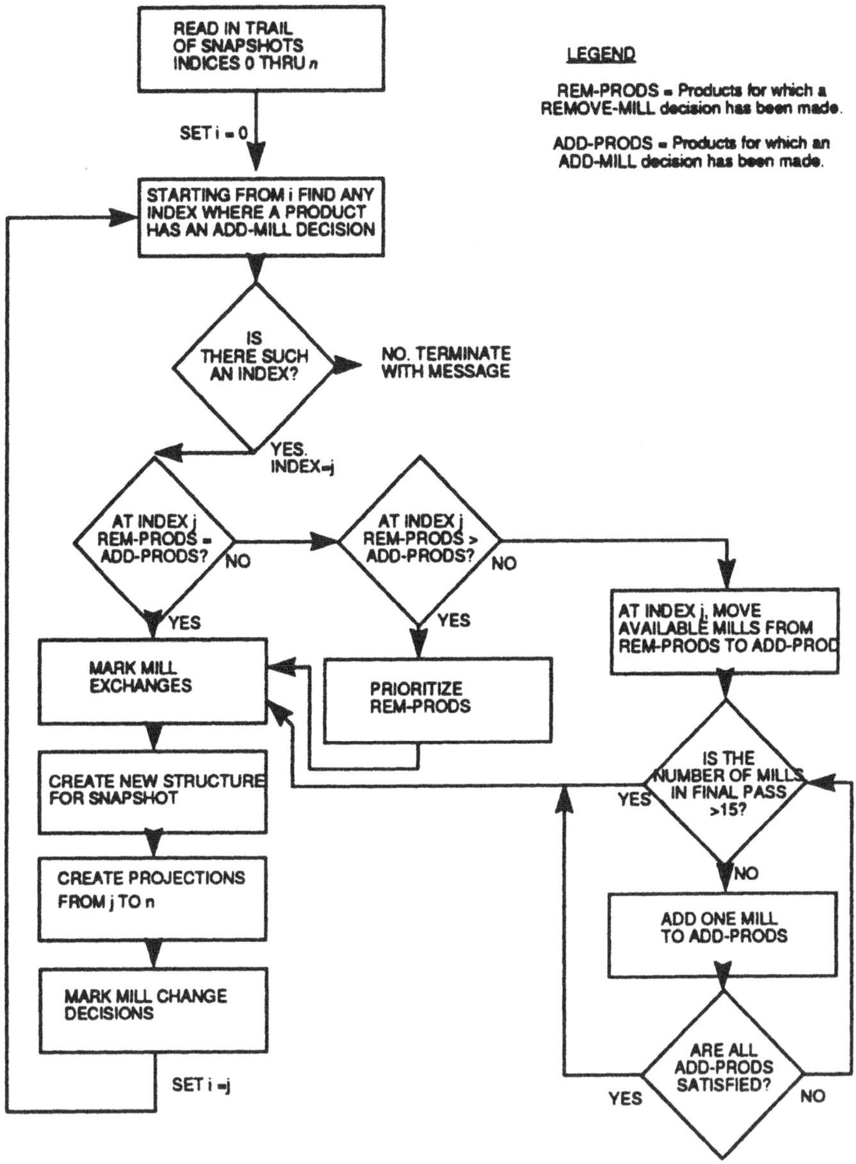

Figure A.3 Backtracking and Re-projecting for Setup Change Decisions

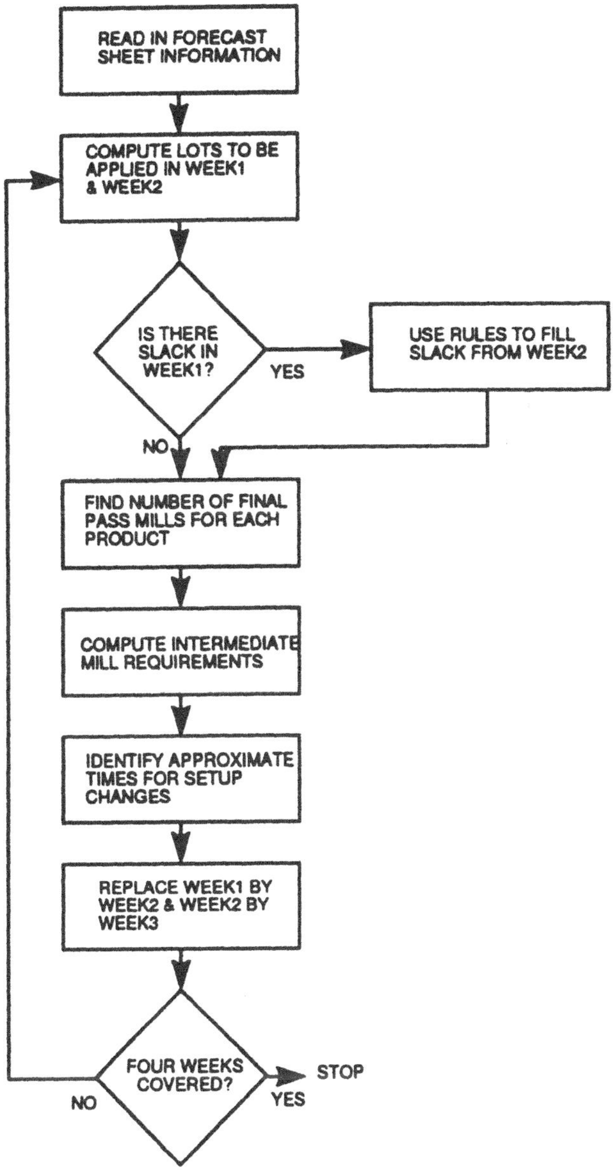

Figure A.4 Decisions for Starting New Material

10

INTEGRATIVE POPULATION ANALYSIS FOR BETTER SOLUTIONS TO LARGE-SCALE MATHEMATICAL PROGRAMS

Fred Glover,
John Mulvey,* Dawei Bai,* Michael T. Tapia**

The University of Colorado at Boulder
Boulder, Colorado 80309
USA

**Department of Civil Engineering and Operations Research*
Princeton University
Princeton, New Jersey 08542
USA

***Department of Management Science and Information Systems*
The University of Texas at Austin
Austin, Texas 78712
USA

ABSTRACT

Integrative Population Analysis unties the learning process called target analysis and a generalized form of sensitivity analysis to yield improved approaches for optimization, particularly where problems from a particular domain must be solved repeatedly. The resulting framework introduces an adaptive design for mapping problems to groups, as a basis for identifying processes that permit problems within a given group to be solved more effectively. We focus in this paper on processes embodied in parameter-based definitions of *regionality*, accompanied by decision rules that extract *representative solutions* from given regions in order to yield effective advanced starting points for our solution methods. Applied to several industrial areas, our approach generates regions and representations that give an order of magnitude improvement in the time required to solve new problems that enter the population and therefore makes the application of large scale optimization models practical in reality.

G. Yu (ed.), Industrial Applications of Combinatorial Optimization, 212–239.

1 INTRODUCTION

Many industrial applications of optimization involve a series of interrelated problems that represent goals associated with different individuals, companies or scenarios. In addition, these interrelated problems are not just "one shot" occurrences but must be addressed over time. Integrative Population Analysis provides a framework that enables successively encountered problems in such applications to be solved more effectively and efficiently. Our approach integrates a refinement of the learning process of target analysis with a generalization of the mathematical programming process of sensitivity analysis.

Integrative Population Analysis characteristically deals with problems that belong to a common class or that arise from a common underlying model. The context for these problems has an importance that is sufficiently encompassing, or sufficiently enduring, to require a significant number of problem instances to be solved. The goal of providing improved methods to operate in such settings is far from new. However, we provide a new design for achieving this goal that proves remarkably effective, based on combining and extending principles from artificial intelligence and mathematical optimization.

Our approach begins by creating special groupings according to a notion of *parametric proximity*. The properties that characterize membership in a particular group are subject to redefinition over time and the parameters that define these properties are a function of both problem data and problem structure.

We may illustrate some of the relevant concerns by reference to several industrial applications. The first application is on flexible manufacturing systems. Consider the situation where products are to be manufactured each day according to specific requirements of customers. The associated machine scheduling problem has to be solved again and again whenever a new product is ordered. These machine scheduling problems differ in the values of their parameters, but otherwise exhibit similar structures. The solutions to some of these problems may resemble each other in various ways, while the solutions to other problems may be unrelated. Another application is on marketing strategy design where the preference patterns of customers need to be uncovered. In this case, we need to look at the population of problems that maximize individual customers' utility functions. Again these problems share similar structures but differ in the specification of preference functions and the available financial resources of individual customers. Finally, we can apply Integrative Population Analysis to product customization. This allows a firm to identify new products that can be created to service different customer segments. Common customer characteristics can be determined so that the a firm's existing strengths can be used to develop products that will attract new customers.

To take appropriate advantage of shared features of problems and solutions, we are prompted to ask the following questions. On what basis can we usefully characterize the individual problems as similar or dissimilar? How can we uncover properties of solutions or solution processes for similar problems that will allow us to solve their problems more effectively? How can we assure a design that will be robust and respond to changing conditions? These issues will be discussed in this paper both in a general context and in the context of the several industrial applications just described. The implications of our approach extend beyond the immediate boundaries of such applications, and are relevant to determining policies for supporting undertakings.

To set the stage for presenting the tools and special features of Integrative Population Analysis, we first examine some of the areas that create an important need for its applications. We then present a description of the key properties of IPA in Section 2, and discuss how it can be applied to enhance the computational efficiency of practical optimization processes in Section 3. We conclude with an examination of future directions in the final section.

1.1 Applications for IPA

IPA can be applied to any problem that has many instances that have to be efficiently solved. These problems can occur in many industrial situations, especially in industrial marketing and design. One area that motivates the need for Integrative Population Analysis is product differentiation, especially in an industrial setting. By an industry we mean a group of firms which produce goods that are close substitutes in consumption. Consumers choose goods from different firms based on combinations of certain product features. If a particular combination of features is highly attractive to customers but unavailable, new enterprises that are capable of providing such features are encouraged to enter the market. For this reason, companies that have monopoly power customarily seek to offer a wide range of product features to block opportunities for new entrants that may otherwise compete with them. The major players in the soft drink industry, for example, provide numerous variations of their basic products to serve people with different tastes. At the same time, a small start-up company that successfully perceives and responds to an opportunity overlooked by the entrenched leaders can sometimes reap considerable rewards. Examples of this abound in markets ranging all the way from ice cream to computer software. It is clearly important for companies, large or small, to analyze the market in order to survive and prosper. Integrative Population Analysis increases the scope and effectiveness of such studies.

Product differentiation [15, 14] can be expressed along many dimensions. For example, Golden Delicious apples differ from Macintosh apples in their juiciness, tartness, sweetness, color, texture, and so forth. When apples on the market neglect to offer a feature that a significant number of people prefer, these individuals will switch to a new brand that more adequately meets their preferences, thus creating an opening for such a brand to emerge. Lancaster [19] was the first to analyze such a situation.

To illustrate, and to establish connections with concerns that will shortly be elaborated in greater detail, we consider goods with only two characteristics, such as apples with characteristics Sw and So (Sweetness and Sourness). The first quadrant in Figure 1 depicts the space of characteristics that represent various types of apples. The quantity of each characteristic is measured along each axis. For example, type C Apple offers three units of sourness for every two units of sweetness. The other rays represent other types of apples which have different So/Sw ratios. Here we have five types.

An efficiency frontier identifies the maximum amounts of characteristics the consumer can obtain with a given level of expenditure. To derive this frontier, we require some standard assumptions that are not entirely realistic in all situations. Nevertheless, they are sufficiently well approximated under many conditions to allow us to rely on them for our preliminary analysis.

The first step is to find out how many units of each variety the consumer would obtain in separate transactions by devoting his total apple budget to buying a single variety at a time. This is found by dividing the apple budget by the price per apple in each case. The result gives a number of units which can be measured as a distance along each of the rays of Figure 1. These are shown as the points a, b, c, d and e in Figure 1.

The efficient frontier then joins successive points that lie on the outer perimeter to produce the convex hull of these points. Therefore b cannot be connected with a and c because it lies strictly inside this outer perimeter. In fact, Type B apple's price is so high that it will not be bought since a combination of A and C will get more units of each characteristic at point w, which lies on the frontier. (The ability to create a "composite good" that corresponds to w is one of the basic assumptions of such analysis.)

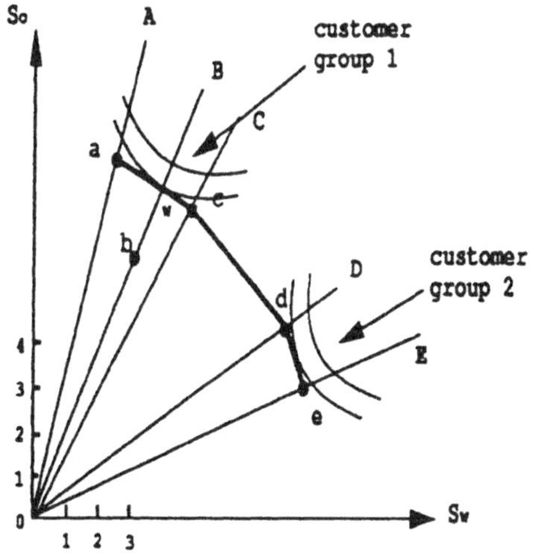

Figure 1 Utility maximization and market segmentation

Individuals possess preference structures which give rise to indifference curves as in Figure 1. Two sets of indifference curves are present, one for each customer group. The utility maximizing choice of consumption can be found by identifying the point at which the highest indifference curve is reached.

Individuals will usually have different preferences for different characteristics. Preferences for certain ratios of characteristics may cluster in one or more parts of the space, leaving other parts of the space largely unoccupied. When this occurs, we say the market *segments* (i.e., becomes segmented). Products outside these segments will find few customers. If the market does not exhibit such segmentation, that is, if customers are allocated uniformly along the efficient frontier, the frontier still provides useful information. We will show how to exploit these classical conditions, and others that are more general, in our subsequent development.

Continuing to lay the foundations that motivate the need for IPA, and that simultaneously introduce certain basic concepts to aid in understanding its characteristics, we turn next to the area of marketing strategy design. The first step is to understand the situation in which products will be accepted by customers, which requires an analysis of the nature of customer demands. This leads us to a characterization that divides the population into neighborhoods or cohorts.

For us, *cohorts* represent groups of customers who share certain common characteristics (see section 2.4). A cohort is constructed in a manner that enables us to evaluate the desirability of various products for *all* customers that lie within it. We may find, for example, that members of a particular cohort will greatly benefit by a new type of home mortgage, but will benefit much less from a traditional mortgage. Such relationships have the advantage of allowing us to design a marketing strategy that matches existing products with the cohort. More particularly, it becomes possible to creat optimization models to exploit market differentiations based on these relationships.

Another potential application area of IPA is in product customization, which is clearly related to product differentiation. For instance, different soft drinks are customized products that serve people with different tastes. An example that merits special consideration concerns investment. In the analysis of product differentiation, if the two attributes underlying the analysis illustrated in Figures 1-3 include elements such as mean and variance of a portfolio, we can then investigate patterns of demand for financial products with different returns and risks. The classical building blocks for generating and exploring efficient frontiers, which we link to customer preferences, can be embedded simply and advantageously in our processes for characterizing cohorts.

Once cohorts are defined via customer preferences, it becomes possible to "engineer" new products based on our analysis of the usefulness of the engineered product. In the investment area, for instance, there is great interest in financial engineering – the creation and trading of novel securities. Integrative Population Analysis provides a vehicle for assisting in the design of new products, by systematically evaluating the improvements for various segmented populations in ways not previously possible.

1.2 Computational Speed of IPA

Optimization problems in many practical settings are large or have attendant complexities such as nonlinearities, combinatorial structures and uncertainties. Size and complexity are often related, as where uncertainties can translate into special nonlinearities whose treatment includes the introduction of numerous variables and constraints. The solution of real world models in such settings typically involves a substantial investment of computation time. We begin by focusing on the goal of reducing the real-time computing burden for solving such problems.

To handle this situation with Integrative Population Analysis, we first face the challenge of mapping different problems into groups so that we can design a way to treat them more effectively based on this differentiation. In our present development, for concreteness, we will devote our attention to a mapping strategy that derives from identifying multiple elite solutions to the problems considered, and grouping problems both in terms of their data parameters and in terms of certain characteristics of these elite solutions. These elite solutions allow us to more efficiently solve similar large-scale mathematical programming problems by providing us with better warm-start solutions. We stress that these characteristics are not defined in advance, and are not simply *descriptive* but *functional.* Specifically, the determination of membership within a common group will depend on the existence of a *community of elite solutions*, derived as a subset of the union of their separate elite solutions, such that problems in the group can be solved more efficiently by utilizing members of this elite community. Note the composition of an elite community rests on several notions that must be clarified. We must specify, for example, what we mean by "utilizing" members of the elite community. Subject to such clarifications, we see that problems of a common group may in fact vary substantially on many descriptive dimensions. Our concern is to unite them at the functional level of identifying ways to exploit them more effectively. At the same time, we anticipate that a connection exists between various problem parameters and such functional concerns, and we will show how to integrate the functional and parametric considerations in our following development.

2 FUNDAMENTAL CHARACTERISTICS OF INTEGRATIVE POPULATION ANALYSIS

2.1 Population Environment

There are three environments for conducting our analysis. In the first situation, data for all the individuals of the population are known a priori. A thorough study can then be performed beforehand. In the second situation, data for the individuals arrive sequentially. We then must determine how to take appropriate account of a growing body of information. In the third situation, data for the individuals are unknown but sampling is possible. This leads to consideration of effective procedures for selecting such samples.

2.2 Problem Formulation

Let P be a population of agents (customers, investors, etc.). Each agent, $p \in P$, has the following optimization problem to solve:

$$\begin{array}{ll} \max & f_p(x_p) \\ \text{s.t.} & x_p \in X_p \end{array} \qquad (10.1)$$

Let x_p^* be an optimal solution to the problem and let $f_p^* = f_p(x_p^*)$ be its optimum objective value. For example, $f(x_p^*)$ represents a consumer's preference function at an optimal solution. We assume that these optimization problems are structurally related, as by representing instances of a particular problem class with different data sets.

For a variety of applications, the feasible region X_p in (10.1) can be defined as a collection of linear inequalities summarized by:

$$A_p x_p \leq b_p \qquad (10.2)$$

where A_p is an $m \times n$ matrix, b_p an $m \times 1$ vector and x_p an $n \times 1$ vector. (These inequalities typically include nonnegativity restrictions for the variables.) The prevalence of such applications leads us to focus on the linear inequality representation to illustrate our development.

The form of f_p and the coefficients of A_p, b_p may be determined by other parameters. For instance, a person's preference, habit, goals all depend on his or her wealth. They also depend on economic factors such as inflation, interest rates and stock prices. In a financial planning model, for example, a coefficient a_{ij} of A may be defined by:

$$a_{ij} = \sum_{t=1}^{T} \frac{1.0}{(1+\rho)^t} \qquad (10.3)$$

indicating the present value of a dollar of income in each period $t = 1, 2, ..., T$. The parameter here is the interest rate ρ.

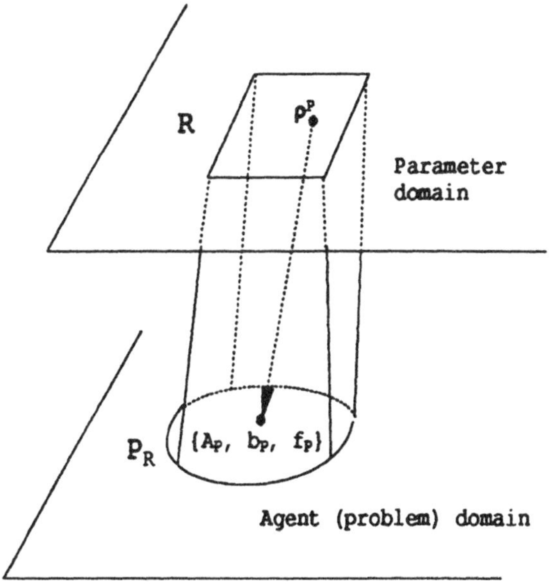

Figure 2 Coefficients are determined by parameters

Formally, we define parameters as data that change independently while coefficients of the optimization problems depend on them. In addition, coefficients that do not depend on any other parameters will themselves be considered parameters. Suppose there are L parameters, $\rho_1, \rho_2, ..., \rho_L$. We stipulate that their special values $\rho_1^p, \rho_2^p, ..., \rho_L^p$ determine the coefficients of A_p, b_p and the form of f_p as follows:

$$a_{ij}^p = h_1^p(\rho_1^p, \rho_2^p, ..., \rho_L^p); \tag{10.4}$$

$$b_i^p = h_2^p(\rho_1^p, \rho_2^p, ..., \rho_L^p); \tag{10.5}$$

$$f_p = h_3^p(\rho_1^p, \rho_2^p, ..., \rho_L^p). \tag{10.6}$$

A triple $\{A, b, f\}$ can be considered as determined by a region in the parameter domain. That is, each instance $\{A_p, b_p, f_p\}$ of the triple is determined by a particular $\rho = \{\rho_1^p, \rho_2^p, ..., \rho_L^p\}$ in the parameter region, as shown in Figure 2. The subset P_R of P identifies an index set for problems whose parameters lie in a region R. That is, we may specify $P_R = \{A(\rho), b(\rho), f(\rho) : \rho \in R\}$. For our present purposes, we will characterize a *region* R of parameters by:

$$\begin{array}{rcl} l_1 \leq & p_1 & \leq u_1 \\ l_2 \leq & p_2 & \leq u_2 \\ \dots & . & \dots \\ l_L \leq & p_L & \leq u_L \end{array} \tag{10.7}$$

where $l_i, u_i : i = 1, ..., L$ are lower and upper bounds for the parameters. Parameters and relationships for linear (and nonlinear) optimization problems such as those specified here can be conveniently represented in modeling languages such as AMPL [3] and GAMS [2].

An important task in our analysis is to characterize the population and its subsets in terms of certain features of their associated optimization problems. For example, we may be interested in the maximum, minimum or average of the objective values for all problems in a set $S \subseteq P$, e.g., in:

$$\max\{f_p^* : p \in S\}, \tag{10.8}$$

We define a *grand objective function* for $S \subseteq P$ as:

$$G_{grand}(S) = G(g(p) : p \in S) \tag{10.9}$$

where the function $g(p)$ evaluates characteristics of an individual agent p while $G_{grand}(S)$ evaluates the collective characteristics of all agents in set S. The function g may not necessarily correspond to the original objective function f. For example, it can identify the amount by which the time to solve problem p is reduced if a particular set of elite solutions is selected to provide starting points for an associated set of strategies. The grand objective function can then be a measure of the total time saved (or average time saved, etc.) by such a design applied to a specified collection of problems. When S is determined by a region R of the parameters, namely, $S = P_R$, we write $G_{grand}(S)$ as G^R_{grand}.

2.3 Generalized Sensitivity Analysis

To provide an approach to exploit the foregoing grand objective function effectively, we propose a generalization of the standard notion of sensitivity analysis to allow it to apply to a *set* of optimization problems rather than to a single optimization problem in isolation. Specifically, we seek to analyze how a grand objective function $G_{grand}(S)$ changes as a function of relaxing or tightening a given constraint for all agents in S, or as a function of expanding or contracting a parameter region along one of its dimensions.

For this purpose, given a parameter region R, we introduce a grand dual variable, $D^i_{grand}(P_R)$, which measures the marginal change in a grand objective function G^R_{grand} with respect to a change in a parameter ρ_i:

$$D^i_{grand}(P_R) = \partial G^R_{grand} / \partial \rho_i \tag{10.10}$$

We can use a deterministic simulation procedure to determine the value of this grand dual variable. This yields an approximation, although one may be able to develop an analytic procedure to determine the true value.

We first obtain the current value for a grand objective function in R and then change the range for ρ_i from

$$l_i \leq \quad \rho_i \quad \leq u_i \tag{10.11}$$

to

$$l_i \leq \quad \rho_i \quad \leq u_i + \Delta_i, \tag{10.12}$$

or

$$l_i - \Delta_i \leq \quad \rho_i \quad \leq u_i, \tag{10.13}$$

with the ranges for other parameters of R fixed.

If the change in the grand objective function is Δ_G, an approximation of the grand dual variable in (10.10) is

$$D^i_{grand}(P_R) \approx \Delta_G / \Delta_i \tag{10.14}$$

It is easy to see that our approach reduces to standard sensitivity analysis in the case where we restrict the domain considered. When a region R shrinks to include only one point, the set it generates will include only one agent, or one problem, say p. At this time,

$$G^R_{grand} = g(p) \tag{10.15}$$

which is a conventional objective function for problem p. If furthermore, we are considering a change in the right hand side of problem p, then

$$D^i_{grand} = \partial g(p)/\partial b_i \tag{10.16}$$

which is a conventional dual variable in LP and NLP.

The concept of a grand dual variable is highly useful to characterize the effects of changes in a population. For example, in a setting where a population consists of consumers and the grand objective function measures their purchasing power, alternative grand dual variables can be used to analyze changes in purchasing power that result when consumers grow a little older, or their incomes become higher or their their savings habits change.

Comparing values for a given grand dual variable in two different populations can reveal population differences. For instance, differing degrees of changed purchasing power that result from specific changes in income or savings habits can be pinpointed in this way. Grand sensitivity analysis, by reference to these notions, yields valuable information about factors that influence populations. Issues of investment, product differentiation and product development are clearly affected by such information.

2.4 IPA Cohorts

We say two agents are "associated" if their corresponding optimization problems are related. There are of course many ways to define the condition of being related. For our present purposes we will rely on a definition that implies the problem parameters lie within a certain bounded region, which we embody in the notion of an IPA cohort. In particular, we define an IPA cohort N^a for an agent a to be a subset of P that contains agents associated with a (including a itself) by the following construction. Denote the parameter set that determines a by $\rho^a = (\rho^a_1, \rho^a_2, ..., \rho^a_L)$. Then the IPA cohort N^a is given by a parameter region R^a that arises by specifying lower and upper bounds for the parameters in a manner we subsequently describe. We will restrict attention to IPA cohorts thus generated from such parameter regions and denote them as N^a_R.

The manner in which we determine IPA cohorts will make use of the grand objective function. To motivate this determination, suppose we stipulate that the grand objective $G_{grand}(S)$ is designed to take on larger values when the parameters that define $S \subset P$ lie within greater proximity to each other. Then we may conceive the problem of searching for a "good" IPA cohort for an agent

a to be formulated as seeking a set S that yields an optimal solution to the following problem

$$\max_{S} \quad G_{grand}(S)$$

$$\text{s.t.} \qquad a \in S \tag{10.17}$$

Clearly, an optimal solution results when the set S includes only the agent a, which is not interesting to us. It is therefore necessary for the grand objective function to encompass other factors, which we consider in following sections. When the grand objective function is appropriately defined, the solution to the problem will be N^a.

Since a set of agents and the coefficients for their optimization problems are determined by a region of parameters, which is in turn calculated as in (10.7), the true variables of interest are actually lower and upper bounds of the parameters. Thus, the preceding formulation can be rewritten as:

$$\max \quad G_{grand}(l, u)$$

$$\text{s.t.} \qquad a \in P_R \tag{10.18}$$

$$P_R = P_R(l, u) \tag{10.19}$$

The last equation requires that the region R be determined by $l = (l_1, l_2, ..., l_L)$ and $u = (u_1, u_2, ..., u_L)$.

2.5 Representative Agents

We link the goal of generating a proper form for the IPA cohorts to the goal of generating a collection of *representative agents*, where we require the union of the IPA cohorts of such special agents to include all agents in P. Clearly, by adjusting the bounds that define IPA cohorts, we can assure that any chosen collection of agents will represent the other agents in this fashion, but we base the determination of representative agents on additional criteria.

We are guided by two interrelated concerns. First, we desire the representative agents to be selected so that the agents in an IPA cohort are associated more strongly than simply by sharing a certain proximity to the agent defining the IPA cohort. For this reason, we will in fact allow a slight distortion by permitting "artificial agents" to be included among those we treat as representative, where these artificial agents are derived by perturbing or interpolating between parameter values of other agents. Such artificial agents are generated by starting from original agents and creating adjustments based on feedback about problem solving efficacy from relying on the current representative agents. Alternatively, we can simply shift the role of representative agent from one original agent to another by such an iterative analysis. In either case, the outcome can lead to more appropriate definitions of IPA cohorts for the goals we seek.

Second, once a set of representative agents is (provisionally) specified, we desire to characterize their IPA cohorts so that the agents within a given IPA cohort will be susceptible to exploitation by information that is specific to this cohort. More precisely, we seek to use this cohort-specific information to solve the optimization problems for the agents in the IPA cohort with significantly increased efficiency. Our design incorporates the principles of the learning approach called target analysis [6, 8, 5, 7]. In the present development, as intimated earlier, we choose to rely on information obtained from elite solutions to problems in the IPA cohort. Still more restrictively, we use these solutions to give advanced starting points for members of the IPA cohort. An interesting discussion of the philosophy and potential applications of focusing on starting solutions, viewed from the perspective of case based reasoning, is given in Kraay and Harker [9]. We emphasize that other decision rules and forms of exploitation are possible within the framework of target analysis, particularly in coordination with tabu search as employed in this study (see, e.g., [8]). We have also found that similar ideas can greatly improve the performance of other methods such as in [1, 17]. Nevertheless, we have found this form of exploitation to be highly effective.

2.6 Computational and Practical Benefits

It may appear from our foregoing abbreviated description that we are investing a great deal of effort to solve and re-solve problems, using different initial conditions. This naturally raises the question of how benefit can be gained from such an undertaking, since we are going well beyond the effort required to obtain solutions to the problems in P. The answer is that P is not a fixed set, but rather is one that grows over time, as new agents or agents with new problems enter the scene. (Alternately, in a static "large population" environment, we may consider P to begin as a subset of a substantially larger P that we desire to deal with.) Thus, we are applying the strategy of intensely focusing on problems already solved in order to improve our ability to solve problems yet to be encountered. By expending a large effort to process a relatively small number of problems, we may save considerably greater amounts of effort in solving a much larger number of problems.

2.7 Choice of Agents and IPA Cohort Design

Our construction of IPA cohorts and critical agents is therefore an anticipatory process, where we seek to organize existing knowledge more effectively to meet future problem solving needs. The process is necessarily dynamic and self-modifying. The organization of our knowledge changes as the knowledge changes.

Often initial conditions are simple, and call for simple measures. This is evidently true in the case of choosing representative agents. To begin, we may simply rely on "conventional wisdom" about a problem, according to the context, to select agents considered typical for different ranges of the underlying parameters.

After making such a selection, we seek intervals about the parameters for these initial representatives so that the resulting IPA cohorts will include all agents. However, this goal must be balanced against the competing goal of creating cohorts sufficiently compact that we can identify a relatively small number of useful elite solutions for agents in one cohort – i.e., solutions that yield good advanced starting points for members of the cohort. Upon analyzing the outcomes of selecting various trial elite solutions (by designs we subsequently describe), we may either add, delete, shift or modify the current representative agents to create a better collection.

We may therefore express the overall plan for our approach in the following outline form:

- Choose a list of representative agents;
- Define a proper cohort for each representative agent and characterize it according to a certain purpose. For example, to improve computation efficiency, we need to identify the associated elite solutions. For marketing strategy design and new product development, careful characterization of the cohorts is needed and sensitivity analysis is required.

- If the current IPA cohorts fail to cover all the important problems, select other representative agents and find other cohorts for the new agents.

3 COMPUTATIONAL EFFICIENCY

We focus on applying integrative population analysis to improve computational efficiency. As emphasized, we seek a proper IPA cohort and its corresponding elite solutions for a representative agent in a manner to permit problems in the cohort to be solved easily. On the other hand, cohort size is also a measure of efficiency: if the IPA cohorts are too small, we have to keep many of them to cover a population and must save a large number of elite solutions. In the extreme case, each cohort would contain only a single problem, requiring us to solve all problems in advance with no advantage derived from prior elite starting solutions. When we speak of elite solutions, we refer not simply to solutions that are actually generated for the problems considered, but also to specially constructed trial solutions that are obtained as "centers" of clusters of other solutions. The final determination of a cohort is a compromise between the following two considerations:

1) The average reduction in computation time that results by starting from appropriately selected elite solutions (obtained from representative problems in the cohort) should be as large as possible.

2) Each cohort should cover a large terrain so that only a few cohorts are needed.

An important element in these considerations is that we will simultaneously be seeking to establish a small and effective set of elite solutions for each cohort generated. With this element implicit, we formulate the problem as follows:

$$\max \quad \theta\, TRR_{grand}(N) + (1-\theta)||N|| \qquad (10.20)$$

where:

TRR : time reduction rate

$||N||$: size of cohort N

θ : weight to balance the two criteria

The time reduction rate TRR is achieved through a five step simulation process discussed in section 3.1. Briefly, we generate random samples of problems in a cohort, solve them with or without elite solutions and get the average time reduction rate. For each sample p:

$$TRR(p) = 1 - \frac{T_e}{T_{ne}} \tag{10.21}$$

where

$$\begin{aligned} T_e &= \text{computing time with elite solutions for problem } p \\ T_{ne} &= \text{computing time without elite solutions for problem } p \end{aligned}$$

The average time reduction rate in a IPA cohort is therefore:

$$TRR_{grand}(N) = \frac{1}{n_N} \sum_{p \in N} TRR(p), \tag{10.22}$$

where n_N is the number of samples in cohort N.

We can also consider time reduction rate in the constraints, in which case we have the following alternative to (10.20):

$$\max \quad ||N|| \tag{10.23}$$

subject to

$$r_1 \leq \quad TRR_{grand} \quad \leq r_2 \tag{10.24}$$

where r_1 and r_2 are the lowest and the highest time reduction rates which we consider to be satisfactory.

The variable in this formulation is the size of a IPA cohort, which is a function of the ranges of the parameters in (10.7). Therefore, we have:

$$||N|| = N(l_1, u_1, l_2, u_2, ..., l_L, u_L) \tag{10.25}$$

or more specifically,

$$||N|| = \sum_{p=1}^{L} w_p(u_p - l_p)^2 \tag{10.26}$$

which is a weighted Euclidian measure.

3.1 Solution Procedure

The procedure for finding a proper IPA cohort to improve computing efficiency consists of the following five steps.

- Determine a list of representative agents;
- For each agent:
  - Define a trial IPA cohort based on conventional observation;
  - Generate a series of samples in the trial cohort;
  - Solve the sample iterates to get a set of optimal or near optimal solutions.
- From the set of solutions, generate points that are (roughly) centered within clusters of these solutions to be treated as trial starting solutions;
- Employ out-of-sample simulations to see if solution time is reduced when the trial solutions are used as advanced starting solutions for problems in the cohort.
- Systematic modification of the IPA cohort:
  - If the time reduction rate is satisfactory, (e.g., between 80% to 90%), stop. The current trial cohort defines a proper cohort for the representative agent whose parameters define the central locus of the cohort. The trial solutions are kept as elite solutions for the cohort.
  - If the time reduction rate exceeds the upper target (say falling above 90%), expand the trial cohort; else, if the time reduction falls below the lower target (say below 80%), shrink the trial cohort. Expanding or shrinking a cohort is guided by grand dual variable information as explained later.

We now discuss each of these steps in detail.

3.2 Choosing Representative Agents

Choosing representative agents is an iterative process. At the beginning, we base the selection on conventional observations as noted earlier. With integrative population analysis, we define an appropriate IPA cohort to encompass each of these choices. We may then find that our initial selection is too coarse (i.e., insufficiently refined) as the population continues to be analyzed. Some important groups of investors may not be covered by any of the cohorts, and additional representatives may be needed. Conversely, we may also find our initial selection is too detailed, as when some of the representatives can be grouped together and share the same set of elite solutions. It should be noted that we do not strive to cover all the investors as it may not be efficient to do so.

3.3 Samples in an IPA Cohort

For each representative agent, we generate random samples according to a multi-dimensional uniform distribution. The similarity of these samples in their financial backgrounds and preferences depend on the size of the IPA cohort.

3.4 Generating Elite Starting Solutions – Cluster Analysis

To generate elite starting solutions, we save each sample's globally optimal (or best known) solutions and those locally optimal solutions which have objective values that are close to the estimated global optimum. (The definition of "close to" is adaptively determined.)

If many of the sample solutions fall within the vicinity of a particular location and have an appropriate objective value, that location is chosen to be a trial solution which may become one of the elite starting solutions. In a parameter input file, we set the maximum number of clusters we need, the minimum number of points to be considered as a cluster, and so on. A diagram of the process is shown in Figure 3.

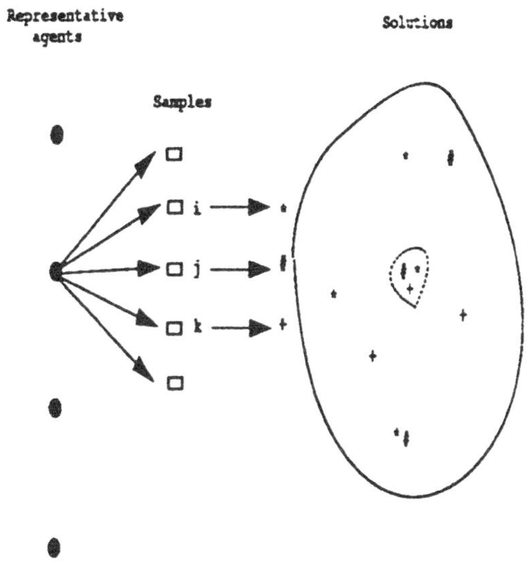

Figure 3 Cluster analysis

3.5 Out-of-sample Simulation

Out-of-sample simulation determines if trial solutions are effective for reducing computation effort. Specifically, new samples are tested with and without the use of elite starting solutions to get the difference in solution time. If the time reduction rate exceeds a certain level, these samples are too similar to each other. We may then need to expand the IPA cohort. On the other hand, if the time reduction rate falls below a desired level, we will decrease the cohort size. However, not all parameters are critical for time reduction in a given cohort. In the next section, we introduce a systematic approach to modify a cohort based on information provided by the grand dual variables.

3.6 Systematic Modification of IPA Cohorts

As in other optimization methods, we employ an iterative procedure to find a desired cohort. At each iteration, we seek a good direction that can substantially improve the objective value. The information on how to choose such a direction is usually provided by the gradient of an objective function. For

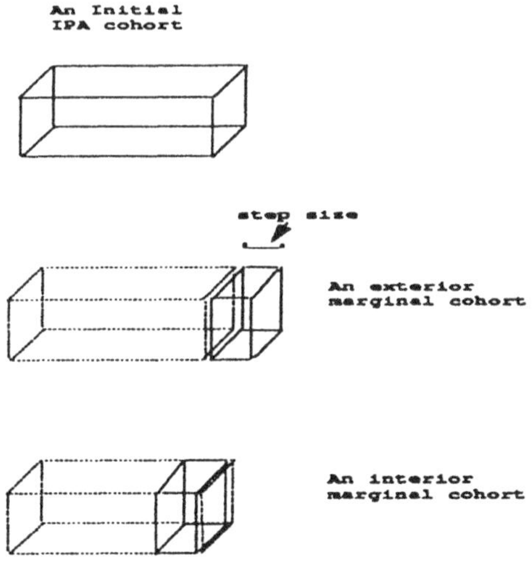

Figure 4 Interior and exterior marginal cohorts

our problem, however, a well-defined gradient is not available because the time reduction rate is not an explicit function of the cohort parameters. We have therefore developed a simulation procedure that can achieve a direction similar to that based on a gradient.

We define the *marginal* problems for a particular cohort to be those that would be added to (dropped from) the cohort if the range of a single parameter is enlarged (reduced) by a specified step size. This step size depends on both the identity of the parameter and the cohort. This is so because each parameter has a specific unit and an unique impact on the benefits derived from the elite starting solutions for the cohort. An adaptive way of chosing the step sizes will be discussed shortly.

Corresponding to marginal problems, we define a marginal cohort to be the region added to (removed from) a given cohort to identify such a set of marginal problems. Moreover, we consider a marginal cohort to result by changing only the lower or upper end of a parameter range; i.e., we modify the two ends of a parameter range separately in order to determine the effects of the two resulting cohort changes. To be more specific, we refer to interior and exterior marginal cohorts (and problems) as those that result, respectively, by decreasing or increasing a parameter's range. Figure 4 illustrates these definitions.

Marginal problems for different parameters have different influences on the average time reduction rates. For example, if the time reduction rate for a parameter mortgage's interior marginal problems is much lower than the average of the cohort, the associated bound for mortgage loan may be too loose. Hence we tighten the bound and reduce the range accordingly. Similarly if the interior marginal problems have much higher time reduction rates than the average, it is possible to expand the range of corresponding parameters. As we have discussed in section 2.3, information about marginal problems yields a sensitivity analysis for comparing time reduction rates to changes in parameters. We can therefore develop a systematic way of modifying a cohort based on this information. To be concrete, we stipulate that a targeted time reduction rate is selected to be 80% in the following discussion. When the time reduction rate is below 80%, we reduce the cohort to achieve a target rate. While the time reduction rate is maintained above 80%, we expand the cohort to cover as many problems as possible. In the following discussion, we consider marginal problems for the upper end of a parameter. The same procedure can be similarly applied to the marginal problems at the lower end of a parameter.

First, we derive a set of trial elite solutions with cluster analysis as described in section 3.4. These solutions offer good starting points for solving problems in the cohort. We keep track of the identities of all samples so that we know which cohort they belong to. This makes it easy for us to drop or add samples when the cohort is modified.

We start the iterative procedure by solving the interior and exterior marginal problems for all the parameters with and without elite starting solutions. Time reduction rates can be determined and summarized as follows.

Table 1 Time reduction rates for the marginal problems

Parameters	1	2	3	4	5	6
TRR (Interior problems)	50%	82%	90%	32%	86%	96%
TRR (Exterior problems)	45%	75%	91%	35%	83%	87%

There are four situations for each parameter according to whether the time reduction rate is above or below the target rate:

1) The time reduction rates for both interior and exterior marginal problems are above the target rate;

2) The time reduction rates for both interior and exterior marginal problems are below the target rate;

3) The time reduction rate for interior marginal problems is above the target rate, while that for exterior marginal problems is below the target rate;

4) The time reduction rate for interior marginal problems is below the target rate, while that for exterior marginal problems is above the target rate.

For a specific parameter and its current range, the time reduction rate for the exterior marginal problems is usually lower than that for the interior marginal problems. This is because the exterior marginal cohort lies farther away from the original cohort for which the elite starting solutions are derived. For this reason, the fourth situation does not occur very often. In case it does occur, the two rates are all close to the target rate because of the similarity in the samples. It is thus reasonable to leave the parameter range unchanged.

The first situation indicates that we can expand the range for the parameter without deteriorating the time reduction rate. Furthermore, the closer the two rates (for interior and exterior marginal cohorts), the more insensitive is the time reduction rate to changes in the parameter. We can therefore expand the range of the parameter accordingly. Let TRR_{in} be the time reduction rate for interior marginal problems and TRR_{ex} for exterior marginal problems. The step size can then be updated from the current value Δ^{now} to a new value Δ^{new} with the following formula:

$$\Delta^{new} = \Delta^{now} + (1.0 - \frac{|TRR_{in} - TRR_{ex}|}{20\%}) * \Delta^{now} \tag{10.27}$$

The value 20% is chosen because it is the highest value that a time reduction rate can change in this situation. When the time reduction rate indeed changes from 100% to 80%, we know the time reduction rate is very sensitive to changes in the parameter and we may expand the parameter range only by the amount of the current step size. Hence the step size stays the same.

In the second situation, we need to reduce the parameter range to improve the time reduction rate. As before, the closer the rates for interior and exterior marginal problems, the more insensitive is the time reduction rate to changes in the parameter. The following formula is then employed to update the step size to reduce the parameter range.

$$\Delta^{new} = \Delta^{now} + (1.0 - \frac{|TRR_{in} - TRR_{ex}|}{80\%}) * \Delta^{now} \tag{10.28}$$

Again, the value 80% is chosen as it is the highest value that a time reduction rate can change in this situation.

Time reduction rates become sensitive to changes in the parameters in the third situation. Here, a change from the interior to the exterior marginal cohort leads to a decrease in the time reduction rate from a value above the target rate to a value below the target rate. At this time, the step size needs to be reduced so that a smaller exterior marginal cohort can have a better time reduction rate. The formula for updating the step size is:

$$\Delta^{new} = (1 - \frac{|TRR_{in} - TRR_{ex}|}{TRR_{in}}) * \Delta^{now} \tag{10.29}$$

With these considerations, we get the following new step sizes based on the data given in Table 1:

Table 2 Updating the step sizes

Parameters	1	2	3	4	5	6
Step size (now)	Δ_1	Δ_2	Δ_3	Δ_4	Δ_5	Δ_6
Step size (new)	$-1.94\Delta_1$	$+0.91\Delta_2$	$+1.95\Delta_3$	$-1.96\Delta_4$	$+1.85\Delta_5$	$+1.55\Delta_6$

The positive or negative sign indicates whether we should expand or reduce the range of the corresponding parameters. The current IPA cohort is then modified according to the new step sizes. That is, the upper end of parameter 1 is reduced by $1.94\Delta_1$, the upper end of parameter 2 is increased by $0.91\Delta_2$, and so on. After the cohort is modified, new elite solutions can be derived from solutions of the samples that are lying within the new cohort. This concludes the first iteration of the procedure.

The new step sizes in Table 2 also provide a promising direction for further investigation of the cohort. For a parameter with a positive step size in the table, we consider its new exterior marginal cohort according to its new step size. The exterior marginal cohort in the previous iteration can be approximately assumed to be its current interior marginal cohort so that the required time reduction rate is readily available. This is only an approximation as the cohort has changed. For a parameter with a negative step size in the table, we consider a new interior marginal cohort. The previous interior marginal cohort becomes approximately its exterior marginal cohort. This saves 50% of the computational effort.

With the same procedure as in the first iteration, we can compare the time reduction rates and decide how to expand or reduce a parameter range. The new step sizes provide a new direction for the next iteration.

When a relatively good IPA cohort is reached, the change in time reduction rates will be in the third situation for all the parameters. That is, when we expand a parameter range a little bit, the time reduction rate will change from above 80% to a value below 80%. Thus we reduce the step sizes for the parameters. The iterative process stops when the step sizes become small compared with pre-determined tolerances or when the maximum iteration limit is reached. As shown in the study of [13], these targeted rates of time reduction are entirely realistic, and in fact, 90% improvements in computation time have been achieved for solving complex financial planning models incorporating nonlinearities and uncertainties.

4 FUTURE DIRECTIONS

We have shown that Integrative Population Analysis is applicable to many industrial engineering areas. The dynamic nature of the industrial world requires operations research models to be flexible and capable of solving new problems promptly and effectively. Many research investigations fail to address the dynamic nature of real world problems. IPA serves to bridge this gap between academic research and the practical needs.

In another paper [13], we have shown that IPA can dramatically reduce execution time for solving multistage financial planning models. These gains make it possible for individuals to conduct systematic analysis of their financial affairs in an interactive fashion. In this context, the optimization capabilities of the financial planning system are critical so that the investor can concentrate on the preference identification and selection process, rather than being compelled to hunt for the best investment strategy. Techniques on carrying out component steps of the solution process are under steady improvement; the planning system can accommodate new algorithms for such steps in a straightforward fashion.

Our future work on Integrative Population Analysis will be devoted to carrying out full scale testing of IPA with populations from a variety of real world settings. A systematic study of issues such as the proper number and composition of IPA cohorts is planned over the next several years. Generalized sensitivity will establish recommendations regarding possible improvement areas – for example, such analysis provides a basis to identify constraints that can be relaxed and thus have a large impact on the results – hopefully in a relatively painless manner.

Another research issue involves enhancing the IPA cohort generation process. We plan to (1) reduce dependence on expensive out of sample tests; (2) develop internal algorithmic tactics for each cohort; and (3) compute generalized dual variables for solutions generated by the adaptive memory process. Finally, we plan to test Integrative Population Analysis in other domains. For example, Benders Decomposition has been the method of choice for financial planning systems without decision rules [1, 17]. We can apply IPA in this setting by substituting elite cuts for the elite solutions. There are also other possible consumer products and services (e.g. travel agents) that could benefit by Integrative Population Analysis, and such applications likewise invite investigation.

Acknowledgements

This paper draws extensively on our development in reference [13], which we extend here to characterize applications of IPA in the realm of industrial engineering.

REFERENCES

[1] D. R. Cariño, T. Kent, D.H. Myers, C. Stacy, M. Sylvanus, A. Turner, K. Watanabe, and W.T. Ziemba, 1994, The Russell–Yasuda Kasai Financial Planning Model, *Interfaces*, 24, 29-49.

[2] A. Brooke, D. Kendrick, and A. Meeraus, 1988, *GAMS: A user's guide*, Scientific Press, Redwood City.

[3] R. Fourer, D.M. Gay, and B.W. Kernighan, 1993, *AMPL: A Modeling Language for Mathematical Programming*, The Scientific Press, San Francisco.

[4] M. R. Anderberg, 1973, *Cluster Analysis for Applications*, Academic Press, New York.

[5] F. Glover and H.J. Greenberg, 1989, New approaches for heuristic search: A bilateral linkage with artificial intelligence, *European Journal of Operations Research*, 39, 119-130.

[6] F. Glover, 1986, Future paths for integer programming and links to artificial intelligence, *Computers and Operations Research*, 13, 533-549.

[7] M. Laguna and F. Glover, 1993, Integrating target analysis and tabu search for improved scheduling systems, *Expert System with Applications*, 6, 287-297.

[8] F. Glover, 1995, Tabu Search fundamentals and uses, University of Colorado, Boulder, CO 80309-0419, Technical report.

[9] D. Kraay and P. Harker, 1995, Case Based Reasoning for Repetitive Combinatorial Optimization Problems, Part I: Framework and Part II: Numerical, Purdue University and the Wharton School, Technical Report(s).

[10] S. Andradottir, 1995, A Stochastic Approximation Algorithm with Varying Bounds, *Operations Research*, 13, 1037-1048.

[11] F. Glover, J. Kelly, and M. Laguna, 1995, Tabu Search and Hybrid Methods for Optimization, INFORMS Conference.

[12] R. Barr and M. Durchholz, 1996, Parallel and Hierarchical Decomposition Approaches for Solving Large-Scale Data Envelopment Analysis Models, to appear, *Annals of Operations Research*. (Technical report available).

[13] F. Glover, J. Mulvey and Dawei Bai, 1996, Improved Approaches To Optimization Via Integrative Population Analysis, Princeton University and University of Colorado, Technical Report SOR-95-25, 1996.

[14] N. Ireland, 1987, *Product Differentiation and Non-Price Competition*, Basil Blackwell, New York.

[15] S. Anderson, A. Palma and J. Thisse, 1992, *Discrete Choice Theory of Product Differentiation*, The MIT Press, Cambridge, Massachusetts.

[16] D. Bai, T. Carpenter, and J. Mulvey, 1994, Stochastic programming to promote network survivability, Department of Civil Engineering and Operations Research, Princeton University, Technical Report SOR-94-14.

[17] G. Infanger, 1994, *Planning under uncertainty: solving large-scale stochastic linear programs*, Boyd and Fraser Publishing Company.

[18] L. A. Cox, W. E. Kuehner, S. H. Parrish, and Y. Qiu, 1993, Optimal Expansion of Fiber-Optic Telecommunication Networks in Metropolitan Areas, *Interfaces*, 23, 35-48.

[19] K. Lancaster, 1966, A new approach to consumer theory, *Journal of Political Economy*, 74, 132-57.

11

LAGRANGIAN RELAXATION FOR FLOW SHOP SCHEDULING

George G. Polak

Department of Management Science and Information Systems
Wright State University
Dayton, Ohio 45435
USA

ABSTRACT

Job shop scheduling encompasses combinatorial problems of extraordinary difficulty. The special structure of a flow shop, in which the machines can be ordered in series, offers some advantages. Employing the disjunctive graph as a model of a flow shop, we proceed to formulate a mixed integer model for the problem of minimizing the makespan. By decomposing the model into unlinked one-machine sequencings without due dates, Lagrangian relaxation introduces subproblems amenable to the greedy algorithm. Moreover, since the subproblems do not have the integrality property, the Lagrangian bound can be stronger than the LP bound.

1 INTRODUCTION

The *makespan* is the time needed to finish all jobs in a shop, and its minimization is often the objective in job shop scheduling. Jobs are processed on machines one at a time with no interruption, and each is of known constant duration (including set-up). A graph with fixed arcs to represent the characteristic sequence of machines for each job, and "variable" arcs for the unknown sequence of jobs on each machine, serves to model the shop. The goal then is to choose the variable arcs to effect a shortest possible critical path through the graph.

Job shop scheduling encompasses combinatorial problems of extraordinary difficulty: not until 1989 did Carlier and Pinson [6] publish a provably optimal solution to a celebrated 10 job by 10 machine problem proposed in 1963 by

G. Yu (ed.), Industrial Applications of Combinatorial Optimization, 240–255.

Fisher and Thompson [7]. Even the one-machine scheduling problem with release and due dates is NP-complete in the strong sense [1], [8], motivating development and testing of heuristics. For the general case, the Shifting Bottleneck Procedure of Adams, Balas and Zawack [1] has met with wide success. At each iteration, the procedure solves a one-machine scheduling problem with release and due dates at a machine deemed the bottleneck in the job shop, then resequences all previous bottlenecks in turn. Balas, Lenstra, and Vazacopoulos [4], and Balas and Vazacopoulos [5] further developed this heuristic.

In the special case of the flow shop, the machines can be ordered in a series followed by all jobs. Sevast'janov [13] derived a polynomially bounded algorithm for obtaining an approximately optimal schedule, along with upper and lower bounds on the optimal makespan. His work is based upon the reduction of the flow shop problem to Compact Vector Summation in a Banach space.

Any decomposition of a general job shop into single machine sequencing problems must incorporate consistent and explicit release and due dates into the latter set. This complication is avoided in the flow shop because of the absence of potential cycles. The difference between these cases is that of complexity: without due dates, the problem of minimizing the makespan for a single machine can be solved greedily; with due dates, it is NP-hard.

Our approach employs the disjunctive graph as a model of a flow shop and formulates a mixed integer model for the problem of minimizing the makespan. We proceed to decompose it into unlinked one-machine sequencings without due dates by relaxing precedences between machines. This application of Lagrangian relaxation introduces subproblems amenable to the greedy algorithm.

Although there is typically a duality gap between an integer optimization and its Lagrange Dual, the latter does provide a bound on the optimal objective value of the former. Such a bound can be useful in implicit enumeration schemes. Moreover, the subproblems of flow shop scheduling do not have Geoffrion's [9] integrality property, hence the Lagrange Dual yields a bound that can be stronger than the LP bound.

Section 2 describes the general job shop model, which is specialized in Section 3 to the case of a flow shop. We proceed to Lagrangian relaxation and the principal results in section 4, followed by a numerical illustration of bounding, and conclusions.

2 THE DISJUNCTIVE JOB SHOP GRAPH

The processing of a given job or item on a particular machine is known as an *operation* and represented as a node $i \in \mathbf{N} = \{0, 1, ..., n\}$, where 0 is a dummy "start," and n a dummy "finish," operation. Directed arc $(i, j) \in \mathbf{A}$ from node i to node j indicates that operation i immediately precedes operation j within a specified job. Associated with this arc is its *length*, the duration d_i of operation i at the head of the arc. In addition to these standard components of a directed graph, a set of *disjunctive* edges $\mathbf{E}$ allows us to model the variable arcs. Specifically, there is an undirected disjunctive edge $[i, j] \in \mathbf{E}_k \subseteq \mathbf{E}$ for each unordered pair of nodes in the set $\mathbf{N}_k \subseteq \mathbf{N}$. That set is known as a *disjunctive clique* and represents all operations on a given machine $k \in \mathbf{K}$, the set of machines.

This disjunctive graph $\mathbf{G} = (\mathbf{N}, \mathbf{A}, \mathbf{E})$ models the job shop (see Figure 1), facilitating formulation of the scheduling problem as a disjunctive optimization by Adams et al. [1]. These authors give an orientation to each disjunctive edge to indicate the sequence of jobs chosen on each: edge $[i, j]$ is given the orientation from i to j to indicate that the job involving operation i is sequenced before that involving operation j.

A *selection* $\mathbf{S}_k$ is a set of directed (conjunctive) arcs, each of which represents a disjunctive edge given an orientation; a *complete selection* $\mathbf{S}$ is the union of selections $\mathbf{S}_k$ over all machines k. An acyclic complete selection represents a feasible schedule, and the makespan is the length of a critical (longest) path through the network on directed graph $\mathbf{D}_S = (\mathbf{N}, \mathbf{A} \cup \mathbf{S})$. An optimal schedule is thus an acyclic complete selection that minimizes the critical path length.

Adams et al. [1] defined a continuous nonnegative epochal variable t_i for each operation i representing the time its processing commences, and formulated the following disjunctive optimization for job shop scheduling:

$$
\begin{aligned}
\textbf{JSP} \quad & \min\ t_n \\
& \textit{subject to} \\
& t_j \geq t_i + d_i \quad \forall (i, j) \in \mathbf{A} && (11.1)\\
& t_j \geq t_i + d_i \ \vee\ t_i \geq t_j + d_j \quad \forall [i.j] \in \mathbf{E}_k,\ \forall k \in \mathbf{K} && (11.2)\\
& t_i \geq 0 \quad i \in \{0, 1, \cdots, n\}.
\end{aligned}
$$

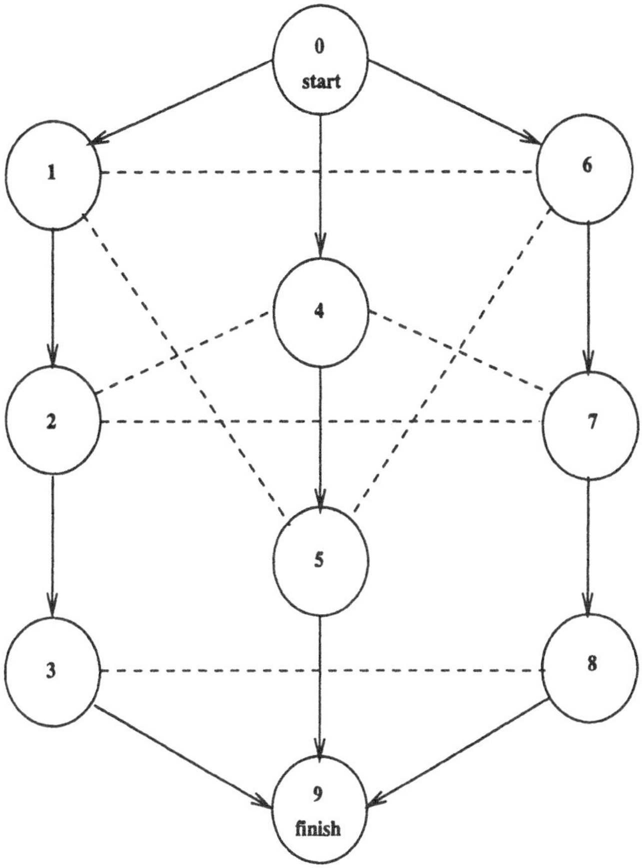

Figure 1 The disjunctive graph for a job shop with 3 jobs consisting of 8 operations on 3 machines. The first job consists of operations 1,2, and 3, the second of 4 and 5, and the third of 6,7, and 8. Operations 1,5, and 6 are performed on machine $\mathcal{A}$,operations 2, 4 and 7 on machine $\mathcal{B}$, and operations 3 and 8 on machine $\mathcal{C}$.

The fixed precedences within jobs are expressed by (11.1) and the variable sequencing precedences by the disjunctive system (11.2). These authors noted that even when $\mathbf{S}_k$ is acyclic for each k, the directed graph $\mathbf{D}_S = (\mathbf{N}, \mathbf{A} \cup \mathbf{S})$ might not be acyclic. In Figure 1, for example, suppose that machine $\mathcal{A}$ is sequenced as 6-5-1 and that machine $\mathcal{B}$ is sequenced as 2-4-7. Thus directed arcs (6,5), (6,1), and (5,1) replace the disjunctive clique $\mathbf{E_A}$, and directed arcs (2,4), (2,7), and (4,7) replace the disjunctive clique $\mathbf{E_B}$. Even though selections $\mathbf{S_A}$ and $\mathbf{S_B}$ are each acyclic, their union contains the cycle 5-1-2-4-5. Such snarls are avoided in a flow shop, which is discussed next.

3 THE FLOW SHOP MODEL

3.1 A Condensed Graph

A flow shop is a job shop in which all work proceeds in a single direction on machines ordered serially [3]. To formalize this notion, we construct the *condensed graph* $\widetilde{\mathbf{G}} = (\widetilde{\mathbf{N}}, \widetilde{\mathbf{A}})$ from the disjunctive graph $\mathbf{G} = (\mathbf{N}, \mathbf{A}, \mathbf{E})$ by replacing each disjunctive clique $\mathbf{N}_k$ by a single condensed node $\widetilde{q} \in \widetilde{\mathbf{N}}$ and deleting all disjunctive arcs. The set of arcs from clique $\mathbf{N}_p$ to clique $\mathbf{N}_q$ in $\mathbf{G}$ is replaced, if nonempty, by a single condensed arc $(\widetilde{p}, \widetilde{q}) \in \widetilde{\mathbf{A}}$. Figure 2 displays the condensed graph for the job shop of Figure 1. A job shop is a flow shop exactly when its condensed graph $\widetilde{\mathbf{G}}$ is acyclic.

Lemma 3 *If selection $\mathbf{S}_k$ is acyclic for each machine $k \in \mathbf{K}$ in a flow shop, then the directed graph $\mathbf{D_S} = (\mathbf{N}, \mathbf{A} \cup \mathbf{S})$, where $\mathbf{S} = \bigcup_k \mathbf{S}_k$ is also acyclic.*

proof. Suppose to the contrary that $\mathbf{D}_S = (\mathbf{N}, \mathbf{A} \cup \mathbf{S})$ contains a cycle, but that each $\mathbf{S}_k$ is acyclic. Then there must be a sequence of directed arcs in $\mathbf{A} \cup \mathbf{S}$ connecting a node $q \in \mathbf{N}_k$ to itself for some machine k. However, this would require a cycle in the condensed graph $\widetilde{\mathbf{G}}$, contrary to the definition of a flow shop. •

The absence of cycles, given any acyclic sequencing decision for each machine, allows us to optimize over a set of directed Hamiltonian tours, one through each disjunctive clique, and replace the disjunctive program **JSP** by the mixed integer model that follows.

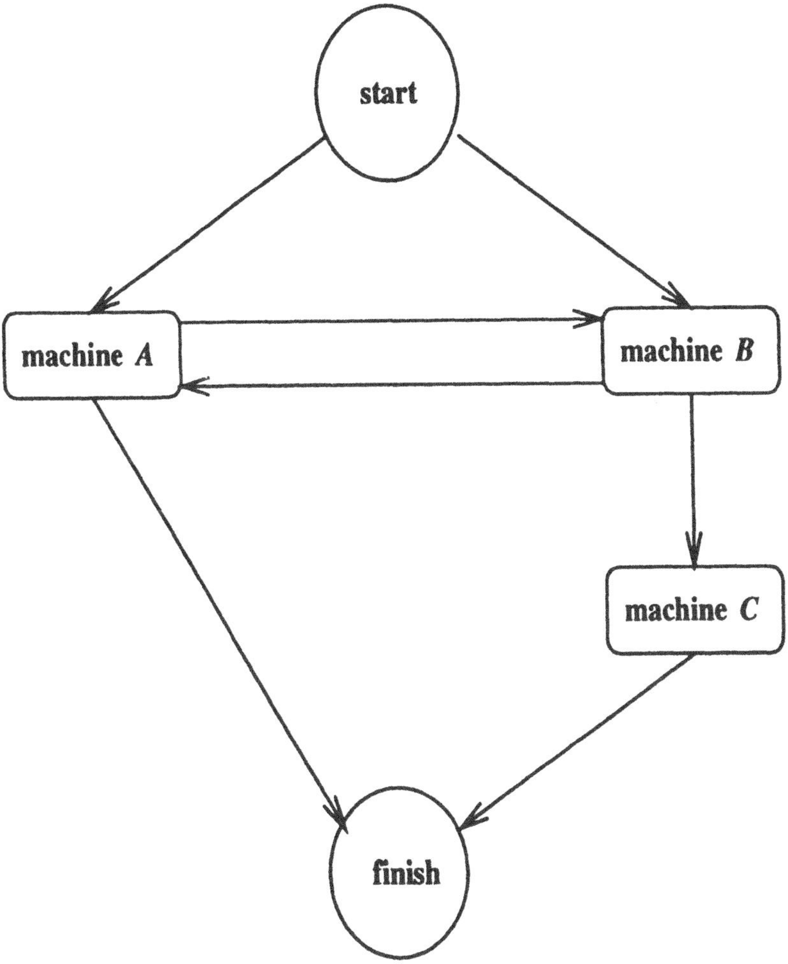

Figure 2 The condensed graph for the job shop in Figure 1. Note the cycle A-B-A.

3.2 A Mixed Integer Model

We treat the disjunctive edges in a slightly different manner than in **JSP**, imposing an orientation on $[i,j]$ from i to j if and only if operation i *immediately* precedes operation j in a given machine's sequence otherwise eliminating the edge completely from an updated graph. For example in Figure 1, if machine $\mathcal{A}$ is sequenced as 6-5-1, only directed arcs (6,5) and (5,1) replace the disjunctive clique $\mathbf{E_A}$. Arc (6,1) is omitted as redundant. Our model of a schedule for machine k is a directed Hamiltonian tour $\mathbf{H}_k$ through the nodes of $\mathbf{N}_k$, rather than a selection $\mathbf{S}_k$. We let $\mathbf{H} = \bigcup_k \mathbf{H}_k$, the union over all machines.

To choose a set of directed Hamiltonian tours that results in a shortest critical path from start to finish is the flow shop scheduling problem. We can formulate it as a mixed integer program in terms of the epochal variables $[t_i]$ as well as a pair of binary variables x_{ij} and x_{ji} for each disjunctive edge $[i,j]$. For convenience, we "close the loop" on each directed Hamiltonian tours by appropriate connections to the dummy start node "0". (No precedence constraints are imposed upon the shop start time t_0.) This allows us to employ familiar traveling salesman problem (TSP) constraints as given by e.g. Ahuja et al [2]. The Flow Shop Program (FSP) is then:

$$\textbf{FSP} \quad \min \; t_n \tag{11.3}$$

subject to

$$t_j \geq t_i + d_i \quad \forall (i,j) \in \mathbf{A}, i \neq 0, \tag{11.4}$$

$$t_j \geq t_i + d_i x_{ij} - M_k(1 - x_{ij}) \tag{11.5}$$

$$\forall i \in \mathbf{N}_k \cup \{0\}, \forall j \in \mathbf{N}_k, \forall k \in \mathbf{K}$$

$$\sum_{i \in \mathbf{N}_k \cup \{0\}} x_{ij} = 1 \quad \forall j \in \mathbf{N}_k \cup \{0\}, \forall k \in \mathbf{K} \tag{11.6}$$

$$\sum_{j \in \mathbf{N}_k \cup \{0\}} x_{ij} = 1 \quad \forall i \in \mathbf{N}_k \cup \{0\}, \forall k \in \mathbf{K} \tag{11.7}$$

$$\sum_{i \in \mathbf{Q}_k \cup \{0\}} \sum_{j \in \mathbf{Q}_k \cup \{0\}} x_{ij} \leq |\mathbf{Q}_k| - 1 \tag{11.8}$$

$$\forall \mathbf{Q}_k \subset \mathbf{N}_k \cup \{0\}, \forall k \in \mathbf{K}$$

$$t_i \geq 0 \quad \forall i \in \{0, 1, \cdots, n\} \tag{11.9}$$

$$x_{ij} \in \{0,1\} \quad \forall i,j \in \{0,1,\cdots,n\}. \tag{11.10}$$

Note that $t_0 = 0$ and $d_0 = 0$. The objective term t_n is the critical path length through the graph with the fixed arcs in $\mathbf{A}$ determining the precedence constraints within jobs (11.4) and the variable arcs as chosen from among the disjunctive edges, determining the sequencing constraints on each machine (11.5). In the latter, the immediate precedence of i before j is enforced exactly when $x_{ij} = 1$; otherwise, $\mathbf{M}_k$ is sufficiently large (see Proposition 3 below) to make (11.5) redundant. The TSP constraints (11.6) - (11.8) ensure a directed Hamiltonian tour through each set of nodes $\mathbf{N}_k$; there, $\mathbf{Q}_k$ is any (proper) subset of $\mathbf{N}_k \cup \{0\}$, and (11.8) is specifically the subtour breaking constraint system.

Proposition 1 *Given any disjunctive graph* $\mathbf{G} = (\mathbf{N}, \mathbf{A}, \mathbf{E})$, *and array of non-negative operational durations* $[d_i]$, *the mixed integer program* **FSP** *is feasible and bounded.*

proof. Any sequencing of machine k yields an Hamiltonian tour $\mathbf{H}_k$ for each machine $k \in \mathbf{K}$, and so $\mathbf{H} = \bigcup_k \mathbf{H}_k$ is a feasible solution. By lemma 3.1, the directed graph $\mathbf{D}_H = (\mathbf{N}, \mathbf{A} \cup \mathbf{H})$ is acyclic, and so a critical path of finite length (the makespan) exists. •

The problem of scheduling steel fabricating jobs in a flow shop with 9 jobs on 8 machines is taken up by Neureuther, Polak, and Sanders [12].

4 LAGRANGIAN RELAXATION

4.1 Decomposition of FSP

In the disjunctive graph $\mathbf{G} = (\mathbf{N}, \mathbf{A}, \mathbf{E})$, the fixed arcs in $\mathbf{A}$ link disjunctive cliques, each containing a one-machine sequencing problem. Lagrangian relaxation of the precedence constraints (11.4) representing these arcs in **FSP** yields the Flow Shop Lagrangian relaxation:

$$\textbf{FSL} \quad \min \left\{ t_n + \sum_{(i,j)\in\mathbf{A}} \lambda_{ij}(d_i + t_i - t_j) : (\mathbf{t}, \mathbf{x}) \in \mathbf{U} \right\} \tag{11.11}$$

where $\mathbf{U} = \bigcup_k \mathbf{U}_k$, and in turn $\mathbf{U}_k = \{(\mathbf{t}^k, \mathbf{x}^k) : \text{(11.5)} - \text{(11.10)}\}$, given any k in $\mathbf{K}$. This relaxation removes the fixed arcs of set $\mathbf{A}$, unlinking **FSL** into $|\mathbf{K}|$ one-machine subproblems:

$$\mathbf{FSL(k)}: \quad \min\left\{\sum_{i\in\mathbf{N}_k}\left(f_i t_i + \sum_{(i,j)\in\mathbf{A}}\lambda_{ij}d_i\right) : (\mathbf{t},\mathbf{x})\in\mathbf{U}_k\right\} \tag{11.12}$$

for given machine $k \in \mathbf{K}$, where

$$f_i = \sum_{\{j:(i,j)\in\mathbf{A}\}}\lambda_{ij} - \sum_{\{j:(j,i)\in\mathbf{A}\}}\lambda_{ji},\ \forall i\in\mathbf{N}, i\notin\{0,n\}. \tag{11.13}$$

The dummy finish operation n is a special case because it is not included in any node set $\mathbf{N}_k$, and only the nonnegativity condition (11.9) applies at that node after the precedences (11.4) are relaxed. The subproblem at node n is

$$\mathbf{FSL(n)} \quad \min\left\{t_n\left(1 - \sum_{(i,n)\in\mathbf{A}}\lambda_{i,n}\right) : t_n \geq 0\right\} \tag{11.14}$$

The *Lagrange function*

$$L(\lambda) = \min\left\{t_n + \sum_{(i,j)\in\mathbf{A}}\lambda_{ij}(d_i + t_i - t_j) : (\mathbf{t},\mathbf{x})\in\mathbf{U}\right\} \tag{11.15}$$

is defined for any vector λ of nonnegative Lagrange multipliers, the best choice of which is the subject of the next result.

Lemma 4 *The Lagrange vector λ that yields the greatest Lagrangian lower bound z_{LD} on the optimal objective value of* **FSP** *is determined by $z_{LD} = \max\{L(\lambda) : \lambda \geq 0\}$. This optimization is known as the* **Lagrange Dual (LD)**, *which is equivalent to the following linear program:*

$$\begin{aligned}
&\mathbf{LD} \quad \max\ w \\
&\quad \textit{subject to} \\
&\qquad w \leq t_n^\pi + \textstyle\sum_{(i,j)\in\mathbf{A}}\lambda_{ij}(d_i + t_i^\pi - t_j^\pi), \quad \forall\pi\in\Pi && (11.16)\\
&\qquad f_i = \textstyle\sum_{j:(i,j)\in\mathbf{A}}\lambda_{ij} - \sum_{j:(j,i)\in\mathbf{A}}\lambda_{ji} \geq 0, && (11.17)\\
&\qquad\qquad \forall i\in\mathbf{N}, i\notin\{0,n\}\\
&\sum_{(i,n)\in\mathbf{A}}\lambda_{in} \leq 1, \quad \forall i\in\mathbf{N} && (11.18)\\
&\qquad \lambda_{ij} \geq 0,\ \forall(i,j)\in\mathbf{A}.
\end{aligned}$$

where Π is the set of all extreme points of the convex hull of $\mathbf{U}$.

proof. That $L(\lambda)$ is a lower bound on the optimal value of **FSP** is established by Nemhauser and Wolsey [11] in Proposition 6.1; its maximum value over nonnegative λ is clearly the greatest such lower bound. The linear programming formulation follows from Corollary 6.3 (op. cit., p. 328): each cut constraint (11.16) corresponds to an extreme point of the convex hull of **U**.

Next we note that if (i) f_i is negative for any $i \in \mathbf{N}_k$, then the objective is unbounded for FSL(k); and similarly, if (ii) $\sum_{(i,n)\in \mathbf{A}} \lambda_{in} > 1$, then the coefficient of t_n is negative, and the objective of FSL is unbounded. Thus, each constraint in the system (11.17) - (11.18) corresponds to an extreme ray of the convex hull of **U**. Finally, it is evident that **FSL** is unbounded if and only if (i) or (ii) holds, so that there can be no other extreme rays. •

Observe that t_n can no longer be the critical path length. In fact, to attain the minimum objective value in **FSL(n)**, t_n can be set to zero in all cases, provided that (11.18) is satisfied. Too, the nonnegativity (11.9) of each t_i obviates the relaxation of any precedence arc $(0, i)$ emanating from the dummy start. **LD** is nonetheless likely to be a large linear program with a great number of constraints, indicating the desirability of subgradient optimization or heuristic techniques.

4.2 On the Lagrange Subproblem FSL

Despite the TSP constraints (11.5) - (11.8) in each one-machine subproblem **FSL(k)**, Lagrangian relaxation introduces subproblem objective functions (11.12) amenable to a simple greedy algorithm. Recall that in a feasible solution to constraints (11.5) - (11.8), each operation starting time t_i represents the epoch at which node i is reached in a directed Hamiltonian tour. Thus for a fixed Lagrange vector λ to minimize (11.12) we "front load" the tour to reach the node with the largest objective coefficient first, the next largest second, and so on. In general, we sort the objective coefficients $\{f_i\}$ in nonincreasing order $i \in \mathbf{N}_k$, with ties broken arbitrarily, then sequence the operations (nodes) in the same order.

For λ feasible in LD, by (11.17) each f_i is nonnegative. The starting time epoch $t_{[1]}$ for the first operation [1] in the sequence must be 0, with the remaining epochs for machine k determined by (11.5); i.e., by the precedences of $\mathbf{H}_k$. Thus we have established the following result:

Proposition 2 *Given a Lagrange vector λ feasible for LD, a directed Hamiltonian tour which visits the nodes in nonincreasing order of the objective coefficients $\{f_i : i \in \mathbf{N}_k\}$ defined by (11.15) is an optimal sequencing of operations for the one-machine Lagrangian subproblem* **FSL(k)**, *for each machine $k \in \mathbf{K}$. Further, an optimal choice of epochs is given by $t_{[m+1]} = t_{[m]} + d_{[m]}, t_{[1]} = 0$, where $[m]$ represents the operation which is m^{th} in the sequence.*

It is well known (see e.g., Theorem 16.9 [2]) that the Lagrangian lower bound z_{LD} on a minimization such as **FSP** is at least as large as the lower bound z_{LP} obtained by solving the linear programming relaxation. Geoffrion [9] showed that if, for every choice of objective coefficients, the Lagrange subproblem has an integer optimal solution even with integrality requirements relaxed for all variables, then the two bounds are identical; i.e., $z_{LD} = z_{LP}$. This property is known as the *integrality property* and is discussed by Ahuja et. al. [2]. Our Lagrange subproblem **FSL** does not have this property, as we show next.

Proposition 3 *The Lagrange subproblem (FSL) does not have the integrality property.*

proof. Given a Lagrange vector λ feasible for **LD** consider the following linear programming relaxation of the single machine integer subproblem, **FSL(k)**:

$$\textbf{FSLP(k)} \quad \min \sum_{i \in \mathbf{N}_k} f_i t_i$$

subject to

$$t_j \geq t_i + d_i x_{ij} - M_k(1 - x_{ij}) \qquad (11.19)$$
$$\forall i \in \mathbf{N}_k \cup \{0\}, \forall j \in \mathbf{N}_k, \forall k \in \mathbf{K}$$

$$\sum_{i \in \mathbf{N}_k \cup \{0\}} x_{ij} = 1, \quad \forall j \in \mathbf{N}_k \cup \{0\}, \forall k \in \mathbf{K} \qquad (11.20)$$

$$\sum_{j \in \mathbf{N}_k \cup \{0\}} x_{ij} = 1, \quad \forall i \in \mathbf{N}_k \cup \{0\}, \forall k \in \mathbf{K} \qquad (11.21)$$

$$\sum_{i \in \mathbf{Q}_k \cup \{0\}} \sum_{j \in \mathbf{Q}_k \cup \{0\}} x_{ij} \leq |\mathbf{Q}_k| - 1, \qquad (11.22)$$

$$\forall \mathbf{Q}_k \subset \mathbf{N}_k \cup \{0\}, \forall k \in \mathbf{K}$$

$$t_i \geq 0, \quad \forall i \in \{0, 1, \cdots, n\}$$

$$x_{ij} \in \{0, 1\} \forall i, j \in \{0, 1, \cdots, n\}.$$

Suppose, to the contrary, that **FSL** has the integrality property. We assume that (t^*, x^*) is an optimal solution to the *integer* program **FSL(k)**, and has a positive objective value; otherwise, the problem is trivial. Recall that M_k is a constant chosen large enough to make (11.19) redundant if $x_{ij} = 0$. Without loss of generality, we can assume that

$$M_k = \frac{1+\sqrt{5}}{2} \cdot \max\{d_i : i \in N_k\}. \tag{11.23}$$

The choice

$$x_{ij} = \begin{cases} d_i/M_k & if x^*_{ij} = 1 \\ 1 - \frac{d_i}{M_k} & if x^*_{ji} = 1 \\ 0 & otherwise \end{cases}$$

is feasible for **FSLP(k)**, since x^* satisfies (11.19) - (11.22). Each constraint (11.19) becomes redundant, yielding an objective value of 0 for **FSL(k)**; it is clear that this value must be an optimal solution to **FSLP(k)** because of the nonnegativity conditions on $\mathbf{t}$, and the nonnegativity of λ. Moreover, no solution $(\mathbf{t}, \mathbf{x})$ with integer x can have an objective value of 0, since the assumed optimal value of the integer program FSL(k) was positive. This contradicts to the assumption that FSL has the integrality property. •

Proposition 3 taken together with Geoffrion's result [9] has the following consequence.

Corollary 1 *The Lagrangian lower bound z_{LD} on* **FSP** *can be stronger, that is, strictly greater, than the LP bound.*

5 A NUMERICAL ILLUSTRATION

Figure 3 represents a flow shop with 3 jobs consisting of 8 operations on 3 machines. The reader can verify that the corresponding condensed graph is acyclic. The displayed set of Lagrange multipliers satisfies (11.17) and (11.18) and therefore is feasible for **LD**. Recall that to solve each independent Lagrangian subproblem **FSL(k)** we sort the objective coefficients $\{f_i\}$ defined by (11.13) in nonincreasing order (with ties broken arbitrarily), then sequence the operations (nodes) in the same order. In Figure 3, we therefore sequence machine $\mathcal{A}$ as 4 - 1 - 6, yielding operational starting epochs $t_4 = 0, t_1 = 9$, and $t_6 = 17$. Similarly we sequence machine $\mathcal{B}$ as 7-2, yielding starting epochs $t_7 = 0$ and $t_2 = 4$.

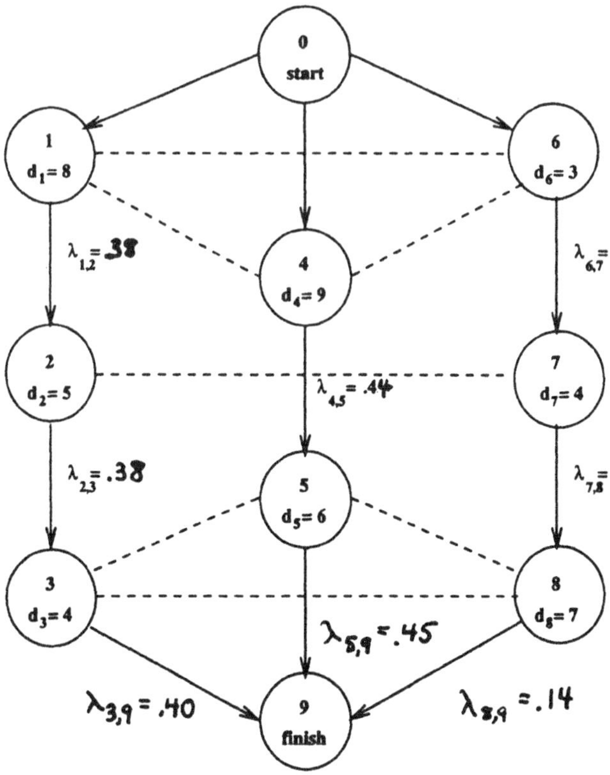

Figure 3 The disjunctive graph for a flow shop with 3 jobs consisting of 8 operations on 3 machines. The duration d_i of operation i appears alongside node i. Operations 1,4, and 6 are done on machine $\mathcal{A}$, operations 2 and 7 on machine $\mathcal{B}$, and 3,5, and 8 on machine $\mathcal{C}$. A feasible set of Lagrange multipliers is displayed.

Machine $\mathcal{C}$ is sequenced as 3 - 5 - 8, yielding starting epochs $t_3 = 0, t_5 = 4$, and $t_8 = 10$. Recall from the discusion following lemma 4 that $t_9 = 0$. Summing the objective function values (11.12) and (11.14) yields (11.11), a lower bound on the optimal makespan for **FSP**. For the Lagrange multipliers given in this illustration that lower bound is computed to be 20.8.

Of course, an *upper* bound on the makespan is given by the length of a critical path through the flowshop network with the precedences (11.4) of each job restored, given some sequencing for each machine. An heuristic, such as the Shifting Bottleneck [1] or Sevast'janov's Algorithm [13], can serve this purpose.

The Lagrange multipliers displayed in Figure 3 were chosen near to their optimal values for the Lagrangian Dual **LD**. For this illustration, the linear program **LD** is as follows:

$$\max \; w$$

$$\textit{subject to}$$

$$w \le \lambda_{1,2}(d_1 + t_1^\pi - t_2^\pi) \; +\lambda_{2,3}(d_2 + t_2^\pi - t_3^\pi) \; +\lambda_{3,9}(d_3 + t_3^\pi) \tag{11.24}$$
$$+\lambda_{4,5}(d_4 + t_4^\pi - t_5^\pi) \; +\lambda_{5,9}(d_5 + t_5^\pi)$$
$$+\lambda_{6,7}(d_6 + t_6^\pi - t_7^\pi) \; +\lambda_{7,8}(d_7 + t_7^\pi - t_8^\pi) \; +\lambda_{8,9}(d_8 + t_8^\pi) \; \forall \pi \in \Pi,$$
$$\lambda_{1,2} - \lambda_{2,3} \le 0, \tag{11.25}$$
$$\lambda_{2,3} - \lambda_{3,9} \le 0, \tag{11.26}$$
$$\lambda_{4,5} - \lambda_{5,9} \le 0, \tag{11.27}$$
$$\lambda_{6,7} - \lambda_{7,8} \le 0, \tag{11.28}$$
$$\lambda_{7,8} - \lambda_{8,9} \le 0, \tag{11.29}$$
$$\lambda_{3,9} + \lambda_{5,9} + \lambda_{8,9} \le 1, \tag{11.30}$$
$$\lambda_{ij} \ge 0, \; \forall (i,j) \in \mathbf{A}.$$

The set Π in (11.24) contains all 72 combined sequencings of machines $\mathcal{A}, \mathcal{B}$, and $\mathcal{C}$. Implementation of the dual simplex method for this instance of **LD** on the Microsoft Excel Solver yielded, as an optimal solution: $\lambda_{1,2} = \lambda_{2,3} = \lambda_{3,9} = .40$, $\lambda_{4,5} = \lambda_{5,9} = .45$, $\lambda_{6,7} = \lambda_{7,8} = \lambda_{8,9} = .15$, with an objective value 21.8.

The actual optimal makespan of this flow shop is 26, resulting from the sequencing 6-1-4 on machine $\mathcal{A}$, 7-2 on machine $\mathcal{B}$, and 8-7-6 on machine $\mathcal{C}$. Thus the duality gap is 4.2, or approximately 16% = [(26 − 21.8)/26] of the optimal makespan. By contrast, the LP bound is simply the critical path length when all machine sequencing precedences (i.e., the disjunctive edges) are ignored; it is 17.

6 CONCLUSIONS

The scheduling of jobs in a flow shop has been posed as a problem of choosing variable arcs in a disjunctive graph which also includes fixed arcs, so as to minimize the critical path length. This graph is acyclic in the case of a flow shop for any set of sequencings of the individual machines. Formulating an integer linear program FSP, we relaxed the constraints implied by the fixed arcs to obtain a Lagrangian subproblem. The latter in turn decomposes into unlinked single machine scheduling problems with no due dates, easily solved to optimality. The subproblem is also unburdened by the integrality property, and so can provide lower bounds stronger than available from linear programming.

The utility of this relaxation in an implicit enumeration approach to **FSP** depends upon the computational tractability of the Lagrange dual. Numerical testing of subgradient optimization and heuristic techniques is a desirable area for future study. Another avenue for exploration is cross decomposition of **FSP**. This scheme alternates between a Benders subproblem (with the x_{ij} as the complicating variables) and the Lagrangian subproblem, solving a master problem (Benders or LD) only if a convergence test fails. (See Holmberg [10] for details of cross decomposition.)

Acknowledgments

The author would like to thank Professors Michael Magazine and Jeffrey D. Camm of the University of Cincinnati, Professor Benjamin T. Smith of the University of Montreal, and an anonymous referee for their helpful comments and suggestions. The author also would like to thank Mr. Sailendra K. Padala of Wright State University for preparing the figures.

REFERENCES

[1] J. Adams, E. Balas, and D. Zawack, 1988, The Shifting Bottleneck Procedure for Job Shop Scheduling, *Management Science*, 34, 391-401.

[2] R. K. Ahuja, T. L. Magnanti, and J. B. Orlin, 1993, *Network Flows: Theory, Algorithms, and Complexity*, Prentice-Hall, Englewood Cliffs, New Jersey.

[3] K. R. Baker, 1992, *Elements of Sequencing and Scheduling*, Hanover, New Hampshire.

[4] E. Balas, J. K. Lenstra, and A. Vazacopoulos, 1995, The One-machine Problem with Delayed Precedence Constraints and Its Use in Job Shop Scheduling, *Management Science*, 41, 94-109.

[5] E. Balas and A. Vazacopoulos, 1996, Efficient Search Procedures Combining Different Neighborhood Structures, 1996 MSOM Conference, Hanover, New Hampshire.

[6] J. Carlier and E. Pinson, 1989, An Algorithm for Solving the Job-Shop Problem, *Management Science*, 35, 164-176.

[7] H. Fisher and G. L. Thompson, 1963, Probabilistic Learning Combinations of Local Job-Shop Scheduling Rules, in J. F. Muth and G. L. Thompson, (Eds.), *Industrial Scheduling*, Prentice-Hall, Englewood Cliffs, NJ.

[8] M. R. Garey and D.S. Johnson, 1979, *Computers and Intractability*, W.H. Freeman and Co., San Francisco, CA.

[9] A. M. Geoffrion, 1974, Lagrangian Relaxations Approaches for Integer Programming, *Mathematical Programming Study*, 2, 82-114.

[10] K. Holmberg, 1990, On the Convergence of Cross Decomposition, *Mathematical Programming*, 47, 269-296.

[11] G. L. Nemhauser and L.A. Wolsey, 1988, *Integer Programming and Combinatorial Optimization*, John Wiley & Sons, New York, NY.

[12] B. Neureuther, G. Polak, and N. Sanders, 1996, A Hierarchical Production Plan for Steel Fabricating. Presentation, Winter Conference, Cincinnati/Dayton INFORMS, Dayton, Ohio.

[13] S. Sevast'janov, 1995, *Mathematics of Operations Research*, 20, 90-103.

12

A HEURISTIC FOR VEHICLE ROUTING AND MANPOWER PLANNING

George A. Bohoris and Jane M. Thomas*

School of Industrial Management
University of Pireaus
185 34, Pireaus
GREECE

**34 Avenue Rd, Ashbey-de-la-Zouch*
Leicestershire LE65 2FE
UNITED KINGDOM

ABSTRACT

This paper discusses the approach taken in solving a specific depot staffing and routing problem. The problem differs from the normal routing exercise in that the route passes through a number of outlets where monetary collections are made. Owing to the ever increasing risks of carrying money there is a requirement for the money to be deposited at banks along the route. Consequently the more traditional routing and scheduling methodologies and software are not easily applied to the problem. This paper details the approach used and highlights some of the good points and the pitfalls in the proposed solution.

1 INTRODUCTION

We are all familiar with the run of the mill routing exercise where a vehicle leaves the depot, visits a number of distribution locations and returns to its starting point. Much work has been done in this area and many methodologies and software packages developed.

We have recently been presented with a more challenging variation of this format. In this instance it is not deliveries but collections that are being made. The collections are of money from vending machines that are located in nu-

G. Yu (ed.), Industrial Applications of Combinatorial Optimization, 256–271.

merous outlets throughout the depot region. Owing to the ever increasing risks associated with carrying large amounts of money, a value limit has been set on the amount which can be carried. Once collectors near that limit then they are required to deposit the money in the nearest bank.

The key aspect to the problem is that collectors do not have sufficient capacity to carry all they collect back to the depot but have to make deposits at various points within the depot area.

Much information is available. The locations of all the banks are known. The location of the depot is known. The locations of all the outlets are known, together with details about their collection values.

The problem presented was to determine depot staffing levels. Traditionally this has been done based on a "rule of thumb" that each depot has 500 machines to be collected per week and each collector can collect from 12 machines per day. A simple division is done to calculate the staffing levels, but this takes no account of the relative distances between outlets or of the amount of money to be collected and counted at each outlet.

A more objective way of determining staffing levels was required. An imbalance of workloads in different depot regions was apparent. It was thus decided to approach this by determining the optimal routes within a depot area and letting the total number of routes determine the staffing level. This was, therefore, a one time exercise, not a daily scheduling task.

The model had to run on a PC. The scope for developing the model was fairly wide. Spreadsheet and database software was available, as was high level programming software (FORTRAN). A further option was to investigate and, where appropriate, purchase commercial software. This paper outlines the approach used to tackle the problem.

Each collector performs the following tasks each day:

(a) Starts from home and follows a route specified by the depot manager.

(b) Takes the money from the vending machines and performs routine tasks at each outlet.

(c) Deposits the money collected at the branches of a specific bank. The collector can visit any of the bank's branches, the constraint being that it must be the branch closest to the current location. There is a set maximum limit for the sum of money that may be carried by a collector at any time, known as the 'Carrying Limit'. This limit must not be exceeded at any time and may necessitate the collector making several deposits per day.

(d) Returns to the depot after the collections for the day have been completed for a debriefing by the manager and to prepare for the following day's collections.

The collection schedule for each collector normally varies from week to week. One of the reasons for this variability is the "Period of Collection." The collection of money from the vast majority of outlets occurs once each week, but there are some outlets where collections are fortnightly or even monthly.

2 PROBLEM REQUIREMENTS

The solution of the problem considered in this paper therefore needed to address the following objectives:

(a) Determine optimal staffing levels.

(b) Distinguish which fortnightly collections are to be made in which week.

The problem was subject to the following constraints:

(i) Money must be deposited at banks once the collection limit is reached.

(ii) Collections from outlets must be scheduled at the same time each week.

(iii) Collections must be scheduled to start at different points (the first outlet on any route) but finish at the depot, ensuring that the route is not circular.

(iv) The time taken to collect the money is variable (depending on the amount to collect from the outlet).

3 LITERATURE REVIEW

The vehicle routing problem is a well-researched area described in many publications over the years. The literature contains details of comprehensive heuristics that take into account different restrictions imposed by realistic operations.

Unfortunately, no single existing heuristic matched our problem, although a number of them seemed to meet some portion of our problem specifications.

Jacobsen and Madsen (1980) for example, used the travelling salesman problem to minimise costs when siting warehouses. It was thought that this approach might have been useful if there were an excess of collectors and the decision was a choice of which collecftors to use.

Having suggested the novel approach of simulated annealing, Golden and Skiscim (1986) concluded that "simulated annealing is a clever, innovative approach but there are probably more effective heuristics available for solving a given combinatorial optimisation problem" and that "simulated annealing turns in disappointing results in terms of accuracy and efficiency."

An appraisal of commercial routing and scheduling software packages currently available on the market was also undertaken to ascertain whether or not any of them could be used for our collectors problem. None of the packages considered would meet all the requirements outlined above. A few of them went some way towards meeting the requirements but could not satisfy the banking and fortnightly collection constraints. It was felt that some of the commercial packages could be tailored to meet our needs but that the cost and effort involved in this would be prohibitive.

Due to the unsuitability of the methods presented in the literature and of the commercial packages considered, it was concluded that a method to determine the staffing levels of collectors should be developed in-house.

4 MODEL DEVELOPMENT

4.1 Assumptions

In developing the first version of the model the following assumptions were made:

(a) An average travelling speed of 30 miles/hour has been used for all types of roads. This, despite being a simplifying assumption, is on average valid for most routes. The depots considered cover a fairly small geographical area. Thus, travel between the outlets, banks and the depot is generally on suburban roads. Few 'A' class roads are used and very rarely motor ways. In the next version of the program a road base system will be built into the system.

(b) All collections are either weekly or fortnightly.

(c) The current time windows for each outlet can be changed as part of an one time exercise.

(d) The amount collected at each outlet is similar from week to week. The collection time for an outlet is calculated by adding the fixed time for the routine tasks to the variable time proportional to the average amount collected and the number of machines.

4.2 The Model

The staffing level was calculated based on determining the minimum number of routes required in order to collect from all of the outlets serviced by a depot. If the minimum number of routes required for a depot is 50, then, for a 5-day working week, ten routes are necessary each day. Therefore, the equivalent of ten full-time collectors is needed.

The method of determining the necessary routes was initially developed by Mistry and Livingstone (1990) under the supervision of the authors. It first assigns outlets to weekly collections and then calculates the collection routes, as follows.

Assignment to Weekly Collections

Using Vogel's Approximation Method

The first stage is to determine which fortnightly collections are to be made in Week 1 and which in Week 2. Over a two-week period the number of fortnightly collections should be equally divided between Weeks 1 and 2. The method used was an adaptation of Vogel's Approximation Method (VAM) (Taha, 1987). VAM was used in a similar manner to assignment problems, where m jobs are to be assigned to n machines and where $m = n$. In this application the aim is to pair the fortnightly outlets together in such a way that the total distance between the pairs is minimised. In each pair of outlets one is assigned to a Week 1 collection and the other is assigned to a Week 2 collection. This reduces the disruption caused by incorporating the fortnightly collections with the weekly collections. Hence in this application:

$$m = n = \text{number of fortnightly collections.}$$

The main reason for using VAM is that, in general, it yields an assignment that is as close to optimum as is possible (that is, it minimises the distance between the paired outlets). In our case, VAM was used as follows:

1. When assigning outlets m to n, an assignment may only take place if outlet m is not equal to outlet n.
2. The row crossed out would represent the assigned outlet m to be collected in Week 1.
3. The column crossed out would represent the assigned outlet n to be collected in Week 2.

Results of the Assignment Process

If the number of fortnightly collections to be made is odd then there is an outlet remaining after VAM has been completed. This outlet is initially assigned for collection in both weeks. It will be removed from the collections from one of the two weeks at a later stage. Hence, the two sets of outlets to be collected from in Weeks 1 and 2 are created. Each set contains:

(a) Fortnightly outlets to be collected, which have been allocated in equal numbers to either Week 1 or Week 2.

(b) Weekly outlets to be collected in both weeks.

(c) (Possibly) A remaining un-allocated fortnightly outlet which is initially to be collected in both weeks.

Calculation of Collection Routes

To determine the minimum number of routes using the sets of outlets obtained in the assignment to weekly collections above, two heuristics have been developed. These are the "Allocation Heuristic" and the "Routing Heuristic."

The Allocation Heuristic

The allocation heuristic is performed once each on the Week 1 and Week 2 data sets obtained from the assignment to weekly collections above. The allocation heuristic is designed in a similar manner to dynamic programming in that it reduces the overall problem into smaller manageable sub-problems.

It operates in two stages:

Stage 1 Assigns outlets to bank regions and calculates the Travelling Salesman route through the bank regions (Steps 1-3).

Stage 2 Determines which outlets form the links between bank regions and calculates the routes through the other outlets within bank regions (Steps 4-9).

Both stages assume that the locations of outlets and banks are known.

Stage 1 - *Assign outlets to regions and calculate travelling salesman route*

Step 1 The numbers indicate the amount collected from machines located within the outlet (Figure 1). For each outlet determine its nearest bank and then assign the outlet to that particular bank (Figure 2).

Step 2 For each bank (B_i) form the set of outlets assigned to it. Subsequently, the whole area serviced by a depot (D) can then be partitioned into, say, k bank regions (Figure 2).

Step 3 Using the banks as nodes, determine the travelling salesman tour starting from and finishing at, the depot (Figure 3).

Stage 2 - *Determine links between bank regions and routes within these regions*

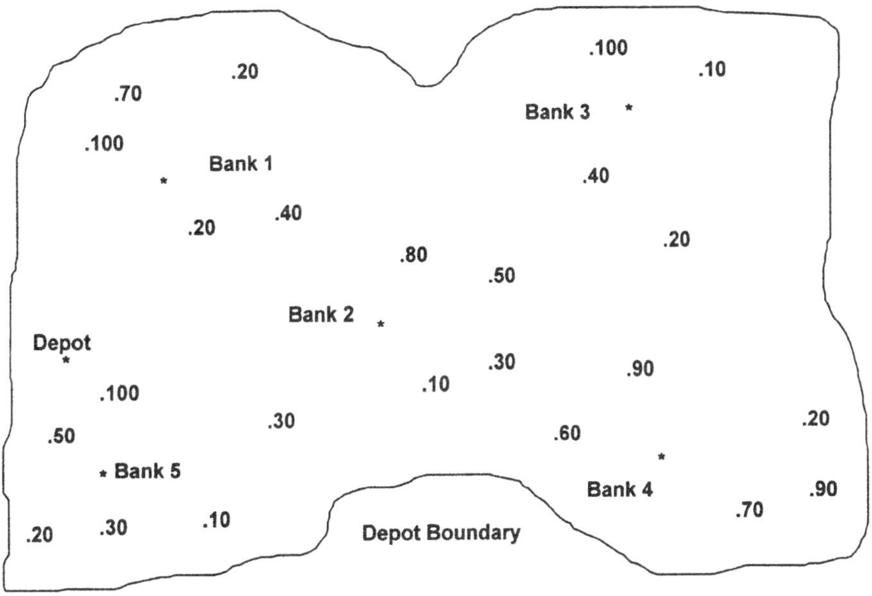

Figure 1 Depot boundary and geographical locations of banks and collection sites (collection sites are labelled with average amount collected).

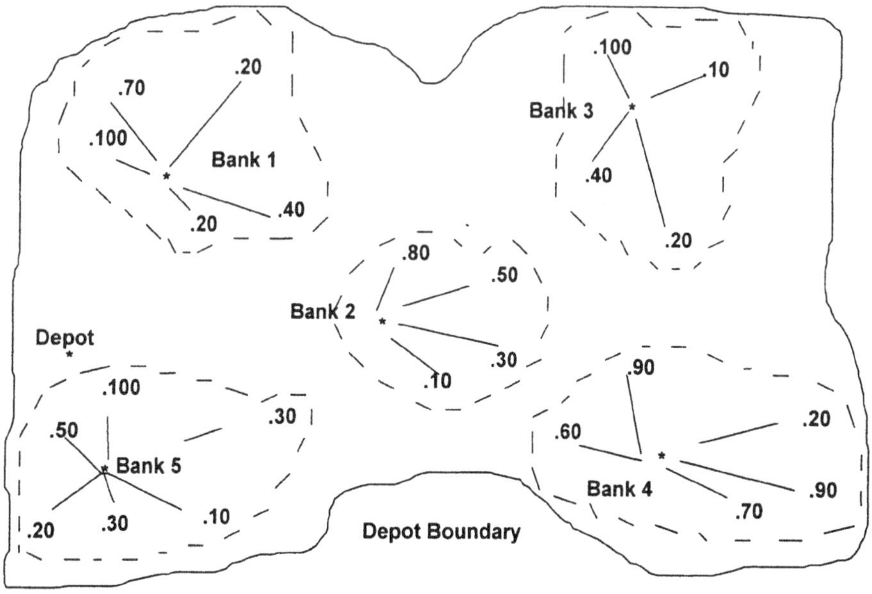

Figure 2 Collection sites assigned to closest bank to create "bank regions."

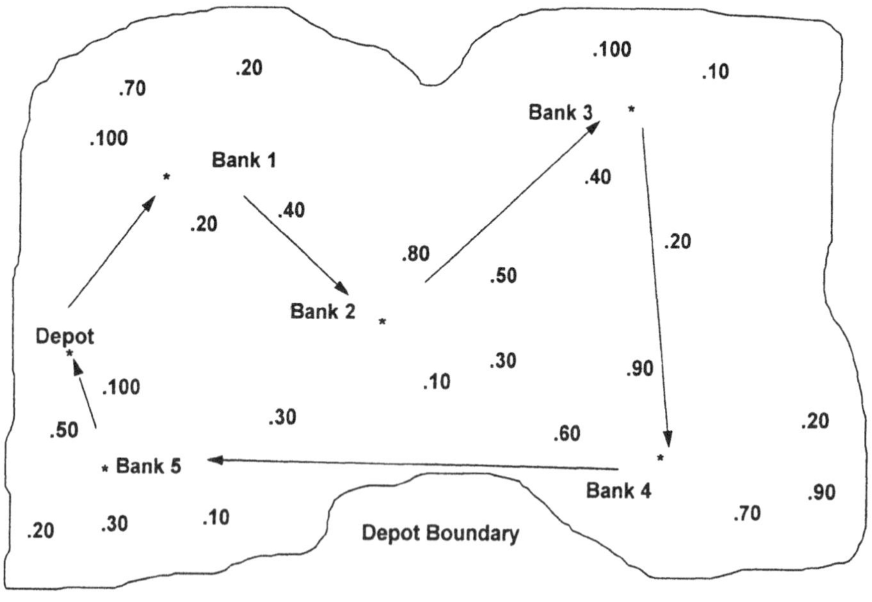

Figure 3 Travelling salesman route connecting depot with banks.

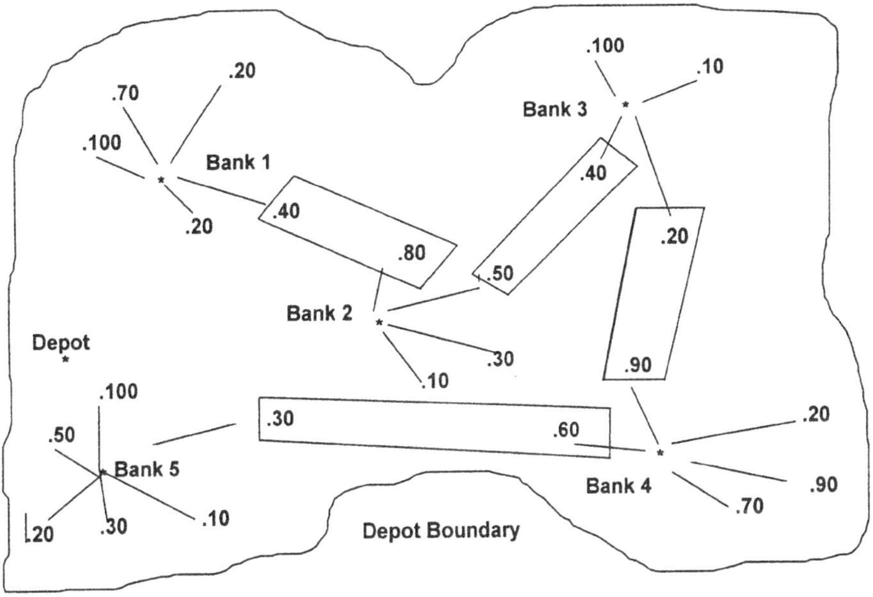

Figure 4 Closest collection sites forming links between bank regions.

Step 4 Between consecutive bank regions along the travelling salesman tour determine the two closest outlets. Remove these outlets from their respective bank regions (Figure 3). These will represent the link through which the collector will travel from one bank region into another.

Step 5 Perform the routing heuristic (described below) within each bank region.

Step 6 Replace those outlets removed in Step 4 (Figure 4). These will represent the link by which to travel from one bank region into another.

Step 7 Following the travelling salesman tour determine how long it takes to travel to the nearest bank from the depot.

Step 8 Follow the route within the bank region (obtained using the routing heuristic). Accumulate the banking and travelling time.

Step 9 If all of the outlets, excluding the one that represents the link into the next bank region, have been collected from, and if the accumulated time is less than a day's shift then:

9-1a Depending on the amount collected from the last outlet either travel along the link into the next bank region, or deposit cash and then travel along the link into the next bank region.
9-1b Return to Step 8 and follow the route within the next region.

Else, if the accumulated time is equal to a day's shift then:

9-2a Note the route used.
9-2b Remove all outlets collected from the data set. Also remove any banks for which all the outlets in their region have been collected.
9-2c Re-run the travelling salesman tour using the depot and remaining banks.
9-2d Return to Step 4.

Else, if the accumulated time is greater than a day's shift then:

9-3a Note the route used (excluding the last outlet that was collected and hence caused the accumulated time to exceed the day's shift length).
9-3b Remove all the outlets collected from the data set (leaving in the last uncollected outlet). Also remove any banks for which all the outlets in their region have been collected.
9-3c Re-run the travelling salesman tour using the depot and remaining banks.
9-3d Return to Step 4.

Repeat Steps 7-9 until all of the outlets to be serviced by the depot have been collected from.

If the number of outlets that are to be collected from on a fortnightly basis is odd then the heuristic is repeated, but this time excluding the unallocated outlet. The unallocated outlet is removed from the week that had the greatest accumulated collection time on the first run of the heuristic (and now becomes allocated to the week that had the lowest accumulated collection time on the first run of the heuristic).

The Routing Heuristic

The routing heuristic is used to obtain a good collection route within a bank region. It has been developed using the principle that no collection routes within a bank region should cross each other. The routing heuristic forms Step 5 of the allocation heuristic. It operates as follows:

Step 1 Within a bank region define the stem distance and direction from the bank to each outlet within its region. Call this the 'stem'.

Step 2 Introduce a link between the two outlets closest together. Check that the amount collected along the link does not exceed the carrying limit and that no stems to other outlets are crossed.

Step 3 Repeat Step 2 until no more links can be introduced.

Step 4 Using the links, calculate the total travelling time, starting from and finishing at the bank, to visit all the outlets within the region.

To improve upon this initial route carry out the following:

Step 5 Check to see if the current solution can be improved.

- **5a** Introduce a link between the closest outlets which were not previously linked. In doing this ensure that no stems are crossed and that the route is feasible.
- **5b** Carry out Step 4 of the routing heuristic.

If the total travelling time is reduced then return to Step 5 in the allocation heuristic with the new feasible route. Otherwise return to Step 5 in the allocation heuristic with the current feasible routes.

5 MODEL VALIDATION

Before the model could be used it had to be checked to ensure that the results produced reflected both what was currently happening in the real world, and produced feasible and workable solutions. This was essential if the chosen methodology was to be given any credence by the depots and operators. Validation concentrated on two areas:

(a) Validation of the data and parameters used.

(b) Validity of the results (routes and staff numbers) produced.

For the purposes of validation a specific depot was selected. The depot currently employs six full time collectors and serves 158 outlets.

The data and parameters used were reasonably straightforward to validate. Locations for depots, outlets and banks were known. Collection times and bank deposit times could be checked by observing working practices. Currently the model uses a straight-line factor and an average speed to determine travel times. These times were sampled and validated using road-based routing software. With subsequent availability of routing software a better approach may be to evaluate all routes more accurately.

Examination of the routes produced showed two significant factors.

1. The first was that the operating statistics from the depot contained inconsistencies. For example, the miles travelled by a collector depended upon the start location (generally starting from home, rather than the depot). The home locations of the current collectors could clearly not be used in the calculation of future staffing levels. Hence, the data had to be carefully reviewed and adjusted.
2. The second factor identified was that the time windows were particularly tight and inflexible. The model is not easily adapted for the inclusion of such tight constraints. Consequently, the model results had to be treated with great care and the limitations of the approach understood.

6 RESULTS

The collection routes determined by our method were mapped out. Visual inspection by the analysts confirmed their feasibility, although the routes were not tested on the road.

Using the set of outlets serviced by the test depot the results indicated that for a week's collection only five collectors were required. This indicated a potential saving of one person (or 17% of collection costs) over the current staffing level of six collectors, although there would still be a need for holidays and sickness cover. None the less there was a reduction in both employment and vehicle costs.

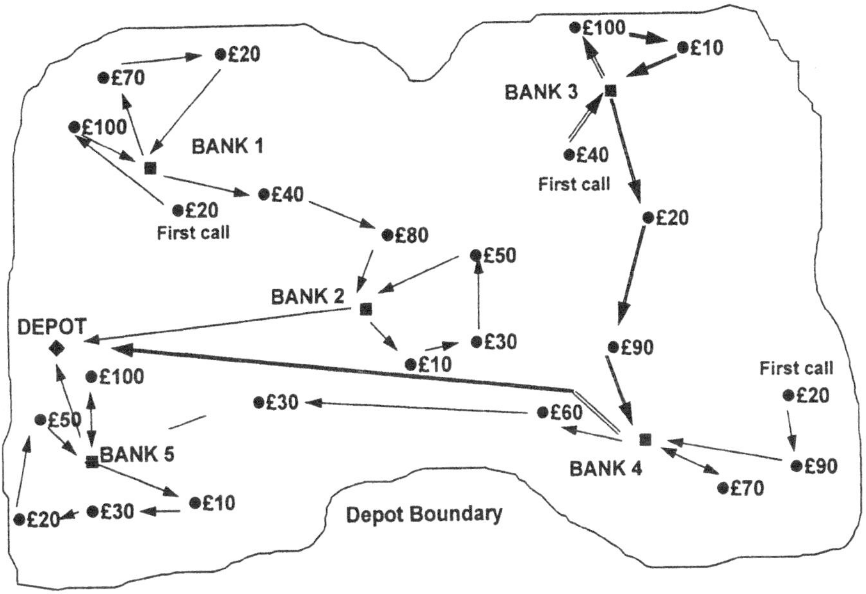

Figure 5 Final collection routes - Start at "first call" and finish at depot.

Running the model using a PC takes only minutes for the size of problem considered - typically six daily collection routes with 200 outlets within the depot area. Different depot areas of similar size would produce similar running times. Although the model has not been tested on larger problems it would be expected that the running time would increase in proportion to the number of outlets and banks and that the increase in time would be additive rather than multiplicative.

Since the methodology is not depot specific it can be applied to determine the number of collectors at any depot, or indeed at any other application of a similar nature.

7 CONCLUSIONS

The methodology devised was a good starting point. The major problem experienced was the time taken to set up the model. With the availability of mapping tools for specifying outlet locations this particular aspect of the exercise could now be speeded up.

Given the size of the existing depots (approximately six collectors at each) and the existence of current routes, the analysts felt that the time and effort required to assess the situation may outweighed the advantages gained. Indeed, the time and effort at existing sites might be better invested in manually assessing the routes and working practices. The issue where the largest benefits can be gained is in determining the location and size of new depots.

It is believed that a modelling approach is appropriate and lends both credence and flexibility to the way depot staffing levels are determined. We feel however that our results can be further improved and hope that this paper will be only the beginning of a fruitful research effort.

Acknowledgements

An important part of the work described in this paper was completed during student projects under the supervision of the authors. These involved M. Mistry and D. Livingstone (Mistry and Livingstone, 1990) and T. Chan, M. Levnajic, M. Newmarch, N. Peters and P. Simonsson (Chan et al, 1991). For confidentiality reasons however their results have not been published and their reports are not unrestrictedly available.

REFERENCES

[1] T. Chan, M. Levnajic, M. Newmarch, N. Peters, and P. Simonsson, 1991, Developing a Methodology for Determining the Staffing Levels at BLMS Depots, Case Study, School of Manufacturing and Mechanical Engineering, University of Birmingham, Birmingham, UK.

[2] M. Desrochers, J. Desrosers, and M. Solomon, 1992, A new optimisation algorithm for the vehicle routing problem with time windows, *Operations Research*, 40, 342-354.

[3] J. Desrosers, M. Sauve, and F. Soumis, 1988, Lagrangian relaxation methods for solving the minimum fleet size multiple travelling salesman problem with time windows, *Management Science*, 34, 1005-1022.

[4] Y. Dumas, J. Desrosers, and F. Soumis, 1991, The pickup and delivery problem with time windows, *European Journal of Operational Research*, 54, 7-22.

[5] B. Golden and C. Skiscim, 1986, Using simulated annealing to solve routing and location problems, *Naval Research Logistics*, 33, 261-279.

[6] S. K. Jacobsen and O. Madsen, 1980, A comparative study of heuristics for a two-level routing location problem, *European Journal of Operational Research*, 5, 378 - 387.

[7] M. Mistry and D. Livingstone, 1990, Bass Leisure Machines Depot Staffing Levels, M.Sc. Dissertation Thesis, School of Manufacturing and Mechanical Engineering, University of Birmingham, Birmingham, UK.

[8] M. Solomon, 1987, Algorithms for the vehicle routing and scheduling problems with time window constraints, *Operations Research*, 35, 264-265.

[9] H. A. Taha, 1987, *Operational Research*, Macmillan Publishing Company, New York.

13

CONCEPTUAL DESIGN AND ANALYSIS OF RAIL CAR UNLOADING AREA

Jonathan F. Bard

Graduate Program in Operations Research
Department of Mechanical Engineering
The University of Texas at Austin
Austin, TX 78712
USA

ABSTRACT

Faced with growing demand at their Alexandria, Louisiana laundry detergent plant, Procter & Gamble asked a design team from the University of Texas to undertake a study of the raw materials unloading area. This chapter reports on the results of that effort and presents the accompanying mathematical model developed to help analyze the material flows. In the first part of the study the design team established daily requirements for the number of raw material rail cars unloaded per day. The related combinatorial optimization problem of assigning rail cars to positions on the platform and unloading equipment to rail cars, was modeled as a mixed integer nonlinear program. The inability of two standard commercial codes to find optimal solutions led to the development of a greedy randomized adaptive search procedure (GRASP). Accounting for the operational and physical limitations of the system, the GRASP was used to determine the maximum performance that could be achieved under normal conditions. The second part of the study called for the conceptualization of alternative designs for meeting an expected 14% increase in demand over the next few years. The analytic hierarchy process in conjunction with a standard scoring model was used to rank the evaluation criteria and to select the most preferred alternative. A worst case analysis of the top candidate confirmed its performance capabilities.

G. Yu (ed.), Industrial Applications of Combinatorial Optimization, 272–300.

1 INTRODUCTION

Scheduling problems in manufacturing start with the issuance of purchase orders and the arrival of raw materials, continue on through production planning, job sequencing, and quality assurance testing, and end with packing, loading, and shipping. In this chapter, we deal with the problem of unloading rail cars containing liquids and powders used in the production of soaps at Proctor & Gamble's Alexandria, Louisiana plant. To give some background, P&G has markets in over 140 countries where it sells such well known brands as Folgers coffee, Pampers diapers, Cover Girl makeup, and Tide laundry detergent. Annual profits exceeded $2 billion during the last fiscal year. In the United States alone, P&G operates 61 plants and eight research facilities. The plant located in Alexandria serves the south central and western regions of the U. S. and is one of four plants in North America that produces dry laundry detergent. Tide is the chief brand manufactured there and accounts for 57% of the plant's total output.

In 1994, the Alexandria plant shipped 28 million cases of dry laundry detergent, up from 10 million cases in 1985. To continue meeting customer demands, the manufacturing process is constantly being improved. Currently, the bottleneck lies in the packaging department. However, P&G is planning to increase production by approximately 14% (by increasing the number of packaging lines from 7 to 8) and is concerned that the raw materials unloading process will not be able to handle the new demands. Therefore, before making the change, P&G asked a design team from the University of Texas to study the underlying issues and to come up with recommendations for implementation. The project consisted of the following tasks:

- analyze the capability of the unloading area
- develop, evaluate, and analyze a redesign of the unloading configuration
- create a computer model of the unloading area to assist in the daily scheduling of rail cars

In the remainder of this chapter, we discuss each of these tasks and present the results of the accompanying analysis. Although our original intent was simply to develop a mathematical model that could be used to assign rail cars to positions on the platform and then assign unloading equipment to each rail car, we found it necessary to expand the scope of the study to accommodate the operational concerns of P&G. This led to the construction of four different design alternatives for the rail yard and a corollary cost-benefit study to select the most preferred. To assess each candidate, we developed a mixed integer nonlinear programming model of the assignment and unloading problem. After linearizing the model we found that it could not be solved with any number of commercial codes, including OSL and CPLEX. Alternatively, we developed a greedy randomized adaptive search procedure – GRASP [4] to find high quality feasible solutions. In all cases, the GRASP outperformed current practice and most likely (though not always verifiable) yielded optimal schedules.

2 RAW MATERIALS UNLOADING PROCESS

The raw materials unloading process begins with the demand from manufacturing for the different components of dry laundry detergent (see Figure 1). As the detergent is being produced, the components flow from their individual storage areas into the manufacturing facility. Silos are used for storing dry powders, and tanks for storing liquids. Replenishment of the storage silos and tanks is done by daily rail car deliveries to the raw material unloading area.

A method based on pressure differential using blowers is employed to unload rail cars containing powders. This type of unloading requires sending low pressure air (less than 15 psi) into the rail car, aerating the powder or granular contents, and sending a mix of the material and air out of the rail car [7]. The material/air mix then enters a *product line*, which transports the material to a storage silo. Product lines are pipes that are hooked to air hoses. Each raw material uses an individually assigned product line which is stationary and located next to a blower drop at the corresponding unloading spot.

Rail cars containing liquid are unloaded using pumps. The liquid is then transported from the rail cars through pipes into storage tanks. Higher viscosity liquids are unloaded with positive displacement pumps while the lower viscosity liquids use centrifugal pumps.

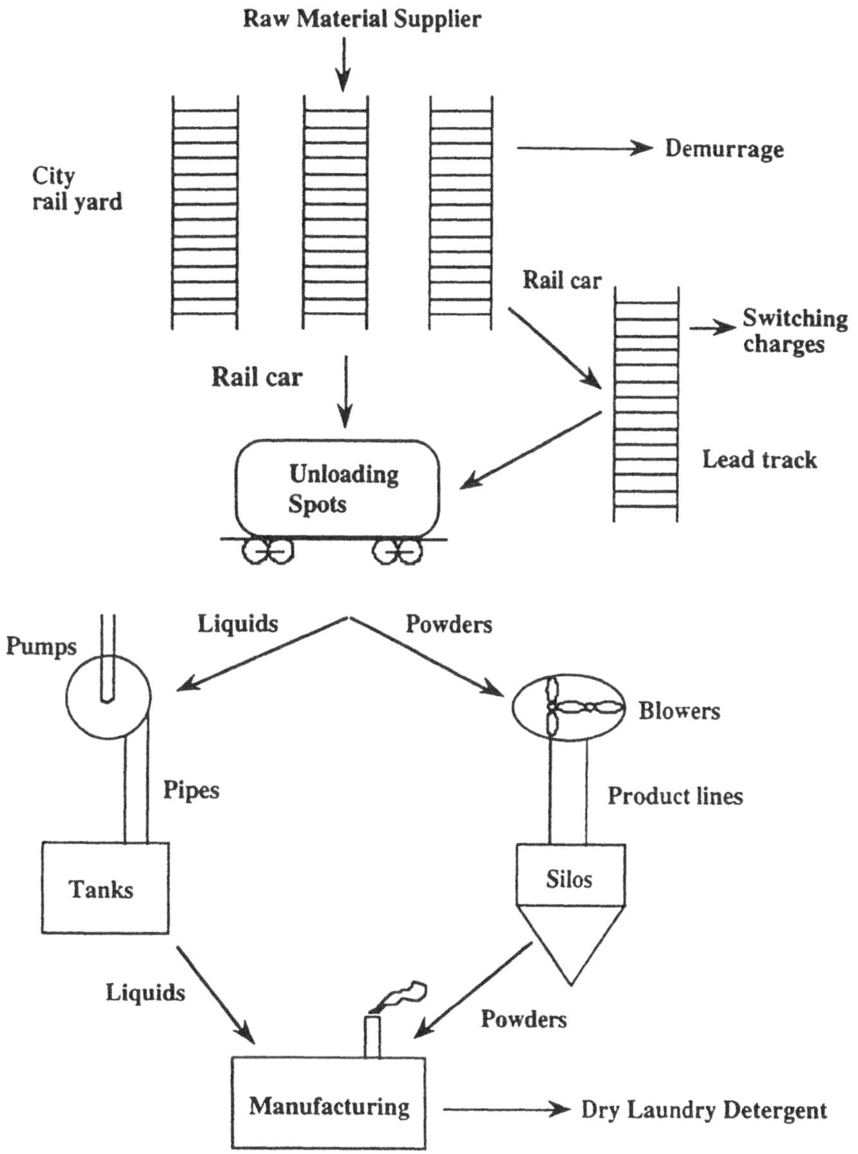

Figure 1 Raw material unloading process

2.1 Unloading Area Issues

The unloading area (see Figure 2) consists of 26 unloading spots distributed along four rail car tracks. Unloading spots are designated according to tracks (A, B, C, and D) and to location numbers (1 through 7). Each raw material (RM) is allocated to certain unloading spots or positions, yet some spots include the allocation of more than one raw material. For example, spot A2 can accommodate only one rail car but this rail car could contain either AC Base (RM Q) or fine soda ash (RM R).

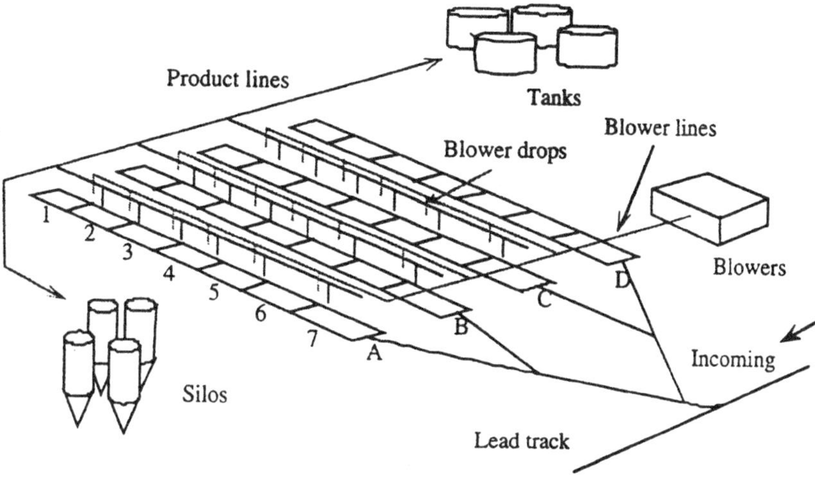

Figure 2 Unloading area schematic

Located between two rail cars is a *blower drop* and/or a pump. Both blower drops and pumps are stationary and can only unload rail cars on either side of the location. For example, the blower drop for blower #3 located between spots A3 and B3 can only unload a rail car located at either of these spots. Although some blowers have more than one blower drop, they can still only unload one rail car at a time. Product lines are stationary and each line transports a single raw material.

A central blower station houses the blowers used to unload the dry powders. Elevated blower pipes leave the station and transport pressurized air to the rail car unloading area. Individual pipes run above and along the unloading area tracks until they drop down to assigned unloading spots. On/off valves along the pipes regulate airflow.

To prepare a rail car for unloading, the raw material unloaders attach one end of a flexible hose to the blower drop connection and the other end to the rail car. Another hose is attached from the rail car to the product line for transport of the material to the silo.

2.2 Scheduling and Switching Issues

Each weekday morning the raw materials planner schedules the delivery ana unloading of the rail cars. Many factors are taken into account, including the rate at which the plant uses each raw material, how much safety stock of each raw material the plant requires, and the amount of time rail cars have been located on the *lead* or in the rail yard. All of these factors must be considered before the scheduler can determine which rail cars should be unloaded and at which spots they should be located. Once the assignments are made, a delivery schedule and a switch list are sent to the railroad company for delivery of the rail cars in the evening when the switching occurs.

The switching operation takes up to 2 hours and is also performed by the railroad company. Once switched, the rail cars remain at their unloading spots until the following evening when the railroad company returns with full replacements.

Because the four tracks in the unloading area dead-end at locations A1, B1, C1, and D1, rail cars can only be moved into and out of the unloading area from one direction. Problems arise when a switching operation is scheduled to be performed and a rail car located at the front of the line has not been completely unloaded. When this occurs, the raw material unloader has to detach the rail car, let the railroad company perform the switching, and reattach the rail car after the empty ones have been switched. This results in a partially empty rail car that must remain on an unloading spot for another day. The full rail car, which was supposed to be switched, must remain idle on the lead or in the city rail yard.

Contributing to the difficulty of the switching operation is the uncertainty in the time it takes to unload the various raw material rail cars. Some materials may take 7 hours on average to unload while others may take 15 hours. Furthermore, substantial differences in unloading times for the same powder material can also occur. These differences result from powder solidifying in the transit due to moisture in the rail cars at the time the powders were initially loaded. The unloader has to manually break the powder up so that it can flow. Rail cars containing liquids are fairly consistent from one material type to another and typically take 5 to 5.5 hours to unload.

Other problems occur when rail cars are not moved onto spots quickly or directly from the rail yard. When they are left in the city rail yard for more than 48 hours, *demurrage* will accumulate. Demurrage can be avoided by placing rail cars on the lead track at the P&G plant site while they await a vacancy. Although time spent on the lead will not result in demurrage, it does occasion unnecessary switching and hence the assessment of switching charges.

3 STUDY REQUIREMENTS

A primary objective of the project was to analyze the throughput of raw material flow from the rail cars to the tanks and silos. Using this information, the design team suggested a redesign of the current unloading area configuration to accommodate an estimated 14% increase in production. A mathematical model of the unloading area was developed to analyze each alternative and to assist in the scheduling and switching of rail cars.

3.1 Throughput Analysis

The throughput study enabled us to evaluate the current unloading process and to determine the requirements of the current system in meeting raw material demands. In this phase of the work, P&G requested that we provide sufficient information to answer following questions:

- *Pump and blower configurations:* How does the allocation of blowers and pumps relative to the unloading spots affect demands on raw materials? What might be a better way to distribute the blowers and pumps?
- *Unloading spots allocations:* How does the allocation of the various unloading spots affect demands on raw materials? What might be a better way to distribute the unloading spots? Is the time that a rail car remains at an unloading spot adequate to deliver the necessary amounts of raw materials (keeping in mind that a switching period occurs once every 24 hours)?

- *Product line replacement:* How does the location of the various raw material product lines affect demands on raw materials? What might be a better way to route the product lines?
- *Inventory levels of raw materials:* Are the inventories of raw materials in the rail cars (located on the lead and in the unloading area) adequate to meet the raw material demands? Do the inventories meet the required safety stock of 2 days production?

3.2 Scheduling

P&G requested the development of an optimization code to assist in scheduling the delivery and unloading of raw materials. The program answered the following questions:

- At which spot should a rail car be unloaded?
- How much raw material is to be unloaded (pounds of material and numbers of rail cars)?
- Which blower should be used to unload each rail car?

4 MATHEMATICAL MODEL FOR RAIL CAR UNLOADING

Problem statement: Rail shipments of raw materials arrive on a daily basis. The rail cars pull up to one of four parallel docks (Figure 3) where specially designed blowers and pumps are used for unloading. This equipment is located on the three center strips on either side of the two middle rails. Each blower and pump can unload only a certain subset of raw material.

On each dock there are a fixed number of spots to which the rail cars can be assigned. Each spot can be serviced by one or two blowers/pumps; a blower/pump can be reassigned after it finishes unloading a car but only to a spot to which it can be attached. It may only be possible to assign blower 1, for example, to spots 3 and 6. In a day, up to 22 hours are available for unloading. The problem is to assign raw materials to spots (this is shorthand for saying that

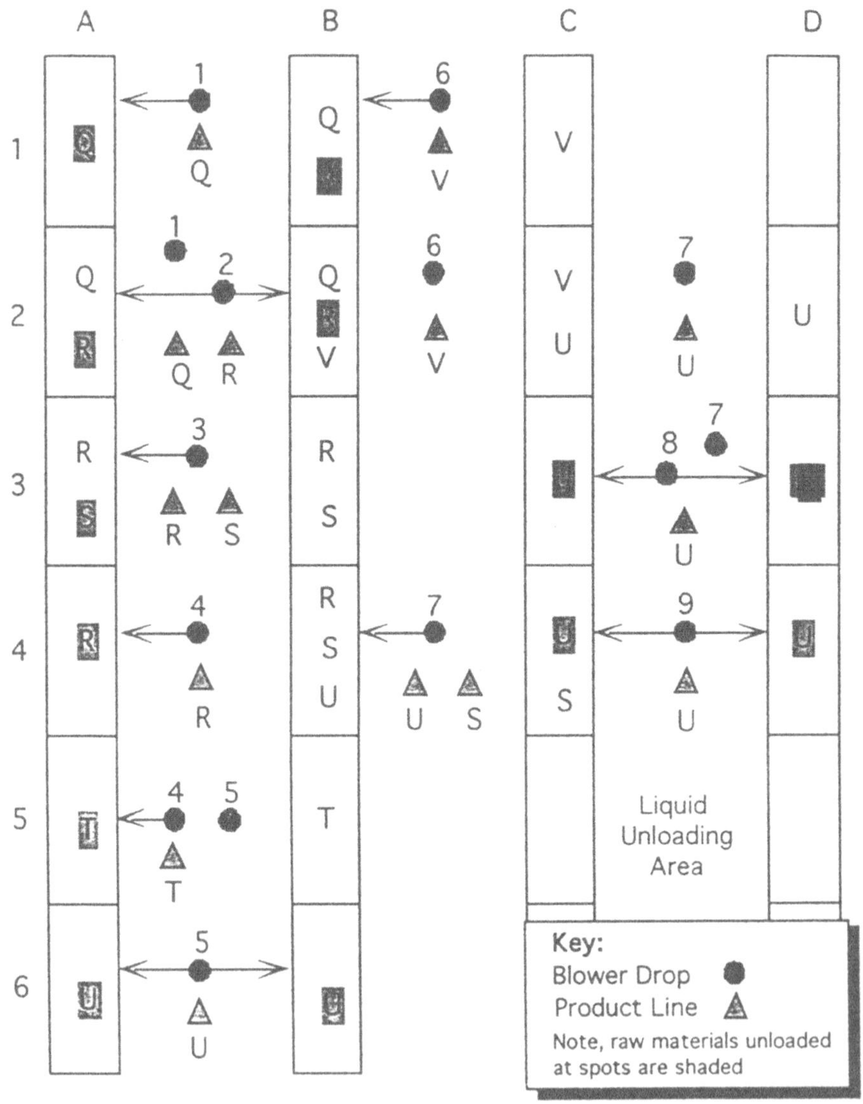

Figure 3 Current unloading area showing maximized configuration

rail cars carrying specific raw materials are assigned to spots), and assign blowers/pumps to spots to unload the raw materials. The objective is to make these assignments so that the daily demand for raw materials is satisfied and that maximum throughput is achieved within the allotted time period.

Indices and sets

i	index for raw material; $i \in I = 1, \cdots, m$
j	index for dock spots; $j \in J = 1, \cdots, n$
k	index for pumps; $k \in K = 1, \cdots, p$
$I(j)$	set of raw material that can be assigned to spot j
$J(i)$	set of spots that are suitable for raw material i (due to the availability of the correct pump or blower)
$J(k)$	set of spots that can be serviced by blower/pump k
$K(i)$	set of blowers/pumps that can unload raw material i
$K(j)$	set of blowers/pumps that can unload spot j

Parameters

c_i	quantity of raw material i in a rail car
d_i	demand for raw material i
t_i	time it takes to unload one rail car of raw material i
n_i	maximum number of spots to which raw material i can be assigned
p_k	maximum number of spots to which blower/pump k can be assigned
p_j	maximum number of pumps that can be assigned to spot j; $p_j \leq 2$

Variables

x_{ij}	(binary) equal to 1 if raw material i is assigned to spot j; 0 otherwise
y_{jk}	(binary) equal to 1 if blower/pump k is assigned to spot j; 0 otherwise
z_{ijk}	fraction of raw material i unloaded at spot j by blower/pump k
τ_{jk}	time spent unloading spot j by blower/pump k

$$\max \quad \Sigma_{i \in I} c_i \Sigma_{j \in J(i)} \Sigma_{k \in K(i)} z_{ijk} \tag{1a}$$

subject to

(spot assignment)
$$\Sigma_{i \in I(j)} x_{ij} = 1 \qquad \forall j \in J \tag{1b}$$

(raw material assignment)	$\Sigma_{j\in J(i)} x_{ij} \leq n_i$	$\forall i \in I$	(1c)
(blower/pump assignment)	$\Sigma_{j\in J(k)} y_{jk} \leq p_k$	$\forall k \in K$	(1d)
(spot assignment)	$\Sigma_{k\in K(j)} y_{jk} \leq p_j$	$\forall j \in J$	(1e)
(feasibility for blower/pump, raw material and spot)	$x_{ij} \leq \Sigma_{k\in K(j)\cap K(i)} y_{jk}$	$\forall i \in I, j \in J(i)$	(1f)
(demand)	$c_i \Sigma_{j\in J(i)} \Sigma_{k\in K(j)} z_{ijk} \geq d_i$	$\forall i \in I$	(1g)
(fraction of raw material)	$\Sigma_{k\in K(j)\cap K(i)} z_{ijk} \leq x_{ij}$	$\forall i \in I, j \in J(i)$	(1h)
(unloading time at j by k)	$\tau_{jk} = \Sigma_{i\in I(j)} t_i z_{ijk}$	$\forall j \in J, k \in K(j)$	(1i)
(total unloading time by k)	$\Sigma_{j\in J(k)} \tau_{jk} y_{jk} \leq 22$	$\forall k \in K$	(1j)
(zero-one variables)	$x_{ij} \in \{0,1\}, y_{jk} \in \{0,1\}$	$\forall i,j,k$	(1k)
(continuous variables)	$0 \leq z_{ijk} \leq 1, 0 \leq \tau_{jk} \leq 22$	$\forall i,j,k$	(1l)

The objective function (1a) sums the quantity of raw material unloaded during the 22 hour period. Constraint (1b) assures that exactly one raw material is assigned to each spot j; its complement (1c) guarantees that raw material i is assigned to at most n_i spots. A lower bound on (1c) can be computed by determining the minimum number of rail cars required to satisfy the demand d_i for raw material i; this value is $[d_i/c_i]$, where $[w]$ is the smallest integer greater than or equal to w.

Constraint (1d) limits to pl_k the number of spots to which blower/pump k can be assigned; (1e) limits to p_j the number of pumps that can be assigned to spot j. Rail cars situated between two docks can be serviced from either side. It may be advantageous to unload a portion of the raw material with one piece of equipment and the remainder with another. Constraint (1f) relates raw material and rail car spots to blower/pump types. It says that if raw material i is assigned to spot j, then a suitable blower/pump must be available. Because the pumps and blowers are in fixed locations and not all equipment can unload all raw materials, it is only necessary for j to range over the set $J(i)$.

The requirement that all demand be satisfied is assured by inequality (1g). It is necessary to introduce the fractional variable z_{ijk} because it may not be possible to unload all the rail cars within the allotted time. If raw material i is assigned to spot j, constraint (1h) puts an upper bound of 1 on the sum of z_{ijk} over the set $K(j)$ of all equipment that can service spot j; alternatively, if raw material i is not assigned to spot j (1h) sets each z_{ijk} to 0. The objective function forces Σz_{ijk} to be as large as possible. Note that if each spot j could only be serviced by one piece of equipment during the unloading period, it would not be necessary to include the subscript k on z_{ijk}. In this case, we would set p_j to 1 and make (1e) an equality. This would guarantee that exactly one blower/pump was assigned to each spot j.

Because pumps can be assigned to more than one spot, it is necessary to keep track of the amount of time that is spent by blower/pump k unloading raw material at spot j. This is the purpose of equation (1i). The next constraint (1j) sums the time blower/pump k is working over all feasible spots j and thus ensures that the total time assigned to blower/pump k is less than or equal to 22 hours. This constitutes that only nonlinearity in the model. We have a continuous variable τ_{jk} multiplied by a binary variable y_{jk}. Note once again that if each spot j could only be serviced by one pump, then the subscript k on τ_{jk} could be eliminated.

Linearization of inequality (1j):

Let $\alpha_{jk} = \tau_{jk} y_{jk} \quad \forall j \in J, k \in K(j)$

Now add the following constraints to the model and delete (1j).

$$\Sigma_{j \in J(k)} \alpha_{jk} \leq 22 \qquad \forall k \in K \tag{2a}$$

$$0 \leq \alpha_{jk} \leq \tau_{jk} \qquad \forall j \in J, k \in K(j) \tag{2b}$$

$$\alpha_{jk} \leq 22 y_{jk} \qquad \forall j \in J, k \in K(j) \tag{2c}$$

$$\alpha \geq \tau_{jk} - 22(1 - y_{jk}) \qquad \forall j \in J, k \in K(j) \tag{2d}$$

In (2), α_{jk} can be thought of as the amount of time blower/pump k spends unloading raw material at spot j. Simple bounds are placed on α_{jk} in (2b). If spot j is not serviced by blower/pump k, then α_{jk} must be 0 as assured by (2c) with $y_{jk} = 0$; in contrast, if blower/pump k services spot j implying that $y_{jk} = 1$, then α_{jk} must equal τ_{jk}. This is guaranteed by (2d) in conjunction with (2b).

5 SOLUTION APPROACH TO UNLOADING PROBLEM

The difficulty in solving the linearized version of model (1) stems from the combinatorial nature of the rail car and unloading equipment assignments. For the data describing the operations of the Alexandria plant, we tried to find solutions using the mixed integer linear codes of both OSL and CPLEX. Lack of success led to the development of a heuristic using the GRASP approach [4].

GRASP was introduced by Feo and Bard [2] for solving airline maintenance base planning and scheduling problems. It combines the power of greedy heuristics, randomization, and local search techniques in two phases: a construction phase and a postprocessing phase. During the construction phase, a feasible solution is built, one element at a time. At each iteration the choice of the next element to be added is determined by ordering a subset of the remaining elements on a candidate list with respect to an appropriate greedy function. This function assesses the benefit of selecting each element. The heuristic is considered adaptive if the benefits associated with every unselected element are updated at each iteration to reflect the changes brought on by the selection of the previous element. The probabilistic component of GRASP is implemented by choosing one element from the candidate list at random.

The most popular way to limit the candidate list is by its cardinality, where say only the top λ elements are included. A different approach is by considering only those elements whose greedy function value is within a fixed percentage ρ of the best move. Sometimes both approaches are employed simultaneously; i.e., only the top λ elements whose greedy function value is within a given percentage ρ of the value of the best move are considered.

The choice of the parameters λ and ρ requires insight into the problem. A balance has to be struck between being too restrictive or too loose. If the criterion used to form the list is too restrictive, only a few candidates will be available. The extreme case is when only one element is allowed. This corresponds to a pure greedy approach, and the same solution will be obtained every time GRASP is executed. The advantage of being restrictive in forming the candidate list is that the greedy objective is not overly compromised. However, the danger of low variability is that the optimum as well as many very good solutions may be overlooked. Lists that are too long, on the other hand, imply a purely random procedure – an approach that is rarely, if ever, effective.

Our implementation of phase I centers on the covering constraint (1g). At each iteration, the assignment of a raw material to a spot and a blower or pump to the raw material is evaluated for feasibility. Assignments that are feasible are ranked according to the amount of unsatisfied demand that they cover. For a particular spot j and blower/pump assignment k, the raw material i with the largest value of c_i/d_i is placed at the top of the candidate list. After a list of size $\lambda = 3$ is constructed, one of the three elements is selected at random and the demand for the corresponding raw material is reduced accordingly. Note that time may not permit the unloading of a full rail car. The procedure continues until all the spots are occupied. If the resulting solution satisfies (1g) we go on to phase 2; if not it is discarded.

Phase 1 is applied N times using different initial seed values to generate a solution (i.e., an assignment of raw materials to spots and blowers/pumps to raw materials) to the problem. Of course, this solution is not necessarily locally optimal with respect to simple neighborhood definitions, even if it is feasible to (1g). Hence a postprocessing phase (phase 2) is applied during which a local search technique is used to improve the current assignments. While such techniques can require exponential running time from an arbitrary starting point, it has been observed that their efficiency significantly improves with the quality of the phase 1 solutions [6]. This effect is conjectured to be a principal factor accounting for the success of GRASP. The result is that many GRASP solutions are generated in the same amount of time required for the local search procedure to converge from a single random start. Furthermore, the best of these GRASP solutions is observed to be significantly better than the solution obtained from a random starting point.

Phase 2 involved two steps. In the first step we tried to remove one rail car and replace it with another. These substitutions continued until no improvement was possible. In the second step, we considered all possible interchanges between rail cars at spots. Proceeding one at a time, these interchanges continued until a local optimum was reached.

During phase 2, the local search was applied to each of the N solutions (N was set to 50) obtained in phase 1. As a consequence, 50 local optima were found (many were duplicates) with the best being kept as the final solution. This approach, where each phase 1 solution is subject to postprocessing, is preferable but only when the computational effort attending the local search is reasonable. For most combinatorial optimization problems it is only practical to apply local search to 1 out of 5 or 10 phase 1 solutions. In the extreme case, phase 2 may be abandoned altogether [3]. The choice of the best strategy depends on a number of factors such as how expensive it is to do the local search and how much improvement results at each iteration.

6 DESIGN METHODOLOGY

The use of a design methodology ensured that the design team remained focused on the task at hand and also provided a structured set of steps for arriving at a solution. This section explains the approach and highlights the results. In all, four alternatives were proposed to accommodate the expected 14% increase in demand over the next few years. A scoring model based on the analytic hierarchy process was used to rank the candidates and make a selection. The performance of each candidate was determined by the GRASP.

Step 1. Analyze current unloading area

The first stage in the design process was to analyze the current unloading area. A throughput study was conducted for this purpose with data obtained from the P&G plant . Factors considered in the study included the raw material demand, rail car capacities, and rail car unloading rates.

Step 2. Determine areas for improvement

After conducting the throughput study of the current unloading configuration, the next step in the design process was to identify the potential areas of improvement. This step included using the GRASP to maximize the raw material output from the current unloading area. Once this was completed, the design team was able to identify the raw materials associated with the bottlenecks in the system.

Step 3. Develop design alternatives

The establishment of areas where the current process could be improved led to four different designs. Each represented various conceptual solutions for improving the current unloading process. An accompanying rationale, details on the advantages and disadvantages of each candidate, and a discussion on how the design improves the current unloading process are included.

Step 4. Evaluate design alternatives

The next step of the design methodology was to determine evaluation criteria on which to rank each design alternative. The design team then evaluated the four designs and determined which one best satisfied the evaluation criteria.

Step 5. Analyze chosen design

After selecting an alternative, the design team analyzed the capabilities of the chosen candidate, analyzed the costs and benefits of implementing the design, and checked the design using the GRASP.

6.1 Analyze Current Unloading Area

The data used in the study consisted of unloading times for the different raw material rail cars, the layout of the unloading area, the number of blowers and pumps, which blowers and pumps can unload which raw materials, silo capacities, and raw material demands from manufacturing. The design team used these data to determine the current requirements of the unloading process. The following assumptions were made.

- Maximum blower operating time is 22 hours per day
- Weekly raw material demands can be divided into a daily demands
- Weekly raw material demands represent typical demands from production
- Silos may be filled to 85% of their maximum capacity
- Average raw material rail car capacities are adequate for calculations

Study findings included:

- *Number of rail cars of each raw material required per day to meet the current demands.* This is probably the most important information obtained. Daily rail car requirements are directly related to meeting production demands.

- *Number of silo volumes of each raw material required per day to meet the current demands.* This represents the flexibility of production in relation to fluctuations of raw material deliveries. A greater demand for silo volumes of material per day means that production will be more sensitive to fluctuations in the delivery of that raw material.
- *Number of blowers and pumps required per day to unload the amount of each raw material required.* This is the ideal minimum number of blowers or pumps required per day by a specific raw material to meet daily demands.
- *Minimum number of each raw material rail car that should be on spots, stored on lead, or in the local rail yard to maintain raw material inventory levels above the required safety stock.* This refers to the number of rail cars available to P&G at the beginning of each day. These results are based on full silos, production demand in silos per day, and rail car capacities.

Table 1 shows the study results for the number of rail cars required per day to meet current demand. Table 2 indicates the corresponding number of blowers and pumps required.

Table 1 Number of rail cars required per day

Material		Rail cars required per day Maximum raw material use	Rail cars required per day Average raw material use
Powders			
AC Base	(RM Q)	0.39	0.35
Fine soda ash	(RM R)	2.58	2.37
NOBS	(RM S)	1.86	1.72
PB-1	(RM T)	0.45	0.32
Soda ash	(RM U)	5.02	4.52
Sulfate	(RM V)	0.32	0.29
Liquids			
Caustic		1.62	1.50
Oleum		1.06	0.98
POLY		0.67	0.59
PEG		0.33	0.29

Table 2 Number of blowers/pumps required per day

Material	Blowers/pumps required per day Maximum raw material use	Blowers/pumps required per day Average raw material use
Maximum unloading time of rail cars		
Powders		
RM Q	0.25	0.22
RM R	1.41	1.30
RM S	1.52	1.41
RM T	0.21	0.15
RM U	1.83	1.64
RM V	0.17	0.16
Liquids		
Caustic	0.40	0.38
Oleum	0.27	0.25
POLY	0.17	0.15
PEG	0.08	0.07
Average unloading time of rail cars		
Powders		
RM Q	0.25	0.22
RM R	1.41	1.30
RM S	1.52	1.41
RM T	0.21	0.15
RM U	1.83	1.64
RM V	0.17	0.16
Liquids		
Caustic	0.40	0.38
Oleum	0.27	0.25
POLY	0.17	0.15
PEG	0.08	0.07

6.2 Determine Areas for Improvement

In light of the throughput study results, we set out to identify areas where the unloading process could be improved. By running the GRASP to maximize the amount of raw materials capable of being unloaded with the current configuration we were able to determine which raw material was the bottleneck. The results provided the foundation for developing conceptual design alternatives.

Recall that the goal of the maximization was to determine the greatest number of rail cars that can be unloaded while still maintaining the proper proportional amounts demanded by manufacturing. A few additional assumptions based on information obtained from analyzing the current unloading process were necessary to round out the model.

- Rail car switching can occur only during the switching period.
- Raw material demands are considered on a one day cycle.
- Average unloading times of rail cars are used for the different raw materials.
- Liquid raw materials are neglected due to specialized unloading spots. High volume materials have their own dedicated unloading area and do not interfere with powders.

With these assumptions in mind, the design team proceeded to determine the maximum number of rail cars that could be unloaded per day given the current configuration. The results are shown in Table 3. The bottleneck constraint was found by dividing the maximum amount of each raw material the process is capable of unloading per day by the current maximum daily production demand for this raw material [8]. Table 3 highlights the results. As can be seen RM S is the bottleneck, permitting only an 8% increase over current maximum production. The next constraint is RM R, capable of handling a 16% increase in production. By removing the RM S constraint, the unloading process can handle an estimated 14% increase in production. Therefore, the design team focused its efforts on removing the RM S constraint in the next phase of the study.

Table 3 Current configuration statistics

Material	Max. no. of rail cars	Maximum capability per day / Current maximum demand
RM Q	1	2.54
RM R	3	1.16
RM S	2	1.08
RM T	1	2.21
RM U	6	1.19
RM V	1	3.14

6.3 Develop Design Alternatives

To eliminate the first bottleneck, three RM S rail cars must be unloaded per day. The design team proposed four conceptual alternatives that complied with the following restrictions:

- maintain the 4 track unloading area layout
- allocate no more than three rail cars per unloading spot
- unload the same number of rail cars as are unloaded in the maximized configuration
- avoid the construction of new unloading spots
- restrict blower operating time to 22 hours or less (except for RMs Q and V)
- avoid the need for additional blowers (use the 9 blowers available)

In the past, when increased production placed greater demands on raw materials unloading, P&G installed new product lines and built new unloading spots. Given the current maximized configuration (see Figure 3), however, we discovered that by simply allocating new unloading spots and installing new product lines or blower drops another rail car of RM S could be unloaded. P&G has also considered adding a fifth track to the unloading area but felt that this would be a costly venture and should be avoided if possible.

The allocation of new unloading spots involves taking an existing unloading spot and upgrading its capability to unload an additional raw material. Ideally, the newly allocated rail car will have a product line located next to it to accommodate the specific raw material. However, to unload three rail cars of RM S, all of the design alternatives required the installation of at least one additional RM S product line. This involves the construction of 200 to 500 feet of pipe running from an unloading spot to a silo. Installation of a new blower drop involves the construction of a short length of pipe connecting an existing blower line to an unloading spot. The cost of installation for the chosen design is discussed in [1].

Design Alternative #1: This design allocates a new RM S rail car at spot B5. Blower 6, which is currently used to unload RM V, has a new drop installed between spots A5 and B5 to accommodate RM S. Blower 1 is then used to unload both RM Q and RM V at spots A1 and B1, respectively. Also required is the installation of a new RM V product line between spots A1 and B1 and a new RM S product line between spots A5 and B5.

Table 4 summarizes the changes associated with alternative #1. The disadvantage of this design is that it calls for the installation of two product lines and one blower drop.

Table 4 Implications for design alternative #1

Changes	Results
Allocate new RM S spot at B5	Allows 3 RM S rail cars to be unloaded
Install new RM S product line between spots A5 and B5	New line accommodates the newly allocated RM S rail car at B5
Install new RM V product line between spots A1 and B1	New line accommodates the RM V rail car at B1
Install new blower drop (blower 6) between spots A5 and B5	New drop accommodates the newly allocated RM S rail car at B5

Design Alternative #2: This design allocates a new RM S rail car at spot B5, and a new RM S product line is installed between spots A5 and B5. Output of the GRASP indicates that blower 5 is then used to unload RM S and RM U at spots B5 and A6, respectively, and blower 7 is used to unload RM U at spot D2. The disadvantage of this design centers on the 22 hour operating time associated with blowers 5 and 7. This leaves no room for error. Table 5 provides a design summary.

Table 5 Implications for design alternative #2

Changes	Results
Allocate new RM S spot at B5	Allows 3 RM S rail cars to be unloaded
Install new RM S product line between spots A5 and B5	New line accommodates the newly allocated RM S rail car at B5

Design Alternative #3: This design allocates a new RM S rail car at spot B5 and a new RM U unloading spot at B3. A new RM S product line is installed between spots A5 and B5 and a new RM U product line is installed between spots A3 and B3. The GRASP determines that blower 5 is then used to unload both RM S and RM U at spots B5 and A6, respectively. The liability of this design is the need to install two product lines, and as with alternative #2, the 22 hour operating time for blowers 5 and 7 does not allow for any mishaps or delays. Table 6 presents the results.

Table 6 Implications for design alternative #3

Changes	Results
Allocate new RM S spot at B5	Allows 3 RM S rail cars to be unloaded
Allocate new RM U spot at B3	Blower 5 used to unload RM U and RM S
Install new RM S product line between spots A5 and B5	New line accommodates the newly allocated RM S rail car at B5
Install new RM U product line between spots A3 and B3.	New line accommodates the newly allocated RM U rail car at B3

Design Alternative #4: This design does not require a new spot allocation. The RM S rail car, currently assigned to spot C4, is accommodated by installing a new RM S product line between spots C4 and D4. Blower 9 is then used to unload RM S and RM U at spots C4 and D4, respectively, and blower 7 is used to unload RM S and RM U at spots B4 and D2, respectively, as indicated by the GRASP. The disadvantages of this design is the installation of one product line and the 22 hour operating time required for blowers 7 and 9. A summary of the changes is provided in Table 7.

Table 7 Implications for design alternative #4

Changes	Results
Install new RM S product line between spots C4 and D4	New line accommodates the currently assigned RM S rail car at C4

Each design meets the goal of unloading three RM S rail cars while satisfying the demand for the five other raw materials. This removes the RM S constraint and allows for a 14% increase in production. Because none of the designs calls for additional tracks or blowers the costs of each is of the same order of magnitude. Table 8 summarizes the advantages and disadvantages of the four candidates.

Table 8 Advantages and disadvantages of four design alternatives

	Advantages	Disadvantages
Alternative #1	Unloading of three RM S rail cars All blowers under 22 hours	Installation of two product lines Installation of one blower drop
Alternative #2	Unloading of three RM S rail cars	Installation of one product line Two blowers each operating 22 hr
Alternative #3	Unloading of three RM S rail cars	Installation of two product lines Two blowers each operating 22 hr
Alternative #4	Unloading of three RM S rail cars	Installation of one product line Two blowers each operating 22 hr

6.4 Evaluation of Design Alternatives

The following criteria were used to evaluate each design alternative. They were arrived at after lengthy conversations with P&G facility planners and operations managers.

Criterion 1. Complexity

Complexity concerns the need for additional unloading spots, product lines, and blower drops. One design is considered more complex than another if it entails the allocation of more unloading spots and the installation of more product lines and blower drops. Greater complexity results in a lower ranking. This criterion was seen to be the most important because of its implications with respect to equipment, construction, and labor costs.

Criterion 2. Blower slack time

Slack time concerns the idle time of a blower after it has finished unloading its assigned rail cars. One design is considered better than another if there is sufficient slack time for each blower, thereby allowing for some mishaps or troubles in the unloading process. A blower scheduled to operate for 18 hours a day has a greater chance of completely unloading its assigned rail cars than a blower scheduled to operate for 22 hours a day.

Criterion 3. Interference with liquids

Liquid rail cars are allocated to unloading spots along tracks C and D. A design is considered better than another if there is less interference with the unloading of liquids. Interference is determined by summing the number of powder raw material rail cars unloaded on tracks C and D and by summing blower operating times.

Criterion 4. Flexibility

Flexibility concerns the choices of unloading spots available that are suitable for a particular rail car. This is important when scheduling the placement of rail cars because it provides the scheduler with a greater number of choices. A design is considered more flexible than another if it provides more spot options for raw material rail cars. For example, in design alternative #1, RM S may be unloaded by blower 3 at either spot A3 or B3. This counts for one spot selection.

Once the evaluation criteria were established, the design team determined the weight of each criterion with the help of the analytic hierarchy process [5]. The results are shown in Figure 4. The weights enabled the design team to evaluate each alternative based on the relative importance of each criterion.

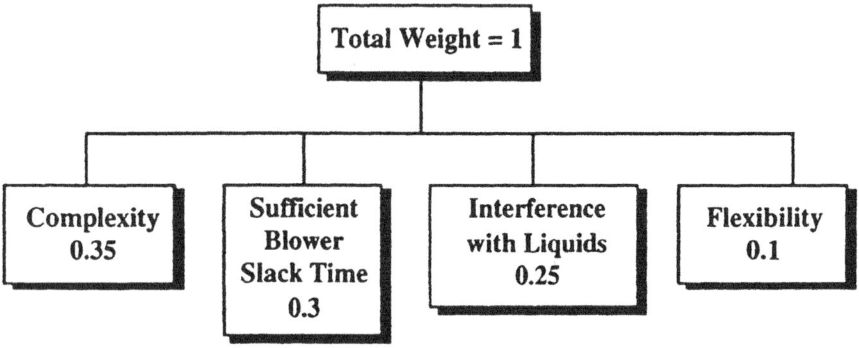

Figure 4 Evaluation criteria weights

Next, each candidate was scored on a 1 to 10 scale depending on how closely they satisfied the respective criteria. Table 9 highlights the results and accompanying rationale.

Table 9 Scoring of design alternatives

	Complexity	Blower slack time	Interference with liquids	Flexibility
	4	10	10	8
Alternative #1	*Reason:* 1 new allocation 2 product lines 1 blower drop	*Reason:* All blowers have at least 2 hours slack time	*Reason:* 4 rail cars along C and D 28 hours blower time	*Reason:* 5 spot selections
	8	8	8	10
Alternative #2	*Reason:* 1 new allocation 1 product line	*Reason:* No slack time for 2 blowers	*Reason:* 5 rail cars along C and D 35 hours blower time	*Reason:* 6 spot selections
	4	8	10	10
Alternative #3	*Reason:* 2 new allocations 2 product lines	*Reason:* No slack time for 2 blowers	*Reason:* 4 rail cars along C and D 28 hours blower time	*Reason:* 6 spot selections
	10	8	6	8
Alternative #4	*Reason:* 1 product line	*Reason:* No slack time for 2 blowers	*Reason:* 5 rail cars along C and D 43 hours blower time	*Reason:* 5 spot selections

The weights and final scores obtained from the analysis are displayed in the decision matrix in Table 10. The total for each alternative is computed by multiplying each criterion score by the respective criterion weight and summing. From the results in the decision matrix, the design team selected alternative #2, which received 8.2 points. The latter was chosen over #4 because it allocates the new RM S unloading spot to a location away from the liquids. The proximity of the new spot to the liquid unloading area required by alternative #4 could lead to conflicts in scheduling. Further analysis, including the determination of implementation costs and capabilities (using the GRASP), confirmed the attractiveness of alternative #2.

Table 10 Decision matrix for evaluation

Design Alternative	Complexity	Blower slack time	Interference with liquids	Flexibility	Total
weight	0.35	0.3	0.25	0.1	1
#1	1.4	3.0	2.5	0.8	7.5
#2	2.8	2.4	2.0	1.0	8.2
#3	2.1	2.4	2.5	1.0	8.0
#4	3.5	2.4	1.5	0.8	8.2

6.5 Recommended Alternative

The differences between alternative #2 and the current configuration are summarized in Table 11. The shaded entries identify the changes. Before a final decision could be made, it was felt that some of the deterministic assumptions in the model should be relaxed and that alternative #2 be evaluated in more detail. This led to a worst case analysis which allowed for fluctuations in daily demand, variability in unloading times, and random equipment breakdowns (see [2] and [3] for the related modification in the code). A final run of the GRASP provided the results shown in Figure 5. Now, a 16% increase in production over current output is possible with RM R being the bottleneck. This exceeds the target 14% increase. Also, blowers 5 and 7 become the critical resource operating for the full 22 hours.

Table 11 Blower and raw material assignments for design alternative #2

Blower #	Raw material rail car assignments	Blower operating time (hours)
1	RM Q (spot A1)	13
2	RM R (spot A2) RM R (spot B2)	20
3	RM S (spot A3)	15
4	RM R (spot A4) RM T (spot A5)	19
5	RM S (spot B5) RM U (spot A6)	22
6	RM V (spot B1)	11
7	RM S (spot B4) RM U (spot D2)	22
8	RM U (spot C3) RM U (spot D3)	14
9	RM U (spot C4) RM U (spot D4)	14

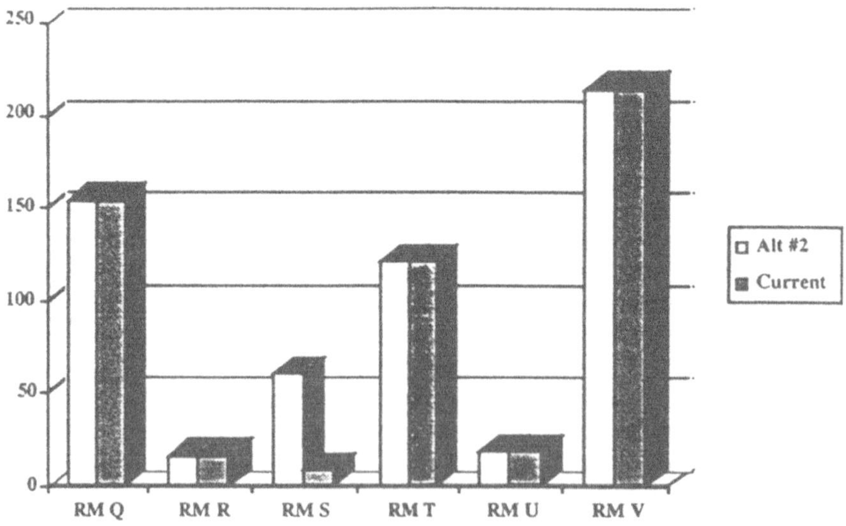

Figure 5 Percent increase in production allowed by design alternative #2

7 SUMMARY AND CONCLUSIONS

The P&G plant in Alexandria, Louisiana receives daily rail car shipments of raw materials used for the production of dry laundry detergent. These rail cars are unloaded on spots in the unloading area. Within the last year interference problems between the scheduling and delivery of raw materials have become more pronounced.

Although the unloading process meets current raw material demands, P&G was concerned that if production increases, this process may become the constraint of the manufacturing system. Consequently, they asked a design team from The University of Texas at Austin to study the unloading area and to develop conceptual designs that allow the unloading process to accommodate a possible 14% increase over current maximum production. The analytic component of the study centered on the development of a GRASP to assign rail cars to spots and blowers to rail cars.

Four alternatives where proposed and evaluated with the help of the GRASP. The results indicated that the current configuration was limited to an 8% increase in production but that a 14% increase could be achieved by adding a single product line. The corresponding costs were minor.

Although a conceptual redesign of the unloading area was completed successfully, it still might be advantageous to perform some additional work to further validate the results. Under certain conditions, for example, alternative #4 which received a close score in the evaluation, might prove highly competitive. Moreover, by removing the assumptions listed below, more insight could be gained.

- *Raw material demands are considered on a one day cycle*

  The elimination of this assumption leads to an analysis that takes into account the excess material unloaded and also considers multiple day cycles. With this in mind, material in the silos will have to be considered as well.

- *Omission of liquid materials in the analysis*

  Although the design team realized that liquid unloading does not interfere with powders, scheduling the two occasionally presents a problem that might get worse as demand increases.

- *Use of average unloading times* Not enough statistics exist on the variability of unloading times. If the range is small, the results should be good;

wide variability, though, could undermine the conclusions of the study. Better data collection would be useful for future analyses.

Incorporating the changes implied by the elimination of these assumptions in the mathematical model (1) while preserving its linearity may be problematic. The advantage of the GRASP is that the modifications would be straightforward. In any case, it would be an easy matter to refine the model to accommodate changes in the unloading area configuration, variations in the product mix, and the addition of product lines and blower drops.

REFERENCES

[1] B. Coates, M. Hutchinson, B. Y. Ko, and S. Melito, 1995, Conceptual design and a mathematical model of a raw materials unloading area, Department of Mechanical Engineering, University of Texas, Austin, Texas.

[2] T. A. Feo and J. F. Bard, 1989, Flight scheduling and maintenance base planning, *Management Science*, 35, 1415-1432.

[3] T. A. Feo, J. F. Bard and S. D. Holland, 1995, Facility wide scheduling of printed wiring board assembly, *Operations Research*, 43, 219-230.

[4] T. A. Feo and M. G. C. Resende, 1995, Greedy randomized adaptive search procedures, *Journal of Global Optimization*, 6, 109-133.

[5] B. L. Golden, W. A. Wasil and P. T. Harker (eds.), 1989, *The Analytic Hierarchy Process*, Springer-Verlag, Berlin.

[6] G. Kontoravdis and J. F. Bard, 1995, A GRASP for the vehicle routing problem with time windows, *ORSA Journal on Computing*, 7,10-23.

[7] M. Kraus, 1980, *Pneumatic conveying of bulk materials, second edition*, McGraw-Hill, New York.

[8] F. E. Meyers, 1993, *Plant layout and material handling*, Prentice Hall, Englewood Cliffs, New Jersey.

14

A STUDY OF THE SPACE SHUTTLE SCHEDULING PROBLEM

Shaukat A. Brah and John L. Hunsucker*

Department of Decision Sciences
National University of Singapore
119260
SINGAPORE

**Department of Industrial Engineering*
University of Houston
Houston, Texas 77204
USA

1 INTRODUCTION

The Space Shuttle Program of the National Aeronautics and Space Administration (NASA) is an extensive strategic and scientific research and development undertaking, with aspirations of flying routinely in space. This program will provide support to various activities such as the installation and manning of the space station, missions to Mars and the Lunar base. It is unique in the sense that most of its components are reusable and there is no precedence for such activity in space. The program continues to grow, in size and complexity, thus increases the need for better planning and scheduling capabilities. The use of such tools will assist in making engineering and management decisions while easing pressures on the whole system.

In this paper, we investigate the space shuttle scheduling problem facing NASA, and propose some solution methodologies. The paper presents a mixed integer programming (MIP) formulation of the scheduling problem, and the results of a simulation study. The mathematical formulation, even though combinatorially explosive, can be used in such capital intensive situations for smaller sized problems. The use of a carefully designed mathematical formulation can help in finding optimal or near optimal solutions in real time and understanding the structure of the problem. The use of heuristics for solving the space shuttle scheduling problem will naturally result in a better understanding of the prob-

G. Yu (ed.), Industrial Applications of Combinatorial Optimization, 301–315.

lem, and allow for a rough-cut analyses which facilitates proper planning of the resources.

The application of this study is not limited to space shuttle scheduling. There are numerous processing facilities, where the setup resembles that of a flow shop with multiple processors, the problem would be less restrictive and hence easier to solve. Brah (1988) presents a mathematical formulation of such problems for optimizing most criteria. Brah and Hunsucker (1991) developed a branch and bound algorithm to solve the makespan minimization problem, while Kochlar and Morris (1988) and Wittrock (1985, 1988) have developed heuristic algorithms.

2 PROCESSING OF THE SPACE SHUTTLE

The Space Shuttle advance through three specific facilities, Orbiter Processing Facility (OPF), Vertical Assembly Building (VAB), and Launch Pad (LP), in this order before being launched into space from the LP at Kennedy Space Center (KSC). Thereafter, it lands at Edwards Air Force Base in California and subsequently it is transported back to KSC on a Shuttle Carrier Aircraft (SCA). The OPF is where basic processing is done on the Orbiter. The VAB is where the Orbiter is mated with its Solid Rocket Boosters and External Tank. The LP is the launch site where most payloads are mounted, and the propellant is loaded. At the VAB, the Orbiter is mounted on the Mobile Launch Platform (MLP), which stays with the Shuttle until it is launched into space. At that time the MLP is brought back and must be processed before it can be made available for the next mission.

The space shuttle goes through a similar processing sequence each time it is preparing to fly a mission in space, making this a flow shop scheduling problem. However, there are multiple processing facilities at the OPF, VAB and LP. The orbiter can practically use any one of them for the processing, which makes this a flow shop with multiple processors scheduling problem. The limitation in this instance, however, is on the number of space shuttles available in the system. In addition, only one space shuttle is allowed in space at any one time, and there is one SCA to bring the shuttle back to KSC after the landing in California. The objective function can be that of minimizing the maximum completion time to finish processing the missions on all processing stages (makespan), average time to complete all missions (mean completion time), or another measure of performance. More detailed work could involve the optimization of multiple objectives or goals.

The space shuttle scheduling problem can also be considered as a restricted flow shop with multiple processors, the restriction being on the number of space shuttles. Furthermore, Brah and Hunsucker (1987) report that the processing times of different types of missions, or missions, such as Standard, Department of Defense, Space Lab and Mix, are proportional. Simplifying that, if one mission type has a larger processing time in one of its facilities, in most conditions, it has a larger processing time in all other facilities as well, except for In-Flight and SCA transportation time.

Fisher and Jaikumar (1978) studied the space shuttle scheduling problem and found that in developing a workable schedule for space shuttle processing, one of the resources was a bottle neck facility. Although, the single machine solution presented by them provides a starting point for the scheduling problem, nevertheless, as the flight rate increases and the processing times decrease, more bottle neck facilities are likely to emerge. The requirement exists for finding a schedule which is feasible and optimal or close to optimal, while considering most of the dominant facilities, such as OPF, VAB and LP, in space shuttle processing. Such a schedule should satisfy the physical constraints of the system as well.

3 PROBLEM DESCRIPTION

The predominant problem of space shuttle processing can be presented graphically as in Figure 1. There is a main queue of incoming missions, and each available mission can be assigned to a space shuttle , which can go to any one of the two OPF's at stage 1, two VAB's at stage 2, two LP's at stage 3. However, only one flight is allowed in space at any one time at stage 4, and there is only one SCA available to bring back the shuttle from the landing site in California to KSC at the last stage. Naturally, for a large problem like the one discussed here, there exist a multitude of other elements of the system which go through different conditions of processing as well. The elements like Solid Rocket Booster, Solid Rocket Motor, Flight Crew, Flight Crew Simulator, and others, are some examples. However, they are mostly processed in parallel, meaning they are not expected to effect the work performed in series at the above central facilities. Further, it is assumed that enough inventory and processing capabilities exist to feed the serial line and therefore such components

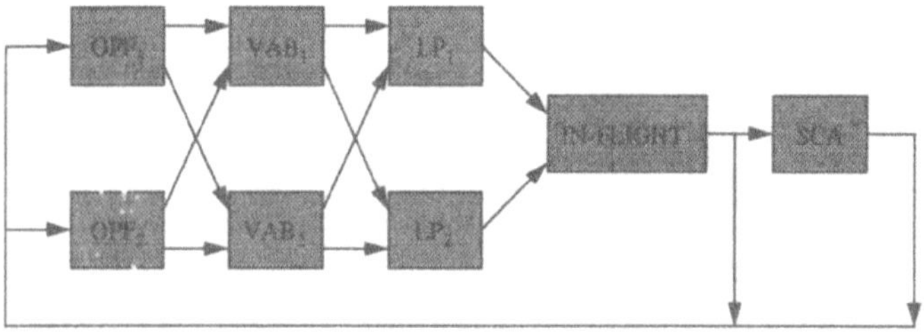

OPF	Orbiter Processing Facility
VAB	Vertical Assembly Building
LP	Launch Pad
In-Flight	Flight of Shuttle in Space
SCA	Shuttle Carrier Aircraft

Figure 1 Schematic view of space shuttle processing

would not create a bottleneck in the planning process. As can be observed, there is a queue at each stage of processing, and hypothetically all missions can be routed to any one of the parallel processing facilities at a stage. When the mission has been processed through the last stage, which is the transportation of the space shuttle to KSC (this step is not necessary if the space shuttle lands at KSC), it is completed and at that point can leave the system.

4 MATHEMATICAL FORMULATION

Table 1 presents the mathematical formulation of the space shuttle scheduling problem making usual assumptions to make the problem tractable. The objective function in this formulation is the minimization of makespan or mean completion time or maximum lateness or mean lateness as shown in (1a), (1b), (1c) and (1d) respectively of Table 1. Slight variations can optimize other criteria such as weighted mean tardiness, earliness, or absolute lateness. Moreover, some or all of these constraints can be used for optimizing, the weighted multiple objective function, or for a goal programming formulation.

The formulation of Table 1 assigns missions to individual processing facilities at each stage. In order to guarantee the assignment of a mission to only one processor at each stage, Equation 2 must hold for any feasible schedule. Similarly, simultaneous processing of the same mission on any two facilities is not

Table 1 Mathematical model of the space shuttle scheduling problem

Minimize Z, Subject to

(Makespan)	$Z \geq C_i$	for all i	(1a)
(Mean completion time)	$Z \geq \Sigma C_i/n$		(1b)
(Maximum lateness)	$Z \geq C_i - d_i$	for all i	(1c)
(Mean lateness)	$Z \geq \Sigma(C_i - d_i)/n$		(1d)
	$\Sigma\, Y_{ijk} = 1$	for all i and j	(2)
	$C_{ij} - C_{i,j-1} \geq \Sigma\, Y_{ijk}\, p_{ijk} + t_{ij}$	for all i and j	(3)
	$Q(2 - Y_{ijk} - Y_{rjk} + X_{irj}) + C_{ij} - C_{rj} \geq p_{ijk}$	for all i, r, j and k	
	$Q(3 - Y_{ijk} - Y_{rjk} - X_{irj}) + C_{rj} - C_{ij} \geq p_{rjk}$	such that i < r	(4)
	$C_{im} - C_{ro} - Q(X_{rim}) = A_{ir}^{+} - A_{ir}^{-}$	for all i and r	(5)
	$\Sigma\, N_{ir} + N_{ri} \leq S - 1$	for all i	(6)
	$A_{ir}^{+} + A_{ri}^{+} - Q(N_{ir}) - Q(N_{ri}) \leq 0$		
	$N_{ir} - A_{ir}^{+} - 0.999 \leq 0$	for all i and r	
	$N_{ri} - A_{ri}^{+} - 0.999 \leq 0$	such that i < r	(7)

where

- p_{ijk} = Processing time of mission i at stage j on facility k, $(1 \leq k \leq M_j)$,
- t_{ij} = Travel time of mission i from stage j-1 to j,
- d_i = Due date for mission i,
- M_j = Number of parallel facilities on stage j,
- C_{ij} = The completion time of mission i at stage j,
- C_i = The completion time of mission i at the last stage,
- $C_{io} \geq r_i$, release time of mission i,
- S = Number of space shuttles,
- Q = A very large number,
- A_{ir}^{+} = The positive part of $C_{im} - C_{ro}$, where $i \neq r$,
- A_{ir}^{-} = The negative part of $C_{im} - C_{ro}$,
- Y_{ijk} = 1 if mission i at stage j is assigned to processor k, 0 otherwise,
- X_{irj} = 1 if i precedes r at stage j on processing facility k, 0 otherwise,
- N_{ir} = 1 when A_{ir}^{+} is greater than zero, 0 otherwise.

possible, or allowed, under the flow shop assumption of the model, the mission must complete processing on one stage before starting on the subsequent stage. However, if such overlap of processing is permitted at any stage, then it can easily be modeled by subtracting the allowable overlay time from the left hand side of Equation 3. In addition, as accounted for in Equation 3, if there is any travel or setup time outside the processing facilities, then it must also elapse before the start of processing at the next level. Next, noninterference is necessary among the missions using a common processing facility at any stage. Meaning, as in Equation set 4, the difference between the processing times of any two missions assigned to the same processing facility must be such that they do not overlap. Furthermore, what differentiates this problem from the general flow shop with multiple processors, is the restriction on the number of available space shuttles in the system. Equations (5), (6) and (7) take care of the above restrictions. The solution of such problems can be obtained by using the standard Linear Programming packages with the option of 0-1 variables.

5 A SIMULATION STUDY

The number of variables and constraints of the mathematical formulation of the space shuttle scheduling problem increases progressively with the size of the problem. The absence of polynomial time algorithms to solve MIP problems, therefore, limits the use of such models. In such situations efficient heuristics, quite often, provide a viable alternative to solve problems in a reasonable amount of computation time.

A representative space shuttle processing model is developed. The computer model is simulated to serve as an experimental vehicle for the analysis of priority rules. It is designed to duplicate the behavior of the system in order to study the performance of priority rules. The model seeks to duplicate the behavior of the system in order to demonstrate the likely effect of various scheduling policies. The effects of priority rules is investigated using regression analysis. In addition, three data sets namely pessimistic, realistic and optimistic processing times are used to evaluate the sensitivity of priority rules to changes in the processing times.

Table 2 lists priority rules investigated in the simulation study of the space shuttle scheduling problem. The basis of the selection of the rules is popularity and performance in similar situations for mean flow time and makespan criteria. The list, however, is not exhaustive, but an obvious set to use. Moreover,

Table 2 Description of priority rules

Rule	*Description*
FIFO	(First In First Out): Select the operation of the mission which was first to enter the queue at that stage,
LIFO	(Last In First Out): Select the operation of the mission which last entered the queue for service,
SPT	(Shortest Processing Time First): Select the operation with the minimum processing time,
LPT	(Largest Processing Time First): Select the operation with the largest processing time,
MTWF	(Most Total Work First): Select the operation associated with the mission having maximum total work,
LTWF	(Least Total Work First): Select the operation associated with the mission having minimum total work,
MWRF	(Most Work Remaining First): Select the operation related with the mission having most work remaining,
LWRF	(Least Work Remaining First): Select the operation related with the mission having least work remaining,
RANDOM	(Random): Select the operation at random.

one can study due-date based rules, to evaluate their performance in a similar situation in order to assist decision makers in satisfying customer related objectives of minimizing mean and maximum tardiness. Nonetheless, this simulation study focuses on resources related criteria of mean flow time and makespan. It is noteworthy that a rule based dispatching is relevant only if there are choices in making a decision. In case of the space program, the planning or scheduling takes place for a set of missions at a time that provide options in making a decision, thus the relevance of the study.

Design of Test Data

The processing times of the missions, as interpreted from the historical data used by Brah and Hunsucker (1987), are generated from the exponential distribution. The OPF, VAB, LP, In-flight, and SCA transportation times are therefore assumed to be following the exponential distribution with means of 40, 5, 17, 7 and 8 days, respectively, for the realistic data set. In addition, two other processing time data sets, optimistic and pessimistic, are studied to observe the sensitivity of the priority rules to the increase or decrease in processing times. The assumption here is that the processing times at each stage are independent of the processing times at other stages. Further, all missions are assumed to be available at the beginning of the scheduling period. A priority or heuristic rule is used to schedule the mission at all stages of processing, namely OPF, VAB and LP, and later In-Flight and thereafter the return journey from Edwards Air Force Base in California to KSC. The analysis for each data set of processing times for a given number of missions or jobs is repeated for all priority rules.

Design of the Experiment

The model is run one hundred times for the processing time data sets and the performance of priority rules is measured for two measures of performance, namely the mean flow time and the makespan. Six sets of missions (5, 10, 15, 20, 30, 50) are studied and each priority rule is tested under each condition. The best rule for the data set under consideration is selected for each criteria and the performance score of the priority rule responsible for obtaining the best solution of each performance measure is incremented by one. In case of ties, the score of all heuristic rules in the tie are incremented. Also, the mean flow time and the makespan values are recorded for the priority rules and averages over all the runs are reported. The procedure is repeated for all the three data sets.

6 FINDINGS OF THE SIMULATION STUDY

The mean flow time and the makespan criteria are studied here for the space shuttle scheduling problem. We considered the number of times the best solution is achieved as one of the indicators of the performance of the rule, and the average value of the measures represent the overall performance. Table 3 presents the results for the average values of the mean flow time over one hundred simulation runs, for all priority rules in the investigation, for the realistic data set, for different numbers of missions or jobs. The numbers in parentheses in the table indicate the frequency of occurrences of the best solution for the mean flow time criteria over one hundred simulation runs. Similarly, Table 4 present the results for the makespan criteria for the realistic data set. The performance of the priority rules for mean flow time and makespan criteria is not observed to be sensitive to any one of the three data sets studied in this research.

Table 3 Simulation study results for the mean flow time criteria

Average values of the mean flow time over one hundred runs									
No. of Missions	*Priority Rule*								
	FIFO	LIFO	SPT	LPT	MTWF	LTWF	MWRF	LWRF	RANDOM
5	104.7 (6)ξ	104.8 (9)	93.9 (71)	129.9 (0)	128.5 (0)	94.2 (63)	128.5 (0)	94.5 (55)	105.7 (7)
10	130.2 (0)	131.2 (0)	105.2 (29)	194.1 (0)	189.0 (0)	104.3 (47)	189.4 (0)	104.7 (42)	129.0 (0)
15	164.1 (0)	165.4 (0)	123.6 (34)	266.3 (0)	254.6 (0)	122.1 (48)	255.9 (0)	125.4 (18)	167.3 (0)
20	190.4 (0)	195.3 (0)	136.6 (50)	338.8 (0)	319.1 (0)	135.9 (38)	321.7 (0)	143.7 (12)	190.6 (0)
30	244.9 (0)	240.4 (0)	163.5 (73)	445.8 (0)	412.6 (0)	167.2 (24)	418.8 (0)	185.5 (3)	244.7 (0)
50	352.2 (0)	348.1 (0)	220.2 (99)	652.0 (0)	582.1 (0)	243.4 (1)	602.4 (0)	279.9 (0)	349.2 (0)
ξ Number in parentheses is number of times each priority rule found best solution.									

Table 4 Simulation study results for the makespan criteria

Average values of the makespan over one hundred runs									
No. of Missions	*Priority Rule*								
	FIFO	LIFO	SPT	LPT	MTWF	LTWF	MWRF	LWRF	RANDOM
5	167.8 (28)ξ	170.0 (27)	172.6 (16)	169.5 (0)	162.7 (39)	175.6 (1)	162.9 (39)	175.5 (2)	166.7 (29)
10	213.3 (19)	217.6 (16)	207.4 (24)	254.6 (2)	232.7 (7)	209.2 (11)	233.1 (9)	210.0 (10)	213.4 (18)
15	265.4 (9)	265.9 (16)	243.6 (36)	359.2 (0)	320.8 (0)	244.6 (24)	321.0 (0)	248.7 (7)	268.3 (8)
20	314.3 (13)	321.3 (6)	282.7 (43)	453.1 (0)	404.6 (0)	284.9 (20)	405.2 (0)	291.7 (12)	313.3 (6)
30	403.2 (5)	397.5 (3)	335.8 (69)	613.9 (0)	536.3 (0)	349.5 (20)	537.4 (0)	368.1 (2)	401.7 (2)
50	591.2 (0)	592.9 (1)	492.9 (81)	936.3 (0)	787.5 (0)	515.6 (17)	792.8 (0)	561.8 (1)	588.0 (0)
ξ Number in parentheses is number of times each priority rule found best solution.									

Table 5 Multiple regression model for the RANDOM strategy

	Mean Flow Time	*Makespan*
Adjusted R^2	0.998	0.999
Constant	80.206	123.268
Jobs	5.425	9.321

Effects of Problem Characteristics

A regression model is developed for the random strategy. The model used number of missions, which is the major determinant of mean flow time and makespan in this study. Table 5 summarizes the regression models for both the mean flow time and makespan. Notice that each model accounted for a very large proportion of the variability in flow times for the two criteria.

Strategy Effects

To determine if a particular strategy resulted in flow times significantly better than or worse than the times predicted by the model, the following approach is used: For the simulation runs for a particular strategy, the actual times are regressed against the times predicted by the regression model for the RANDOM strategy. Assuming that the adjusted R^2 is high, verifying that the model is a good predictor, one could interpret the slopes and intercepts to determine the impact of the strategy. If the slope of the line relating actual times to predicted times is not significantly different from 1.0 and the intercept is not significantly different from 0.0, then the strategy is not significantly different from the RANDOM strategy. A slope less than 1.0 indicated a strategy that is faster than the RANDOM strategy; a slope greater than 1.0 indicated a slower strategy. If the slope is not different from 1.0 but the intercept is non-zero, it indicated a constant amount of time added to or subtracted from the predicted time. A negative intercept indicated a faster process, a positive intercept a slower process. If the slope is significantly different from 1.0 and the intercept is significantly different from 0.0, the intercept had a more complex interpretation, which will be discussed on a case by case basis when the results are shown. If the adjusted R^2 values are low, it would mean that for that particular strategy, the model is simply poor.

In fact, for all strategies the model is very good, with adjusted R^2 values ranging from 0.991 to 0.999 for the mean flow time and from 0.995 to 0.999 for the makespan criterion. Table 6 lists the slopes, intercepts, and adjusted R^2 values for each strategy under both criteria. The strategies are listed in rank order of performance. For the mean flow time criterion, the slope for SPT, LTWF, LWRF and LIFO are significantly less than 1.0. and so are about 48%, 38%, 23% and 1% faster than RANDOM strategy, respectively. However, the intercepts for SPT and LTWF are significantly larger than 0.0. The remaining strategies, namely FIFO, MTWF, MWRF and LPT had slopes that are significantly greater than 1.0 which quantified to be about 2%, 85%, 93%, and 113% greater time than RANDOM strategy, respectively. However, intercepts for MTWF, MWRF and LPT are significantly less than 0.0. Similarly, for the makespan criterion, SPT, LTWF and LWRF are 25%, 19% and 8% faster than RANDOM strategy as compared to MTWF, MWRF and LPT which is found to be 50%, 50% and 83% slower than RANDOM strategy, respectively. The intercepts for SPT and LTWF are significantly larger than 0.0 as compared to the intercepts for MTWF, MWRF and LPT which are significantly less than 0.0. Looking at both criteria, it became apparent that for larger number of missions, SPT, LTWF and LWRF will definitely outperform MTWF, MWRF and LPT and this will not be as clear for smaller number of missions.

Table 6 Actual times regressed on RANDOM stragegy model

Strategy	*Adjusted R^2*	*Slope*	*Intercept*
	Mean Flow Time Criterion		
SPT	0.999	0.519	37.777
LTWF	0.993	0.617	22.525
LWRF	0.991	0.775	2.405
LIFO	0.997	0.990	1.774
FIFO	0.999	1.015	-3.035
MTWF	0.987	1.846	-50.793
MWRF	0.991	1.931	-62.440
LPT	0.993	2.132	-83.705
	Makespan Criterion		
SPT	0.996	0.754	43.911
LTWF	0.997	0.808	33.622
LWRF	0.995	0.923	9.258
LIFO	0.998	1.000	2.234
FIFO	0.999	1.008	-2.079
MTWF	0.995	1.492	-77.778
MWRF	0.995	1.504	-80.377
LPT	0.997	1.826	-129.558

Analysis of Variance

The regression analysis indicated which strategies are significantly faster or slower than RANDOM. However, it did not provide a way of comparing the strategies with each other. For that purpose, a one way analysis of variance is performed. The analysis of variance is done using actual simulated values as the dependent variable and strategy as the independent variable. Least significant difference (LSD) and Scheffe paired comparison tests are performed to compare the strategies. Table 7 summarizes the analysis of variance results.

Table 7 Group means and paired comparisons

Mean Flow Time		*Makespan*	
Strategy	*Group Mean*	*Strategy*	*Group Mean*
SPT	140.500	SPT	289.167
LTWF	144.517	LTWF	296.567
LWRF	155.617	LWRF	309.300
LIFO	197.533	LIFO	327.533
FIFO	197.750	FIFO	325.867
RANDOM	197.750	RANDOM	325.233
MTWF	314.317	MTWF	407.433
MWRF	319.350	MWRF	408.733
LPT	337.817	LPT	464.433
Homogeneous Subsets ($p < 0.05$)			
LSD Test	*Scheffe Test*	*LSD Test*	*Scheffe Test*
Subset 1:		*Subset1:*	
SPT, LTWF, LWRF, LIFO, FIFO, RANDOM	SPT, LTWF, LWRF, LIFO, FIFO, RANDOM, MTWF, MWRF, LPT	SPT, LTWF, LWRF, RANDOM, FIFO, LIFO, MTWF, MWRF, LPT	SPT, LTWF, LWRF, RANDOM, FIFO, LIFO, MTWF, MWRF, LPT
Subset 2:		*Subset 2:*	
LIFO, FIFO, RANDOM, MTWF, MWRF			
Subset 3:		*Subset 3:*	
MTWF, MWRF, LPT			

For both criteria we rejected the null hypothesis that all strategies have the same means. For mean flow time, F Ratio = 2.6929 with F Prob. = 0.0164 and df = 8, 45; for the makespan F Ratio = 0.6695 with F Prob. = 0.7153 and df = 8, 45. For the mean flow time criterion with $p < 0.05$, the least significant differences test has grouped SPT, LTWF, LWRF, LIFO, FIFO and RANDOM in a homogeneous subset. Therefore, these strategies are not significantly different from each other although some of them, namely SPT, LTWF, LWRF and LIFO are significantly faster than RANDOM. Using the least significant differences test, LIFO, FIFO, RANDOM, MTWF and MWRF formed a homogeneous subset. Under the same test, LPT, MTWF and MWRF also formed a subset. These results implied that LPT being the slowest strategy is also significantly different from the rest of the strategies except MTWF and MWRF. Furthermore, MTWF and MWRF are also shown to be significantly different from SPT, LTWF and LWRF. However, the result implied by the scheffe test is that all nine strategies are not significantly different from each other since they are placed in the same homogeneous subset.

Makespan criterion with $p < 0.05$ for both tests also produced similar results as the scheffe test for mean flow time criterion. None of the strategies are found to be significantly different from each other, although SPT, LTWF and LWRF are significantly faster than the remaining strategies. Therefore, all nine strategies are placed in the same homogeneous subset for both the least differences test and scheffe test.

7 SUMMARY AND CONCLUSION

The space shuttle scheduling problem is one of the major issues faced by the National Space Transportation System of NASA. The problems will continue to grow in size and complexity as the program continues to develop. Although the high variability among the processing times of the missions will continue for a long while - which incidentally is typical of most high-tech programs - the need exists for developing an operational framework to make them predictable. Once stability in predicting the processing times is achieved, then mathematical tools can be readily used to optimize the system. The mathematical models however, can also be used in stochastic modeling of the system or by building allowances into the processing times of the deterministic models. Naturally, as experience grows, such allowances in similar deterministic models can be reduced. Although the mathematical model formulation presented is combinatorial in nature, it is useful for capital intensive programs like the space shuttle. On the other hand, the simulation study can be usefully employed as a planning tool for the rough-cut analysis.

The findings of the simulation study suggests that the primary determinant of the two flow time criteria is the number of missions. This is true for either the mean flow time criterion or the makespan criterion. Simple regression models for both criteria accounted for more than 99% of the variability in flow time. The regression models showed that flow time increased in a linear fashion with number of missions. When actual simulated processing times are regressed on the RANDOM strategy model, SPT, LTWF and LWRF are found to be faster than the remaining strategies for both criteria. However, only the analysis of variance using the least significant differences tested for the mean flow time criterion showed strong evidence that SPT, LTWF and LWRF are found to be significantly superior to the other strategies.

Acknowledgements

This research was partially supported by the National Aeronautics and Space Administration (NASA) through Grant No. 9-BC-19-6-1P.

REFERENCES

[1] S. A. Brah and J. L. Hunsucker, 1991, Branch and Bound Method for Flow Shop with Multiple Processors Scheduling, *European Journal of Operational Research*, 51, 88-99.

[2] S. A. Brah, 1988, Scheduling in a Flow Shop with Multiple Processors, Dissertation Abstracts International, 50, 1587B. (University Microfilms No. 89-12667).

[3] S. A. Brah and J. L. Hunsucker, 1987, An Analysis of the Flight Rate Capability of NASA's Space Shuttle Program, *Logistics Spectrum*, 21, 7-12.

[4] M. L. Fisher and R. Jaikumar, 1978, An Algorithm for the Space-Shuttle Scheduling Problem, *Operations Research*, 26, 166-181.

[5] S. Kochlar and R. J. T. Morris, 1987, Heuristic Methods for Flexible Flow Line Scheduling, *Journal of Manufacturing Systems*, 6, 299-314.

[6] R. J. Wittrock, 1985, Scheduling Algorithms for Flexible Flow Lines, *IBM Journal of Research and Development*, 29, 401-412.

[7] R. J. Wittrock, 1988, An Adaptable Scheduling Algorithms for Flexible Flow Lines, *Operations Research*, 32, 445-453.

15

MODELS AND IMPLEMENTATION TECHNIQUES FOR LOCAL ACCESS TELECOMMUNICATION NETWORK DESIGN

Geon Cho and Dong X. Shaw*

Department of Business Administration
Chonnan National University
Kwang Ju 500-757
KOREA

**School of Industrial Engineering*
Purdue University
West Lafayette, IN 47907
USA

ABSTRACT

This paper presents models, algorithms and implementation techniques for Local Access Telecommunication Network (LATN) Design and Expansion problems. In particular, we develop detailed algorithms for implementing the so-called Limited Column Generation approach. We first formulate both problems into a tree-partitioning problem with an exponential number of variables. A dedicated column generation procedure is developed so that the master tree partitioning linear program can be solved in just n pivots, where n is the number of nodes in the network. As only a limited number of columns are generated, we call such a procedure Limited Column Generation. Prior to each pivot, an entering variable is selected by detecting the Locally Most Violated (LMV) reduced profit (reduced cost), which is equivalent to solve a subproblem in pseudo-polynomial time. In particular, for the LATN Design problem, the corresponding subproblem is the Tree Knapsack Problem. As dual variables are updated at each pivot, the reduced profits have to be computed in an on-line fashion. An efficient implementation is developed to execute such a task so that the LATN Design problem can be solved in $O(n^2H)$, and the LATN Expansion problem can be solved in $O(n^2\delta H)$ time, where δ is the largest existing cable capacity and H

G. Yu (ed.), Industrial Applications of Combinatorial Optimization, 316–344.

is the maximum capacity of concentrators. Our computational experiments indicate that our algorithm delivers outstanding performance. For instance, the LATN Design and Expansion problems with 200 nodes can be solved in approximately 90 seconds and 484 seconds, respectively, on a SUN SPARC 1000 workstation.

1 INTRODUCTION

The telecommunication industry is undergoing rapid and fundamental change. Technological innovation creates new market opportunities and growing customer demand for enhanced products and services such as multimedia applications. The legislative regulatory reform is reshaping the structure and business practice of major industry segments. The federal government has challenged the private sector to participate in an effort to modernize the nation's entire information infrastructure by building the "information super-highway." As a result, many major telecommunication service providers are in the process of upgrading and expanding their facilities and services. The planning and design of various types of telecommunication networks plays an important role in this process. It is well known that optimization techniques can be used to improve telecommunication network design. In this paper, we study a portion of the telecommunication system, the Local Access Telecommunication Network (LATN).

Most existing LATNs have a tree structure. Each customer node associates with a demand representing the required number of circuits from that node to the switching center. This demand can be satisfied by either connecting the node through a cable along the unique path to the switching center, or routing it first to a concentrator which compresses the incoming traffic into a higher frequency signal that requires fewer outgoing lines. The objective of the LATN Design and Expansion problem is to make a trade-off between cable expansion and concentrator installation to minimize the total cost.

Recently, many researchers have been working on the LATN Design and Expansion problems, which are NP-complete problems since they include the Knapsack Problem as a special case. A discussion of the modeling issues can be found in Balakrishnan *et al.* (1991), and a Lagrangian relaxation based approach has been developed by Balakrishnan, Magnanti and Wong (1991). A related but different model was formulated by Cook (1990), and a solution procedure and related knapsack-type problem have also been studied by Bienstock (1993a, b).

A so-called Limited Column Generation procedure was developed by Shaw (1994). The LATN Design and Expansion problem is first formulated as a tree–partitioning problem with an exponential number of variables, each of which corresponds to a subtree. Chvatal (1975) and Gavril (1974) have shown that the polyhedron of its linear programming relaxation is integral. Hence, the tree–partitioning problem can be simply reduced to a linear program (with an exponential number of variables, though). Then, a dedicated column generation procedure is developed so that the linear program can be solved in just $n+1$ pivots ($n+1$ is the number of nodes in the tree). Prior to each pivot, the Locally Most Violated (LMV) pivoting rule is applied to select the entering variable.

Our computational experiments indicate that the Limited Column Generation is a very efficient approach for these models. The LATN Design and Expansion problems with 200 nodes can be solved in approximately 90 seconds and 484 seconds, respectively, on a SUN SPARC 1000 workstation. A problem with 41 nodes given by Balakrishnan, Magnanti, and Wong (1995) , which took 879 seconds on a Macintosh II by a Lagrangian relaxation based approach to reach a solution with 7% gap, can be solved by our algorithm optimally in 135 seconds and 306 seconds if the largest concentrator capacities are chosen to be 5,000 and 10,000, respectively.

Our computational results might suggest that the Limited Column Generation is an effective alternative other than the polyhedron/facet-cutting approach, which has proved to be very successful in solving some large-scale combinatorial optimization problems such as Traveling Salesman Problem, (see Padberg and Rinaldi (1986), and Grötschel and Holland (1991) among others), and Network Design Problem with connectivity constraints (see Grötschel, Monma and Stoer (1992)). For the LATN Expansion problem, some valid-inequalities are derived by Balakrishnan, Magnanti, and Wong (1995). However, it is interesting to notice that the complete valid inequality (row-wise) descriptions for the polyhedron of the LATN Design and Expansion problems are unknown, while their complete column-wise descriptions are known in form of tree-partitioning formulation. Hence, for such a problem, a column-wise approach would be viable.

For the LATN Design problem, the LMV reduced profit can be obtained by solving n subproblems, each of which is the Tree Knapsack Problem (TKP) and has been studied by Cho and Shaw (1994), Johnson and Niemi (1983), Lukes (1974), Shaw (1994), and Shaw, Cho and Chang (1995). Johnson and Niemi (1983) solved the TKP by the so–called "left–right" dynamic programming algorithm with a running time of $O(nC^*)$, where C^* is the optimal value of the TKP. Recently, Cho and Shaw (1994) proposed the so-called "depth-first" dynamic programming algorithm with a running time of $O(nH)$ which can be regarded as a refinement of Johnson and Niemi (1983), where H is the maximum concentrator capacity.

A straight-forward approach for combining Limited Column Generation with a low complexity TKP solver such as the "depth-first" dynamic programming algorithm developed by Cho and Shaw (1994) leads to an $O(n^3H)$ algorithm for the LATN Design problem because $n+1$ pivots are required and prior to each pivot, n CTKPs have to be solved, each of which requires $O(nH)$ time. However, as reduced profits are related to the dual variables, which are updated at each iteration, the LMV reduced profits have to be computed in an on-line fashion. One of the main results of this paper is to develop an efficient on-line implementation which recursively computes all the LMV reduced profits in the total of $O(n^2H)$ time for the LATN Design problem. As a result, we obtain an algorithm for the problem with the overall complexity of $O(n^2H)$.

For the LATN Expansion problem, all the LMV reduced profits can be computed by solving n subproblems ETKP, namely the Extended Tree Knapsack Problem with arc- expansion, each of which can be solved by a depth-first dynamic programming procedure similar to the one for the TKP in $O(n\delta H)$ time, where δ is the largest existing cable capacity. Consequently, the LATN Expansion problem can be solved in $O(n^2\delta H)$ time.

This paper is organized as follows. In Section 1 we present the fixed-charge network formulations for the LATN Design and Expansion problems, and then we describe the Limited Column Generation procedure and its implementation in Section 2. In Section 3 we present an on-line recursive procedure for computing all the LMV reduced profits. This is the main result of this paper. The implementation details for the LATN Design problem are presented in Section 4. In Section 5 we discuss the LATN Expansion problem. Finally, we present our computational test results in Section 6.

2 PROBLEM DESCRIPTION

In this section, we describe the LATN Design and Expansion problems in detail. The LATN is a tree rooted at the switch center, and each non-root node is a customer node and is associated with a demand, which represents the number of circuits required from the customer node to the switching center. This demand can be satisfied by either connecting a cable from the customer node to the switching center or routing the circuits to a concentrator located at another customer node which compresses the incoming traffic into a higher frequency signal that requires fewer outgoing cables. A variety of electronic devices can perform traffic compression through frequency division or time division such as concentrators, multiplexers, remote switches and fiber optical terminals. Since these devices perform essentially equivalent functions from our modeling point of view, we collectively refer them as concentrators. In this paper, we assume that the compressed signal requires a dedicated cable point–to–point (home–run) routing to the switching center. The LATN Expansion problem is to make tradeoff between installing concentrators and expanding cable capacities that already exist along the arcs in order to minimize the total cost. The LATN Design problem is its special case, where all of the existing cable capacities along the arcs are equal to zero.

Let $\hat{T} = (V, E)$ be an undirected tree rooted at node 0, and $V = \{0, 1, 2, ..., n\}$. Then, without loss of generality, we assume that all nodes in the rooted tree $\hat{T}$ are labeled by the Depth–First–Search (DFS) order. Then, for any subtree T of $\hat{T}$, the *root* of subtree T, $r(T)$, is the node in T with the smallest label. Let $P[i, j]$ be a sequence of nodes on the unique *path* in tree $\hat{T}$ from node i to node j. We denote p_j^i as the *predecessor* of node j with respect to root i, which is defined as the first node other than node j on the path $P[j, i]$. For convenience, we denote p_j^0 as p_j for $j \in V$. The node subset $T(i) = \{j | i \in P[j, 0]\}$ is called the *complete subtree* rooted at i. As nodes are labeled in DFS order, we define $LAST(i) = \max\{j | j \in T(i)\}$, the last node in the subtree$T(i)$, because $T(i) = \{i, i+1, ..., LAST(i)\}$.

A directed graph $\overrightarrow{T}$ is derived from $\hat{T}$ by defining 2 directed arcs for every edge. By convention, the arcs (p_i, i), i.e., the out-tree arcs, are labeled i, and the arcs (i, p_i), i.e., the in-tree arcs, are labeled $-i$. As $\hat{T}$ has $n + 1$ nodes and n edges, $\overrightarrow{T}$ has $n+1$ nodes and $2n$ arcs. Let A be the $n \times 2n$ node–arc incidence matrix of the directed graph $\overrightarrow{T}$ which excludes the row corresponding to node 0.

We denote d_i as the demand at node i for $i = 1, 2, \cdots, n$. The total investment on the LATN involves two categories of costs: the concentrator installation and cable expansion.

Suppose there are m different capacities for concentrator modules. A concentrator with capacity h^t can be installed at node i with the fixed cost $\hat{F}_i^t$, which represents the concentrator purchase cost and installation costs as well as other infrastructure investment, and there is also a variable concentrator cost $\hat{c}_i$ representing the operating expenses, where $i = 1, 2, \cdots, n$, and $t = 1, 2, \cdots, m$. We assume $h^1 < h^2 < \cdots < h^m$. Several concentrators with different capacities may be installed at one node to compress the incoming signals. Decision must be made concerning the combination of concentrators at every node. The cost for compressing traffic amount h at node i is $\hat{F}_i(h) = \min\{\sum_{t=1}^m \hat{F}_i^t y^t | \sum_{t=1}^m h^t y^t \geq h, y^1, \cdots, y^m \geq 0 \text{ are integers}\} + \hat{c}_i h$, which can be obtained by solving a reversed Knapsack Problem through, say, a branch-and-bound method. Since m is a very small number around 5 in practice, these costs can be easily computed and treated as a part of preprocessing. Hence, to simplify the model and to discuss our problem efficiently, we regard the concentrator as an "aggregate concentrator" and assume that at most one concentrator can be installed at each node. More general models are discussed in Shaw (1995).

The existing cable capacity on arc i is b_i for all $i = \pm 1, \pm 2, \cdots, \pm n$, and $b_i = b_{-i}$, as all cables are two-way. If the cable requirement on arc i exceeds the existing capacity b_i, a fixed cable expansion cost F_i as well as a marginal variable cable expansion cost of c_i occurs, where $i = \pm 1, \pm 2, \cdots, \pm n$. The fixed cost represents the expenses for digging trenches and the cable purchasing and laying pipes on arc i, while the variable cost represents the cable purchasing and maintenance cost on arc i.

Now we introduce the decision variables as follows. Let $\hat{x}_i$ be the load of the concentrator at node i, $i = 0, 1, 2, \cdots, n$ and x_i be the cable requirement on arc i, $i = \pm 1, \pm 2, \cdots, \pm n$. Then we set

$$x_i = x_i' + x_i'',$$

where $0 \leq x_i' \leq b_i$, and x_i'' is the amount of the cable expansion beyond existing capacity, for $i = \pm 1, \pm 2, \cdots, \pm n$.

We also define binary variables as follows:

$$y_i' = \begin{cases} 1 & \text{if } x_i' > 0 \\ 0 & \text{otherwise}, \quad i = \pm 1, \pm 2, \cdots, \pm n \,, \end{cases}$$

$$y_i'' = \begin{cases} 1 & \text{if } x_i'' > 0 \\ 0 & \text{otherwise}, \quad i = \pm 1, \pm 2, \cdots, \pm n \,, \end{cases}$$

and

$$\hat{y}_i^t = \begin{cases} 1 & \text{if } h^t \text{ is used at node } i \\ 0 & \text{otherwise}, \quad i = 0, 1, 2, \cdots, n \text{ and } t = 1, 2, \cdots, m. \end{cases}$$

In practice, to reduce the complexity of network planning, management and maintenance, some restrictions are imposed on the routing patterns by planners. To simplify the problem, we also make additional assumptions. A discussion of these assumptions can be found in Balakrishnan *et al.* (1991a) and Balakrishnan *et al.* (1991b). These restrictions can be summarized as follows:

1. Only one-level traffic compression is allowed (i.e., all demands can be compressed at most once before reaching the switching center).
2. The compressed signals are point-to-point routed to the switching center through a dedicated cable (for example, fiber optics).
3. The *non-bifurcated routing*, that is, all circuits from one customer node must follow the same routing pattern.
4. The *contiguity restriction*, that is, if node j's traffic is compressed by a concentrator located at node i, then traffic from all nodes on path $P[i,j]$ is compressed by the same concentrator located at node i. In particular, if a concentrator is installed at node i, then the demand at node i should be routed to that concentrator.

Based on the above discussion, the LATN Expansion problem can be formulated as the following special fixed–charge network flow problem (E):

$$(E) \quad \min \quad c^T x'' + F^T y'' + \hat{c}^T \hat{x} + \sum_{t=1}^{m} (\hat{F}^t)^T \hat{y}^t \tag{15.1}$$

$$\text{s.t.} \quad A(x' + x'') + \hat{x} = d \tag{15.2}$$

$$0 \leq x' \leq b * y' \tag{15.3}$$

$$0 \leq x'' \leq My'' \tag{15.4}$$

$$0 \leq \hat{x} \leq \sum_{t=1}^{m} h^t \hat{y}^t \tag{15.5}$$

$$y'_i + \sum_{t=1}^{m} \hat{y}_i^t + \sum_{j \in \{j | p_j = i\}} y'_{-j} = 1, \quad i = 1, 2, \cdots, n \tag{15.6}$$

$$y' \geq y'' \tag{15.7}$$

$$x', x'' \in R^{2n}, \hat{x} \in R^n, y', y'' \in \{0,1\}^{2n}, \hat{y}^t \in \{0,1\}^n \tag{15.8}$$

where $M = \sum_{k=1}^{n} d_k$ and "*" denotes the component by component product.

The objective function (15.1) seeks to minimize the sum of the cable and the concentrator installation costs. Constraints (15.2) specify the flow conservation constraints, that is, the amount of traffic coming into node i minus the amount of traffic going out from node i plus the load of concentrator located at node i should be exactly equal to the demand d_i at node i. Constraints (15.3) and (15.4) are simply the upper bounds for the existing cable capacity and for the cable expansion requirement, respectively. Constraints (15.6) imply $\sum_{t=1}^{m} \hat{y}_i^t \leq 1$, which means that only one type of concentrator modules can be used at each node. Constraints (15.6) and (15.7) together state that the demand at node i is satisfied from either one of the incoming arcs or a concentrator installed at that node, which reflect the non-bifurcated routing and contiguity assumptions. Constraints (15.7) simply state that if the existing cable capacity on arc i is not used, then cable expansion should not occur on arc i.

In the LATN Design problem, the existing cable capacity b_i on arc i is zero for all $i = \pm 1, \pm 2, \cdots, \pm n$. Therefore, we eliminate constraints (15.3) and (15.7) in (E), and set $x' = 0$, $x = x''$, $y' = 0$ and $y = y''$. Then we have $y_i + \sum_{t=1}^{m} \hat{y}_i^t + \sum_{j \in \{j | p_j = i\}} y_{-j} = 1$. The following is the fixed-charge network flow formulation for the LATN Design problem (D);

$$
\begin{aligned}
(D) \qquad & \min \quad c^T x + F^T y + \hat{c}^T \hat{x} + \sum_{t=1}^{m} (\hat{F}^t)^T \hat{y}^t \\
& \text{s.t.} \quad Ax + \hat{x} = d \\
& \qquad\quad 0 \leq x \leq My \\
& \qquad\quad 0 \leq \hat{x} \leq \sum_{t=1}^{m} h^t \hat{y}^t \\
& \qquad\quad y_i + \sum_{t=1}^{m} \hat{y}_i^t + \sum_{j \in \{j | p_j = i\}} y_{-j} = 1, \quad i = 1, 2, \cdots, n \\
& \qquad\quad y \in \{0,1\}^{2n}, \ \hat{y}^t \in \{0,1\}^n
\end{aligned}
$$

where $M = \sum_{k=1}^{n} d_k$.

In the next section, we introduce a tree partitioning formulation and the Limited Column Generation procedure for the LATN Design and Expansion problems developed by Shaw (1994 and 1996).

3 TREE PARTITIONING FORMULATION AND LIMITED COLUMN GENERATION PROCEDURE

Because of the contiguity assumption, the set of nodes allocated to the same concentrator forms a subtree of $\hat{T}$. Hence, we can formulate the LATN Design and Expansion problem as a tree partitioning problem.

Let $d_0 = \hat{c}_0 = \hat{F}_0 = 0$. We define the *routing cost* c_{ij} from node j to a concentrator location i for the LATN Design problem as follows : for $i = 0, 1, 2, \cdots, n$ and $j = 1, 2, \cdots, n$,

$$c_{ij} = \begin{cases} d_j(\hat{c}_i + \sum_{k \in K_{ij}} c_k) + F_j & \text{if } j \notin P[i,0] \\ d_j(\hat{c}_i + \sum_{k \in K_{ij}} c_k) + F_{-p_j^i} & \text{if } j \in P[i,0] \setminus \{i\} \\ d_j \hat{c}_j & \text{otherwise,} \end{cases} \tag{15.9}$$

where K_{ij} is the set of arcs on the path $P[i,j]$ from node i to node j.

The set of nodes allocated to the same concentrator forms a subtree T of $\hat{T}$. We denote by k the root of such subtree (i.e., the node with smallest label), and we denote by i the center of such a subtree, i.e., the node where the concentrator is located. Note that if k is the switching center (the root node 0), then $k = i$.

$$c_T^i = \sum_{j \in T} c_{ij} + \hat{F}_i^{t^*}, \tag{15.10}$$

where $t^* = \min \{t \mid \sum_{j \in T} d_j \leq h^t,\ t = 1, 2, \cdots, m\}$.

Similarly, we can define the total cost c_T^i of T for the LATN Expansion problem as

$$\begin{aligned} c_T^i &= \sum_{j \in T} \hat{c}_i d_j + \hat{F}_i^{t^*} + \sum_{j \in T \setminus P[i,k]} e_j \Big(\sum_{v \in T \cap T(j)} d_v \Big) \\ &+ \sum_{j \in P[i,k] \setminus \{k\}} e_{-j} \sum_{v \in T \cap T^c(j)} d_v \end{aligned} \tag{15.11}$$

where $T^c(i)$ stands for the complement of $T(i)$, and the cable expansion cost on arc j is defined as

$$e_j(z) = \begin{cases} F_j + c_j(z - b_j) & \text{if } z > b_j \\ 0 & \text{otherwise.} \end{cases} \tag{15.12}$$

Let $c_T = \min_{i \in T} c_T^i$ be the smallest cost of assigning all nodes in T to a common concentrator.

Then, the LATN Design and Expansion problem can be reformulated as follows:

$$\begin{aligned} (P) \quad & \min \quad c^T \xi \\ & \text{s.t.} \quad G\xi = 1 \\ & \qquad\ \ \xi_T \in \{0, 1\} \quad \text{for all subtrees } T\ , \end{aligned}$$

where $c = (c_T)$, $\xi = (\xi_T)$ and G is a node subtree incidence matrix.

As the intersection graph of G is a chordal graph and, of course, a perfect graph (for detail, see Chvatal (1975), Gavril (1974), Golumbic (1980), and Lovasz (1972)), the linear programming relaxation of (P) has all integral vertices. Shaw(1994) presents an elementary proof that the linear programming relaxation of (P) always has an integer optimal solution. Therefore, (P) can be written as

$$
\begin{aligned}
(P) \qquad & \min \quad c^T\xi \\
& \text{s.t.} \quad G\xi = 1 \\
& \qquad\quad 0 \le \xi_T \le 1 \quad \text{for all subtrees } T.
\end{aligned}
$$

Since the number of columns in G is exponential in n, a natural way to solve this problem is to use a column generation technique, which was first proposed by Gilmore and Gomory (1961). However, because of the special structure of our problem, much better results can be achieved. Shaw (1994) has shown that there exists a pivot rule which selects the so–called Locally Most Violated (LMV) reduced profit and solves (P) in just $n+1$ pivots.

Let B be the $(n+1) \times (n+1)$ basis for the system of constraints in (P) and π_i be the dual variable associated with the i-th constraint (corresponding to node i) in (P) for $i = 0, 1, 2, \cdots, n$. Then the reduced profit of a variable ξ_T is

$$
\gamma_T = \sum_{i \in T} \pi_i - c_T.
$$

The Locally Most Violated (LMV) reduced profit γ_k^* is defined as

$$
\begin{aligned}
\gamma_k^* &= \max_{r(T)=k} \gamma_T \\
&= \gamma_{T_k^*},
\end{aligned}
$$

where $T_k^* = \arg \max_{r(T)=k} \gamma_T$ and $r(T) = \min\{k | k \in T\}$ is the root of T as all nodes in $\hat{T}$ are labeled in DFS order.

Initially, we partition $\hat{T}$ into singleton sets $\{k\}$ for $k = 0, 1, 2, \cdots, n$ (i.e., $\hat{T} = \cup_{k=0}^{n}\{k\}$). This can be interpreted as the case that every node has its own "concentrator." In such a case, the basis B is an $(n+1) \times (n+1)$ identity matrix I and the dual variable π_k is

$$
\begin{aligned}
\pi_k &= c_{\{k\}} \\
&= d_k \hat{c}_k + \hat{F}_k^{t_k^*},
\end{aligned}
$$

where $t_k^* = \min\{t | d_k \le h^t,\ t = 1, 2, \cdots, m\}$ and $k = 0, 1, \cdots, n$.

Let $1_{T_k^*} = (\nu_j) \in R^{n+1}$, where $\nu_j = \begin{cases} 1 & \text{if } \; k \in T_k^* \\ 0 & \text{otherwise.} \end{cases}$

Now, the Limited Column Generation procedure developed by Shaw(1994) can be formally described as follows:

Algorithm 1. Limited Column Generation;

begin

{**comment:** Initialization}

$B := I$;

for $k := 0$ up to n **do** $\pi_k := c_{\{k\}}$;

{**comment:** Main Loop}

for $k := n$ down to 0 **do**

begin

Find_γ_k^*; {**comment:** $\gamma_k^* = \max_{r(T)=k} \gamma_T = \gamma_{T_k^*}$ }

if $(\gamma_k^* > 0)$ **then**

replace k–th column of B by $1_{T_k^*}$;

$\pi_k := \pi_k - \gamma_k^*$;

Update_γ_k^*;

end if

end

opt_value $:= \sum_{k=0}^{n} \pi_k$;

Opt_Solution;

end

Theorem 1. (Shaw 1994 and 1996) *If the LMV reduced profit is given at each pivot, then the LATN Design and Expansion problem can be solved by Algorithm 1 which essentially takes $n+1$ simplex pivots.*

The proof of Theorem 1 can be found in Shaw (1994 and 1996). Because of Theorem 1, a critical step for solving the LATN Design and Expansion problem is to find the LMV reduced profits by the procedure **Find_γ_k^*** at each pivot. Since γ_k^* depends on $\{\pi_i | i \in T(k)\}$, which is updated at each iteration, we show in the next section how the set $\{\gamma_k^* | k = n, n-1, \cdots, 0\}$ can be computed recursively in an on-line fashion. The procedures **Find_γ_k^*** and **Update_γ_k^*** are given in the next section.

Finally, we present the procedure **Opt_Solution** which is a tree search algorithm and can be solved in linear time. We define an array $SOLUTION(j)$ as follows: $SOLUTION(j) = k$ if and only if node j is covered by a subtree rooted at k.

Procedure **Opt_Solution**;

begin

$k := 0$; $STACK := \emptyset$;

for $j := 0$ up to n **do**

begin

if $(j > LAST(k))$ **then**

keep pick k from $STACK$ until $LAST(k) \geq j$;

end if

if $(B_{jk} = 0)$ **then** {**comment:** $B = (B_{ij})$ is the optimal basis}

put k to $STACK$;

$k := j$;

end if

$SOLUTION(j) := k;$

end

end

One remark we would like to make here is that the center of the subtree at k can be specified through an array $LOCATION(k)$ which is defined as follows: $LOCATION(k) = i$ if and only if the subtree rooted at k is served by a concentrator located at i. Then $LOCATION(SOLUTION(j))$ specifies the location of concentrator that serves node j.

Lemma 2.1. *If we exclude the computational time for computing* **Find_**γ_k^* *and* **Update_**γ_k^*, *then Algorithm 1 can be terminated in* $O(n^2)$ *time.*

Proof: Since replacing a column of B takes $O(n)$ time and there are $n+1$ iterations in Algorithm 1, the overall complexity is $O(n^2)$. □

4 COMPUTING THE LMV REDUCED COST

Let π_i be the dual variable associated node i, $i = 0, 1, 2, \cdots, n$, in (P). Then we have

$$\begin{aligned}
\gamma_k^* &= \max_{r(T)=k} \Big(\sum_{j \in T} \pi_j - c_T\Big) \\
&= \max_{r(T)=k} \Big(\sum_{j \in T} \pi_j - \min_{i \in T} c_T^i\Big) \\
&= \max_{r(T)=k} \max_{i \in T} \Big(\sum_{j \in T} \pi_j - c_T^i\Big) \\
&= \max_{i \in T(k)} \max_{\substack{T \ni i \\ r(T)=k}} \Big(\sum_{j \in T} \pi_j - c_T^i\Big) \\
&= \max_{i \in T(k)} \gamma_k^i,
\end{aligned}$$

for $k = 1, 2, ..., n$, where

$$\gamma_k^i = \max_{\substack{T \ni i \\ r(T)=k}} \Big(\sum_{j \in T} \pi_j - c_T^i\Big). \tag{15.13}$$

for $i, k = 1, 2, ..., n$, and $\{T | T \ni i, r(T) = k\}$ represents all subtrees that include node i and are rooted at node k.

Given node k, let $i \in T(k)$. For the LATN Design problem, we define the subproblem $(S_k^i(h))$

$$
\begin{aligned}
(S_k^i(h)) \qquad \max \quad & \sum_{j \in T(k)} \bar{c}_{ij} z_j \\
\text{s.t.} \quad & z_{p_j^i} \geq z_j, \quad j \in T(k) \setminus \{i\} \\
& \sum_{j \in T(k)} d_j z_j \leq h \\
& z_k = 1 \\
& z_j \in \{0, 1\},
\end{aligned}
$$

where $\bar{c}_{ij} = \pi_j - c_{ij}$ and c_{ij} is defined in (15.9).

Similarly, for the LATN Expansion problem, we define the subproblem $(\tilde{S}_k^i(h))$

$$
\begin{aligned}
(\tilde{S}_k^i(h)) \qquad \max \quad & \sum_{j \in T(k)} \bar{c}_{ij} z_j - \sum_{j \in T(k) \setminus P[i,k]} e_j(u_j) - \sum_{j \in P[i,k] \setminus \{k\}} e_{-j}(u_{-j}) \\
\text{s.t.} \quad & z_{p_j^i} \geq z_j, \quad j \in T(k) \setminus \{i\} \\
& \sum_{j \in T(k)} d_j z_j = h \\
& u_j = \sum_{v \in T(j)} d_v z_v, \quad j \in T(k) \setminus P[i,k] \\
& u_{-j} = \sum_{v \in T(k) \setminus T(j)} d_v z_v, \quad j \in P[i,k] \setminus \{k\} \\
& z_k = 1 \\
& z_j \in \{0, 1\},
\end{aligned}
$$

where $\bar{c}_{ij} = \pi_j - \hat{c}_i d_j$.

It is interesting to notice that in $(\tilde{S}_k^i(h))$ we use constraint $\sum_{j \in T(k)} d_j z_j = h$, instead of the inequality constraint, because, unlike the $(S_k^i(h))$, the optimal objective value of $(\tilde{S}_k^i(h))$ is not monotonic with h.

We use $P_{T(k)}(i, k, h)$ and $\tilde{P}_{T(k)}(i, k, h)$ to denote the optimal value of $(S_k^i(h))$ and $(\tilde{S}_k^i(h))$, respectively. (See $(CTKP)$ in Section 4 and $(ETKP)$ in Section

5 for detailed discussion about how to obtain these values.) Then, it follows from (15.9)-(15.11) and (15.13) that γ_k^i for the LATN Design problem can be obtained as follows:

$$\gamma_k^i = \max_{1 \leq t \leq m} \{P_{T(k)}(i, k, h^t) - \hat{F}_i^t\}. \tag{15.14}$$

For the LATN Expansion problem, we can compute γ_i^k accordingly. The detailed discussions are given in section 5. Hence, a critical step in computing γ_i^k for the LATN Design and Expansion problem is to solve the subproblems $(S_k^i(h))$ and $(\tilde{S}_k^i(h))$, respectively. When $i = k$, $(S_i^i(h))$ is the TKP discussed by Cho and Shaw (1994), Johnson and Niemi (1983), Lukes (1974), Shaw (1994), and Shaw *et al.* (1995). In particular, Cho and Shaw (1994) proposed an efficient Depth–First dynamic programming algorithm that runs in $O(|T(k)|h)$ time to solve the TKP. In general, for $i \in T(k)$ (or $k \in P[i, 0]$), we call $(S_k^i(h))$ the Centered Tree Knapsack Problem (CTKP). In the following section, we show that for fixed i, the CTKP $(S_k^i(h))$ can be solved recursively from an optimal solution of $(S_{p_k^i}^i(h))$ in $O(|T(k) \setminus T(p_k^i)|h)$ time for all $k \in P[i, 0] \setminus \{i\}$. Consequently, we are able to compute γ_k^i through (3.2) and finally obtain γ_k^*. However, all $(S_k^i(h))$ are related to $\{\pi_i | i = 0, 1, 2, \cdots, n\}$, which are updated in Algorithm 1 in the reverse order of the node labeling. Therefore, all $(S_k^i(h))$ with $i \in T(k)$ have to be solved in an on–line fashion. The following two procedures, **Find_**γ_k^* and **Update_**γ_k^*, resolve these difficulties. Technically, we define $\underline{d} = \min\{d_j | j \in V \setminus \{0\}\}$.

Procedure **Find_**γ_k^*;

begin

compute $P_{T(k)}(k, k, h)$ for all $h = \underline{d}, \underline{d} + 1, \cdots, H$;

if $(k \neq 0)$ **then**

for $(i \in T(k) \setminus \{k\})$ **do**

begin

compute $P_{T(k)}(i, k, h)$ starting from $P_{T(k)}(i, p_k^i, h)$ for all $h = \underline{d}, \underline{d} + 1, \cdots, H$;

end

$\gamma_k^* := \max_{i \in T(k)} \max_{1 \le t \le m} \{P_{T(k)}(i, k, h^t) - \hat{F}_i^t\};$

end if

end

Procedure **Update_γ_k^*;**

begin

$w := p_0^k;$

for $(i \in T(w))$ **do** $\bar{c}_{ik} := \bar{c}_{ik} - \gamma_k^*;$

for $(i \in T(k))$ **do**

begin

for $h := \underline{d}$ up to H **do** $P_{T(k)}(i, k, h) := P_{T(k)}(i, k, h) - \gamma_k^*;$

end

end

Consequently, we have the following main results.

Theorem 2. *Algorithm 1 solves the LATN Design problem (D) in $O(n^2H)$, where $H = h^m$.*

Proof: The correctness of Algorithm 1 is given in Theorem 1. Because of Lemma 2.1, we only need to estimate the total complexity for the procedures **Find_γ_k^*** and **Update_γ_k^***. We will show in Lemma 4.1 that for a given k, $P_{T(k)}(i, k, h)$ can be computed in $O((|T(k)| - |T(p_k^i)|)H)$ time by starting from $P_{T(k)}(i, p_k^i, h)$ for $h = \underline{d}, \underline{d} + 1, \cdots, H$. Therefore, each **Find_γ_k^*** takes $O(|T(k)|H) + O(\sum_{i \in T(k) \setminus \{k\}} (|T(k)| - |T(p_k^i)|)H) + O(nm)$ time. Hence, the total complexity of Algorithm 1 for performing the procedure **Find_γ_k^*** is

$$\begin{aligned} & O\{\sum_{k=0}^{n}(|T(k)|H + \sum_{i\in T(k)\setminus\{k\}}(|T(k)| - |T(p_k^i)|)H)\} \\ = & O\{\sum_{i=0}^{n}(|T(i)|H + \sum_{k\in P[i,0]\setminus\{i\}}(|T(k)| - |T(p_k^i)|)H)\} \\ = & O\{\sum_{i=0}^{n}(|T(i)| + |T(0)| - |T(i)|)H\} \\ = & O(n^2H). \end{aligned}$$

Moreover, since each **Update_γ_k^*** takes $O(|T(p_0^k)| + |T(k)|H) = O(nH)$, the total time taken by Algorithm 1 for performing the procedure **Update_γ_k^*** is $O(n^2H)$. Therefore, the overall complexity taken by Algorithm 1 for solving the LATN Design problem is $O(n^2H)$. □

Similarly, for $i \in T(k)$ (or $k \in P[i,0]$), we call $(\tilde{S}_k^i(h))$ the Extended Tree Knapsack Problem (ETKP) with arc expansion costs. Since the ETKP $(\tilde{S}_k^k(h))$ can also be solved by a depth–first dynamic programming algorithm described in Shaw, Cho and Chang (1995) in $O(|T(k)|\delta h)$ time, in general, the ETKP $(\tilde{S}_k^i(h))$ can be computed in $O((|T(k)| - |T(p_k^i)|)\delta h)$ time by starting from $(\tilde{S}_{p_k^i}^i(h))$. As a result, we have the following theorem.

Theorem $2'$. *Algorithm 1 solves the LATN Expansion problem* (E) *in* $O(n^2\delta H)$, *where* $H = h^m$.

Proof: Similar to the proof of Theorem 2. □

5 SOLVING THE SUBPROBLEM FOR THE LATN DESIGN PROBLEM

Let T be a subtree of $T(k)$ rooted at k and $i \in T$ be the center of the tree T. For any $v \in T$, we define

$$(CTKP) \qquad P_T(i,v,h) = \max \sum_{j\in T} \bar{c}_{ij}x_j \tag{15.15}$$

$$\text{s.t.} \quad x_{p_j^i} \geq x_j, \quad j \in T\setminus\{i\} \tag{15.16}$$

$$\sum_{j \in T} d_j x_j \leq h \tag{15.17}$$

$$x_v = 1 \tag{15.18}$$

$$x_j \in \{0, 1\}. \tag{15.19}$$

Then our objective is to find the optimal value $P_{T(k)}(i, k, h)$ of the Centered Tree Knapsack Problem $(S_k^i(h))$ using a dynamic programming algorithm.

We can solve $(S_k^i(h))$ by the depth-first dynamic programming algorithm by applying the following recursive rules which are similar to ones for the TKP discussed by Cho and Shaw (1994);

1. (Initialization)

$$P_{\{i\}}(i, i, h) = \begin{cases} \bar{c}_{ii} & \text{if } h \geq d_i \\ -\infty & \text{otherwise} \end{cases}$$

2. (Forward move to expand T)
   For $v \notin T$ and $p_v^i \in T$,

$$P_{T \cup \{v\}}(i, v, h) = P_T(i, p_v^i, h - d_v) + \bar{c}_{iv} \quad \text{if} \quad h \geq \sum_{j \in P[i,v]} d_j$$

3. (Backward move to visit p_v from v)
   For $v \in T \setminus P[i, k]$,

$$P_T(i, p_v, h) = \max\{P_{T \setminus T(v)}(i, p_v, h), P_T(i, v, h)\}.$$

Suppose that we have solved a TKP $(S_i^i(h))$ on $T(i)$. It is important to observe that, for all $k \in P[i, 0] \setminus \{i\}$, the problem $(S_k^i(h))$ can be solved from an optimal solution of $(S_{p_k^i}^i(h))$ as follows:

Step 1) Perform a 'forward move' from node p_k^i to node k, by applying

$$P_{T(p_k^i) \cup \{k\}}(i, k, h) = P_{T(p_k^i)}(i, p_k^i, h - d_k) + \bar{c}_{ik} \quad \text{if} \quad h \geq \sum_{j \in P[i,k]} d_j$$

Step 2) Apply the depth-first dynamic programming algorithm by using the above recursive rules on $T(k) \setminus T(p_k^i)$ to find $P_{T(k)}(i, k, h)$.

To find the optimal solution for $(S^i_k(h))$, we need to define the *index* $I_T(i,v,h)$ corresponding to $P_T(i,v,h)$ for all $v \in T(k)$ as follows:

1. (Initialization)
$$I_{\{i\}}(i,i,h) = 1 \quad \text{if} \quad h \geq d_i$$

2. (Forward move to expand T)
For $v \in P[i,k]$ and $p_v \notin T$,
$$I_{T\cup\{p_v\}}(i,p_v,h) = \begin{cases} 1 & \text{if} \quad h \geq \sum\limits_{j\in P[i,p_v]} d_j \\ 0 & \text{otherwise} \end{cases}$$

3. (Backward move to visit p_v from v)
For $v \in T \setminus P[i,k]$,
$$I_T(i,v,h) = \begin{cases} 1 & \text{if} \quad P_{T\setminus T(v)}(i,p_v,h) < P_T(i,v,h) \\ 0 & \text{otherwise.} \end{cases}$$

We now present an algorithm which solves the Centered Tree Knapsack Problem $(S^i_k(h))$ recursively, starting from $(S^i_{p^i_k}(h))$. It is important to notice that if both the "forward move" and the "backward move" follow the specific order described in Algorithm 2, the value $P_T(i,v,h)$ can be uniquely determined by (i,v,h), that is, T can be uniquely determined by v. Therefore, we can omit T from the notation and simply use $P(i,v,h)$ and $I(i,v,h)$ in the dynamic programming procedure. The pseudo-code of Algorithm 2, including the related procedures, are given in the Appendix.

From the algorithm, we see that the 'forward move' visits nodes in the Depth-First Search order in Algorithm 2. This is the reason we refer to this dynamic programming procedure as "Depth-First." Due to such ordering, the value $P_T(i,v,h)$ can be uniquely determined by (i,v,h), that is, T can be uniquely determined by v, the current node being visited by the algorithm. Let $T^{(v)}$ be the current set of nodes being visited. If v has not been visited by a 'backward move', then $T^{(v)} = \{k, k+1, \cdots, v\}$, where k is the root of subtree $T^{(v)}$. Otherwise, $T^{(v)} = T^{(u)}$, where u is the most recent successor of v from which a 'backward move' visits v. This result comes from the nature of the depth–first dynamic programming procedure.

With the above procedures, we can prove the following theorem.

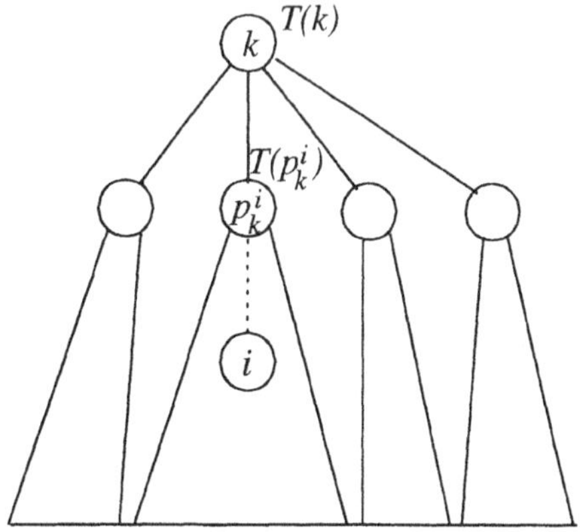

Figure 1 $T(k) = T(p^i_k) \cup (T(k) \setminus T(p^i_k))$

Theorem 3. *If $i = k$, then $(S^i_i(H))$ can be solved in $O(|T(i)|H)$ time, where $|T(i)|$ is the number of nodes in the tree $T(i)$. If $i \neq k$, then the problem $(S^i_k(H))$ can be solved by Algorithm 2 in $O((|T(k)| - |T(p^i_k)|)H)$, provided that the optimal value $P(i, p^i_k, H)$ of $(S^i_{p^i_k}(H))$ is given.*

Proof: If $i = k$, then $(S^i_i(H))$ is the Tree Knapsack Problem and therefore it can be solved in $O(|T(i)|H)$ time (see Cho and Shaw (1994)). When we reroot the tree to node i, the set of nodes in $T(k)$ can be considered as the set of nodes extended from $T(p^i_k)$ by attaching a subtree $T(k) \setminus T(p^i_k)$ as shown in Figure 1. Hence, the correctness of this algorithm is obvious from the correctness of the algorithm for the TKP centered at the root (see Cho and Shaw (1994)). As **Forward_Move(·)** and **Backward_Move(·)** require $O(H)$ time, respectively, we only need to count the number of 'forward moves' and 'backward moves'. Clearly, every node in $T(k) \setminus T(p^i_k)$ is visited by **Forward_Move(·)** and if a 'forward move' moves into a node $v(\neq k)$, then there must be a 'backward move' which moves out from the node v. Hence, the total number of 'forward moves' and 'backward moves' is $O(|T(k)| - |T(p^i_k)|)$. Therefore, the overall complexity of Alorithm 2 is $O((|T(k)| - |T(p^i_k)|)H)$. ◻

We now can rewrite the procedure **Find**_γ_k^* for finding the LMV reduced profit (the negative of the reduced cost) γ_k^* for the LATN Design problem by using the algorithm **CTKP**(i,k,v) to find $P(i,k,h)$ for all $h = \underline{d}, \underline{d}+1, \cdots, H$. The detailed pseudo-code is provided in the Appendix.

Note that we can reduce the space complexity of the algorithm by using two 2–dimensional arrays $P'(i,h)$ and $P''(v,h)$ to represent the three-dimensional array for the optimal value $P(i,v,h)$. The idea can be described as follows. Within the procedure **CTKP**$(i,k,pred)$, we use $P''(v,h)$ to represent $P(i,v,h)$. This is possible because the concentrator location i is fixed. At the end of this procedure, the optimal value $P''(k,h) = P(i,k,h)$ is obtained. Then, we pass this value to $P'(i,h)$, which stores $P(i,k,h)$, where k is the most recent node on the path $P[i,0]$ visited by the Limited Column Generation procedure, Algorithm 1. Therefore, we use $P'(i,h)$ to represent $P(i,k,h)$ outside the procedure **CTKP**$(i,k,pred)$.

6 SOLVING THE SUBPROBLEM FOR THE LATN EXPANSION PROBLEM

Let T be a subtree of $T(k)$ rooted at k and $i \in T$ be the center of T. For any $v \in T$, we define the Extended Tree Knapsack Problem (ETKP) as follows.

$$
\begin{aligned}
(ETKP) \quad \tilde{P}_T(i,v,h) = \quad \max \quad & \sum_{j \in T} \bar{c}_{ij} x_j - \left(\sum_{u \in T \setminus P[i,k]} e_u(z_u) + \sum_{u \in P[i,k] \setminus \{k\}} e_{-u}(z_{-u}) \right) \\
\text{s.t.} \quad & x_{p_j^i} \geq x_j, \quad j \in T \setminus \{i\} \\
& \sum_{j \in T} d_j x_j = h \\
& z_u = \sum_{j \in T \cap T(u)} d_j x_j, \quad u \in T \setminus P[i,k] \\
& z_{-u} = \sum_{j \in T \setminus T(u)} d_j x_j, \quad u \in P[i,k] \setminus \{k\} \\
& x_v = 1 \\
& x_j \in \{0,1\},
\end{aligned}
$$

where $e_u(\cdot)$ is defined in (2.4). The LMV reduced profit can be obtained as

$$\gamma_k^* = \max_{i \in T(k)} \max_{0 \leq h \leq H} \{\tilde{P}_{T(k)}(i, k, h) - F_i^{t(h)}\}, \tag{15.20}$$

where $t(h) = \min\{t | 1 \leq t \leq m, h^t \geq h\}$.

Therefore, the LATN Expansion problem can be solved in a way similar to the LATN Design problem, except that each column generation subproblem in LATN Expansion problem is the ETKP, instead of the TKP. The detailed algorithm for the ETKP is provided in Shaw, Cho and Chang (1995). Each ETKP can be solved by the dynamic programming algorithm in $O(n\delta H)$, where $\delta = \max_{1 \leq j \leq n} b_j$ is the largest existing cable capacity. (See (15.12)). Therefore, we can show that the overall complexity for the LATN Expansion problem is $O(n^2 \delta H)$.

7 COMPUTATIONAL RESULTS AND DATA STRUCTURE

In this section, we report the computational results of the Limited Column Generation procedure for the LATN Design and Expansion problems which incorporate the depth-first dynamic programming algorithm presented. Our Limited Column Generation algorithm LCG is coded in C.

We use two lists, the predecessor p_i and the last node $LAST(i)$ in subtree $T(i)$, to represent the topology of the tree. The 2–dimensional array $B(i, j)$ is used to store the optimal basis and the 3–dimensional arrays $P(i, v, h)$ and $I(i, u, h)$ are used by the dynamic programming procedure for the LATN Design problem. Notice that $P(i, v, h)$ can be replaced by two 2–dimensional arrays. For the LATN Expansion problem, extra memory space is required to store $D'(\cdot, \cdot, \cdot)$ and $D''(\cdot, \cdot, \cdot)$. To save memory space, a dynamic memory allocation technique is required to store these arrays for large scale problems.

A 41–node problem presented in Balakrishnan, Magnanti, and Wong (1991) , which took 879 seconds on a Macintosh II by a Lagrangian relaxation based approach to reach a solution with 7% gap, can be solved by our algorithm to optimality in 135 seconds and 306 seconds if the largest concentrator capacities are chosen to be 5,000 and 10,000, respectively. We have also tested our algorithm on a set of randomly generated problems. To generate a tree randomly, we first specified the total number n of nodes. Starting from the root node, we randomly generated the number of successors of each node from an interval $[0, \log_2 n]$ in Breadth First Search order until the total number of nodes was met.

Two types of concentrator capacities $h^1 < h^2 < \cdots < h^m = H$ were generated randomly in the interval $[H/2, H]$. For each node i, we randomly generated $\hat{c}_i \in [1, 50]$, $\hat{F}_i^1 < \hat{F}_i^2 \cdots < \hat{F}_i^m \in [1, 1000]$, and $d_i \in [1, 50]$ if $H = 500$, and $d_i \in [1, 100]$ if $H = 1000$. For each arc i, we also randomly generated $c_i, F_i \in [1, 50]$ and $b_i \in [1, 500]$.

We fixed the number of concentrator types to $m = 3$, and tested eight problems generated from the ranges defined above. We averaged the CPU time over those eight randomly generated test problems and reported the worst and the best CPU time obtained in each case. The CPU time reported here includes input/output time and is measured by the sum of the user time and the system time. The results are shown in Tables 1 and 2. All these tests run on a SUN SPARC 1000 workstation, and the CPU time is measured in seconds.

Table 1 CPU time of LCG for the LATN Design problems

n	H	LATNE		
		worst	average	best
20	500	1.06	0.72	0.43
	1000	1.94	1.31	0.77
30	500	2.28	1.41	1.00
	1000	4.25	2.41	1.61
50	500	6.94	3.95	2.68
	1000	12.28	9.75	5.87
80	500	17.48	9.09	6.14
	1000	11.29	9.20	6.38
100	500	26.61	13.79	8.18
	1000	50.44	33.94	23.36
150	500	63.72	32.16	14.77
	1000	112.29	66.92	22.20
200	500	120.04	50.09	24.40
	1000	126.28	89.14	51.74

Table 2 CPU time of LCG for the LATN Expansion problems

n	H	LATNE		
		worst	average	best
20	500	3.12	2.83	2.48
	1000	5.66	3.94	3.03
30	500	7.56	6.10	4.72
	1000	13.69	10.82	9.16
50	500	26.59	26.34	25.93
	1000	50.86	50.31	49.45
80	500	67.08	60.01	30.00
	1000	138.06	95.13	33.21
100	500	109.66	102.76	83.12
	1000	195.24	142.75	92.72
150	500	285.61	217.35	93.16
	1000	627.57	339.14	179.17
200	500	476.35	298.76	195.76
	1000	878.48	483.93	202.83

Table 3 Results of standard test problems for the LATN Expansion problems

problems	n	m	H ($h^1/h^2/h^3$)	Optimal Value	CPU time	
					LCG 1.0	CPLEX 3.0
Exp5a	5	1	41	559	0.04	0.25
Exp9a	9	1	51	1305	0.16	0.83
Exp9b	9	1	70	559	0.22	0.55
Exp9c	9	1	340	25847	0.38	1.67
Exp15a	15	3	390/416/451	86717	1.26	1024.85
Exp15b	15	3	387/430/474	65693	0.81	1648.07
Exp20a	20	1	140	23382	0.89	177.33
Exp20b	20	3	374/422/443	65859	2.24	7446.87
Exp20c	20	3	890/916/951	162936	6.13	22664.50*
Exp30a	30	1	340	93238	4.46	1911.32
Exp30b	30	3	390/416/451	86717	6.56	76720.11*

*the CPU time given here was obtained when CPLEX met the branch and bound node limit of 500,000. CPLEX could not solve them within that node limit.

As shown in the tables, the CPU time increases approximately at a quadratic rate with respect to n and at a linear rate with respect to H. Because of the nature of dynamic programming, the complexity of our algorithm is actually $\Omega(n^2 H)$ for the LATN Design problem, and we can see from Table 1 that the running time for the worst case in each category is less than twice that of the average case, whereas the running time for the best case is about one half of that of the average case. Comparing with the LATN Design problem, the LATN Expansion problem is about 4–5 times slower for small size problems (for example, $n \leq 30$) and 6–7 times slower for large size problems. This phenomenon reflects that the complexity of the LATN Expansion problem is about $O(\delta)$ times of that of the LATN Design problem. We don't have a comparison with results from the literature since no other algorithm has been implemented to solve problems of more than 41 nodes. Therefore, we created 11 standard test problems to check the efficiency of the Limited Column Generation procedure as well as the correctness of our code LCG for the LATN Expansion problem by comparing with results from CPLEX 3.0. As shown in Table 3, the Limited Column Generation procedure LCG is clearly superior to CPLEX. The huge CPU time required by CPLEX even for the small size problems (15–30 nodes) is one of the indications that the problems we have studied in this paper are very difficult.

Our computer code LCG 1.0, together with data of 11 small test problems described above and a random problem generator used in our experiments, are now available through WWW. To obtain these computer programs, one can simply access the home-page "http://palette. ecn.purdue.edu/~shawdx2/" and select the item "Computer Programs."

8 CONCLUSIONS

In this paper we developed an efficient implementation of the Limited Column Generation approach for the Local Access Telecommunication Network (LATN) Design and Expansion problems which was originally proposed by Shaw (1994). Our computational results indicate that the Limited Column Generation is a very effective and efficient method for this kind of problems. The LATN Design and Expansion models have a very special combinatorial structure. For example, an appropriate Lagrangian relaxation approach can yield a bound exactly equal to the optimal value of the integer program (see Shaw (1995)). In general, these two models are the special cases of the general tree–partitioning problems (see Shaw (1996)), or the generalization of the capacitated economic lot-sizing problem (see Shaw and Wagelmans (1995)). The Limited Column Generation method is a special-purpose algorithm for this class of problems, which essentially decomposes the problem into a series of subproblems. Our results might suggest a column- wise approach to such integer programming problems rather than the row-wise (cutting plane) approach. In our LATN Design and Expansion problems the corresponding subproblems are Centered Tree Knapsack Problems (CTKP) and Extended Tree Knapsack Problems (ETKP) with arc expansion cost, respectively. These subproblems not only have their *own* special structure, but also have close relationships to each other. One of the main results of this paper is the utilization of these relationships to develop an efficient implementation.

Because of the NP-complete nature of the original problem, the subproblems themselves are NP-complete discrete optimization problems. Usually, the convex hull of the integer feasible solutions can be described either by its facets through valid inequalities or by its vertices directly. In this paper, we adopt the latter approach through dynamic programming. Such an approach is natural and straightforward, but proven to be very effective. As a Knapsack Problem is a special case of the Tree Knapsack Problem (TKP), the complexity of $O(nH)$ for the TKP (and CTKP) might be the best we can expect. However, the question as to whether the bound of $O(n\delta H)$ for ETKP can be further improved is certainly worth investigating in future research.

REFERENCES

[1] A. Balakrishnan, T. L. Magnanti, A. Shulman, and R.T. Wong., 1991a, Models for planning capacity expansion in local access telecommunication networks, *Annals of Operations Research*, 33, 239-284.

[2] A. Balakrishnan, T. L. Magnanti, and R.T. Wong, 1991b, A decomposition algorithm for expanding local access telecommunication networks, Manuscript, Sloan School of Management, M.I.T., Cambridge, Massachusetts.

[3] D. Bienstock, 1993a, A lot-sizing problem on trees, related to network design, *Mathematics of Operations Research*, 18, 402-422.

[4] D. Bienstock, 1993b, Computational experience with an effective heuristic for some capacity expansion problems in local access networks, *Telecommunication Systems*, 1, 379-400.

[5] G. Cho and D. Shaw, 1994a, A Depth-First Dynamic Programming Procedure for Tree Knapsack Problem, Technical Report, School of Industrial Engineering, Purdue University.

[6] V. Chvatal, 1975, On certain polytopes associated with graphs, *J. Combin. Theory B*, 18, 138-154.

[7] W. Cook, 1990, Integer programming solutions for capacity expansion of the local access network, Manuscript, Bell Communication Research.

[8] F. Gavril, 1974, The intersection graphs of subtrees in tree are exactly the chordal graphs, *J. Combin. Theory B*, 16, 47-56.

[9] P. C. Gilmore and R.E. Gomory, 1961, A linear programming approach to the cutting-stock problem, *Operations Research*, 9, 849-859.

[10] M. C. Golumbic, 1980, *Algorithmic graph theory and perfect graphs.* Academic Press, New York.

[11] M. Grötschel and O. Holland, 1991, Solution of large-scale symmetric traveling salesman problems, *Mathematical Programming*, 51, 141-202.

[12] M. Grötschel, C. L. Monma, and M. Stoer, 1992, Computational results with a cutting plane algorithm for designing communication networks with low-connectivity constraints, *Operatons Research*, 40, 309-330.

[13] D. S. Johnson and K.A. Niemi, 1983, On Knapsacks, partitions and a new dynamic programming technique for trees, *Mathematics of Operations Research*, 8, 1-14.

[14] L. Lovasz, 1972, A characterization of perfect graphs, *J. Combin. Theory B*, 13, 95-98.

[15] J. A. Lukes, 1974, Efficient algorithm for the partitioning of trees, *IBM Journal of Research and Development*, 18, 214-224.

[16] M. W. Padberg and G. Rinaldi, 1986, Optimization of a 532-city symmetric traveling salesman problem by brach-and-cut, Manuscript.

[17] D. Shaw, 1994, Reformulation and Column Generation for Several Telecommunication Network Design Problems, *Proceedings of the 2nd International Telecommunication Conference*, Nashville, Tennessee.

[18] D. Shaw and G. Cho, 1994, A Branch-and Bound Procedure for the Tree Knapsack Problem, Technical Report, School of Industrial Engineering, Purdue University.

[19] D. Shaw and A. Wagelmans, 1995, An Algorithm for Single-item Capacitated Economic Lot Sizing with Piecewise Linear Production Costs and General Holding Costs, Technical Report, School of Industrial Engineering, Purdue University.

[20] D. Shaw, 1995, Local Access Network Design and Its Extensions, Technical Report, School of Industrial Engineering, Purdue University.

[21] D. Shaw, G. Cho, and H. Chang, 1995, The Extended Tree Knapsack Problem for Local Access Telecommunication Network, Working Paper, School of Industrial Engineering, Purdue University.

[22] D. Shaw and G. Cho, 1996, Limited Column Generation for Local Access Network Expansion and Its Extensions — Formulations, Algorithms and Implementations, Technical Report, School of Industrial Engineering, Purdue University.

[23] D. Shaw, 1996, A Note on Tree Partitioning Problem, Technical Report, School of Industrial Engineering, Purdue University.

APPENDIX A

PSEUDO-CODE OF ALGORITHM 2

Algorithm 2. **CTKP**$(i, k, pred)$;

{**comment:** $i = center,\ k = root,\ pred = p_k^i$}
begin
if $(k \neq 0)$ **then**
$d\_path := \sum_{i \in P[i,pred]} d_i;$
else
$d\_path := 0;$
end if
if $(i \neq k)$ **then**
Forward_Move$(i, pred, k)$;
end if
$j := k + 1;$
while $(j \leq LAST(k))$ **do**
begin
if $(j \neq pred)$ **then**
Forward_Move(i, p_j, j);
if $(j = LAST(j))$ **then** {**comment:** node j is a leaf node}
$w := j;$
do
Backward_Move(i, w, p_w);
$w := p_w;$
while $(LAST(w) = j$ and $w \neq k)$
{**comment:** w has no successor t such that $t > j$ and $w \neq k$}
end if
$j := j + 1;$
else
$j := LAST(pred) + 1;$
end if
end
end

Procedure 2.1 **Forward_Move**(i, j, k);

```
begin
d_path := d_path + d_k;
for h := d up to d_k − 1   do   P(i, k, h) := −∞;
for h := d_k up to H   do
begin
if (d_path ≤ h)   then
P(i, k, h) := P(i, j, h − d_k) + c̄_ik;
if (p_j = k)   then
I(i, k, h) := 1;
end if
else
P(i, k, h) := −∞;
end if
end
end
```

Procedure 2.2 **Backward_Move**(i, j, k);

```
begin
d_path := d_path − d_j;
for h := d up to H   do
begin
if (P(i, k, h) ≥ P(i, j, h))   then
I(i, j, h) := 0;
else
P(i, k, h) := P(i, j, h);
I(i, j, h) := 1;
end if
end
end
```

Procedure 1.1′ **Find_**γ_k^*;

```
begin
CTKP(k, k, −1);      {comment: as p_k^k is undefined, we set p_k^k = −1}
if (k ≠ 0)   then
pred := k + 1;      {comment: pred = p_k^i}
```

```
for i := k + 1 up to LAST(k) do
begin
if (i > LAST(pred)) then
pred := i;
end if
CTKP(i, k, pred);
end
```

$$\gamma_k^* := \max_{i \in T(k)} \max_{1 \le t \le m} \{P(i,k,h^t) - \hat{F}_i^t\};$$

```
  end if
end
```

Applied Optimization

1. D.-Z. Du and D.F. Hsu (eds.): *Combinatorial Network Theory.* 1996 ISBN 0-7923-3777-8
2. M.J. Panik: *Linear Programming: Mathematics, Theory and Algorithms.* 1996 ISBN 0-7923-3782-4
3. R.B. Kearfott and V. Kreinovich (eds.): *Applications of Interval Computations.* 1996 ISBN 0-7923-3847-2
4. N. Hritonenko and Y. Yatsenko: *Modeling and Optimimization of the Lifetime of Technology.* 1996 ISBN 0-7923-4014-0
5. T. Terlaky (ed.): *Interior Point Methods of Mathematical Programming.* 1996 ISBN 0-7923-4201-1
6. B. Jansen: *Interior Point Techniques in Optimization.* Complementarity, Sensitivity and Algorithms. 1997 ISBN 0-7923-4430-8
7. A. Migdalas, P.M. Pardalos and S. Storøy (eds.): *Parallel Computing in Optimization.* 1997 ISBN 0-7923-4583-5
8. F.A. Lootsma: *Fuzzy Logic for Planning and Decision Making.* 1997 ISBN 0-7923-4681-5
9. J.A. dos Santos Gromicho: *Quasiconvex Optimization and Location Theory.* 1998 ISBN 0-7923-4694-7
10. V. Kreinovich, A. Lakeyev, J. Rohn and P. Kahl: *Computational Complexity and Feasibility of Data Processing and Interval Computations.* 1998 ISBN 0-7923-4865-6
11. J. Gil-Aluja: *The Interactive Management of Human Resources in Uncertainty.* 1998 ISBN 0-7923-4886-9
12. C. Zopounidis and A.I. Dimitras: *Multicriteria Decision Aid Methods for the Prediction of Business Failure.* 1998 ISBN 0-7923-4900-8
13. F. Giannessi, S. Komlósi and T. Rapcsák (eds.): *New Trends in Mathematical Programming.* Homage to Steven Vajda. 1998 ISBN 0-7923-5036-7
14. Ya-xiang Yuan (ed.): *Advances in Nonlinear Programming.* Proceedings of the '96 International Conference on Nonlinear Programming. 1998 ISBN 0-7923-5053-7
15. W.W. Hager and P.M. Pardalos: *Optimal Control.* Theory, Algorithms, and Applications. 1998 ISBN 0-7923-5067-7
16. Gang Yu (ed.): *Industrial Applications of Combinatorial Optimization.* 1998 ISBN 0-7923-5073-1
17. D. Braha and O. Maimon (eds.): *A Mathematical Theory of Design: Foundations, Algorithms and Applications.* 1998 ISBN 0-7923-5079-0

KLUWER ACADEMIC PUBLISHERS – DORDRECHT / BOSTON / LONDON

GPSR Compliance

The European Union's (EU) General Product Safety Regulation (GPSR) is a set of rules that requires consumer products to be safe and our obligations to ensure this.

If you have any concerns about our products, you can contact us on ProductSafety@springernature.com

In case Publisher is established outside the EU, the EU authorized representative is:

Springer Nature Customer Service Center GmbH
Europaplatz 3
69115 Heidelberg, Germany

Batch number: 10782423

Printed by Printforce, the Netherlands